# Cioffari's Experiments in College Physics

**9**TH EDITION

## Dean S. Edmonds, Jr.

Boston University and
The University of Western Ontario

D. C. HEATH AND COMPANY

Lexington, Massachusetts

Toronto

*Cover:* Photograph of light pattern of a transverse wave in a vibrating cord by Fundamental Photographs/Richard Megna.

International Standard Book Number: 0-669-28570-6

1  2  3  4  5  6  7  8  9  0

discussions and suggestions and in fact originated some of the experimental procedures presented herein. The same applies to Professors Donald Moorcroft, Graham Rose, and Patrick Whippey of the Physics Department at the University of Western Ontario, who also made available laboratory space and material with which to try the new experiments. And finally, I would like to thank Mr. Gerald Ezcurra of the Pasco Scientific Company for his help in obtaining and operating some of his company's equipment.

*Dean S. Edmonds, Jr.*

# Preface

The ninth edition of *Cioffari's Experiments in College Physics* is, like its predecessors, intended for use in the laboratory segment of either a calculus- or algebra-based freshman or sophomore physics course. Some of the more advanced experiments can be used in later laboratory work, but an effort has been made to keep things in the present revision simple. To that end experiments such as the photoelectric effect and the famous Franck-Hertz and Millikan oil drop experiments, which appeared in earlier editions, have been omitted as being too long and tricky for the elementary laboratory and requiring equipment of too expensive or specialized a nature. In addition, several experiments have been rewritten to improve the clarity of the theoretical presentation and data-taking instructions, and some new problems have been added to the Questions sections to give instructors a wider choice when assigning such work. The fact that calculus is needed in a few Questions is clearly indicated in those cases so that they can be omitted from assignments in non-calculus courses.

One major change is the rewriting of Experiment 10 on rotational motion to use the Pasco model ME-9341 apparatus rather than the Cenco catalog no. 75290 specified in earlier editions. One reason for this is the difficulty of obtaining the latter, despite its continued listing in the Cenco catalog. Another reason is the fact that the Pasco system places the rotating assembly in the horizontal plane so that various objects may be loaded on the basic disk to vary the assembly's moment of inertia. As these objects have standard, easily measured shapes, their moments of inertia may be calculated from their dimensions for a further comparison with experiment. The Pasco apparatus also provides a driving drum with three diameters so that the torque may be varied by changing both the radius on which the string tension acts and the weight on the string. If the auxiliary disk, the steel ring, and the rectangular plate are all used in turn on the basic disk and runs are made with the various possible torques, the experiment may get too long for the laboratory period. However, the write-up has been organized so that an instructor can assign whatever number of sections he or she deems to be consistent with the allotted time.

Another change is the revision of Experiments 48, 49, and 50 on radioactivity. In earlier editions a count-rate meter was specified, but the radioactive sources sold for use in teaching laboratories without the requirement of a license turned out to be insufficiently intense to give good results with such a meter. The present versions make use of a counter-timer unit that provides a great improvement in statistics even when reasonably short counting intervals are chosen. Intervals that should prove satisfactory are given in the text, but an instructor can clearly vary these to give the best compromise between good statistics and time available in the laboratory.

Finally, Experiment 35, which formerly involved both an introduction to the cathode ray oscilloscope and a study of ac circuits and was therefore considered overly long, has been split into two (Experiments 35 and 36). The first of these is an introduction to the oscilloscope providing extra practice with this instrument and thus promoting better familiarity with it before the business of making actual measurements on ac circuits is attempted.

The ninth edition is also accompanied by an updated *Instructor's Guide* that contains more sample data and answers to the new and revised Questions. It is again hoped that the purposes both of instructors charged with setting up new laboratories and veterans presiding over well-established ones may be advantageously served.

## ACKNOWLEDGMENTS

The ninth edition reflects ongoing feedback from many users of previous editions of *Cioffari's Experiments in College Physics* whose reviews and evaluations were helpful in making improvements. For their valuable contributions to the ninth edition, I would like to acknowledge the following users: James J. Coughlin, Northwestern Michigan College; William J. DeBuvitz, Middlesex County College; Allan Fryant, Jamestown College; Frederick W. Oliver, Morgan State University; Arnold Walker, Southeastern Oklahoma State University.

In addition, Professors Wolfgang Franzen, Bernard Chasan, George Zimmerman and the late J. Gordon Stipe of the Boston University Physics Department deserve special mention for their contributions. They were always ready with helpful

iii

# Contents

because physics gives them some fundamentally special status but because we can express all other units in terms of them. Thus velocity is measured in meters per second (m/s), acceleration in meters per second per second (m/s²), and force in kilogram-meters per second per second (kg-m/s²). In the case of a unit like force, which is an unwieldy combination of fundamental units, a special name is given to represent that combination. It has become customary to honor famous scientists by using their names for this purpose; thus our unit of force is called the newton (after Isaac Newton) and is defined as a kilogram-meter per second per second.

The meter, the kilogram, and the second suffice to give us all the units we need for measurements in mechanics. But electricity introduces a new physical quantity, electric charge, for which we need an additional fundamental unit. Actually, because a given amount of charge is hard to determine precisely, the International Conference agreed instead on the size of the ampere, the unit of electric current. Clearly, since by definition of current an ampere is a unit of charge going by in a second, and since a second has already been defined, a precise definition of the unit of electric charge is immediately obtainable. It is called the coulomb after the French physicist Charles Augustin Coulomb, just as the name of the unit of current honors André Marie Ampere.

Although the meter, the kilogram, and the second are convenient in size for measurement of the lengths, masses, and times encountered in everyday life, both smaller and larger units are needed in many areas of physics. Accordingly, a system of prefixes indicating multiples by powers of ten is used to express such quantities. These prefixes are listed in Table III at the end of the book. Some common examples are the centimeter ($10^{-2}$ m), the millimeter ($10^{-3}$ m), the kilometer ($10^{3}$ m); and for time, the millisecond ($10^{-3}$ s), the microsecond ($10^{-6}$ s) and, in this day of electronic circuitry, the nanosecond ($10^{-9}$ s) and the picosecond ($10^{-12}$ s). Note that the standard unit of mass is the kilogram, not the gram, although this presents no particular problem, a standard gram being a mass of $10^{-3}$ kg. Conversion to other units in common use but not in the agreed-upon system, such as the foot, the mile, the minute, and the hour are given in Table II at the end of the book, which points out that in the so-called English system the unit of length is now defined by making the inch *exactly* 2.54 cm.

The system agreed upon at the 1971 International Conference and described earlier is called the SI system, an abbreviation for the full French title "Système Internationale d'Unités." Because the meter, the kilogram, the second, and the ampere (and hence the coulomb) are its fundamental units, the SI system is essentially identical with what was previously called the MKS system (for meter-kilogram-second). The only difference, in fact, lies in the precise definitions of the meter and second described above. These definitions superseded earlier, less precise ones, but the number of wavelengths of the krypton light making up a meter and the number of cycles of the cesium frequency making up a second were chosen to agree with the earlier definitions of these units within the precision already established. Thus existing secondary standards did not have to be altered following the 1971 conference. For our purposes, the terms MKS system and SI system are synonymous, but we shall use the presently preferred designation SI. We will also follow current practice in using the SI system in this book; however, there will be some exceptions. For example, meter sticks, vernier calipers, and micrometer calipers are calibrated in centimeters, and recording the reading of such instruments in centimeters should certainly be permissible. But conversion to meters will usually be advisable if not essential when the data is used in calculations.

One other difficulty that needs to be overcome right from the start is the confusion between mass and weight. *It is imperative to remember that weight is a force, namely, for our purposes in a laboratory on the earth's surface, the force with which objects are attracted to the earth by gravity.* Mass, on the other hand, is an intrinsic property of all objects. Since the force of gravity goes inversely with the square of the distance separating the attracting objects, an object can be made weightless by taking it far away from the earth and all other heavenly bodies, but it never becomes massless. Confusion arises because an object's mass is apparently responsible for its gravitational attraction to other objects, and in fact the gravitational force is proportional to the masses involved. Thus a mass $m$ will be attracted to the earth with a force $W$ (its weight) that is proportional to $m$. As the proportionality constant is a known number (the acceleration of gravity $g$), a measurement of the weight of an object also determines its mass. As a result, scales and balances, which actually measure weight, are often calibrated in grams or kilograms, which are units of mass. Thus it is all too easy to say that a certain object "weighs so many grams." Great care must be taken not to do this but to say rather that the object's mass is so many grams, even though the instrument making the measurement is actually measuring the force of gravity on the object. The distinction between mass and weight will be carefully observed throughout this book, although sometimes a seeming contradiction may appear. Thus it is common parlance to speak of "a set of weights," meaning objects of calibrated mass to be used in balances to produce known forces of gravity. We shall adopt such usual terms rather than the somewhat forced "set of calibrated masses," but the distinction between mass and weight must be kept in mind at all times, even when confusing situations such as those just mentioned arise.

## PLOTTING OF CURVES

Graphs are of particular importance in physics because they display the relationship between pairs of inter-dependent quantities in a readily visualized form. Thus if two quantities $x$ and $y$ have the linear relation

$$y = ax + b \qquad (1)$$

a graph of $y$ against $x$ will be a straight line whose slope is $a$ and whose $y$ intercept (the value of $y$ at which the line crosses the $y$ axis, that is, at $x = 0$) is $b$. Conversely, if a given theory predicts that a certain physical quantity $y$ depends linearly on another physical quantity $x$, this conclusion can be tested experimentally by measuring corresponding values of $x$ and $y$ and plotting these results. The plotted points will readily show whether a straight line can be drawn through them, even if various errors cause them to have a "scatter" instead of all lying right on a line. If there is a scatter, the straight line that represents the best average should be drawn as shown in Fig. 1. There are numerous rules for obtaining the line that is a true "best fit" to a given set of experimental points, but a simple determination by visual inspection using a transparent (plastic) straight edge is usually good enough and is all that will be required in these experiments. If the scatter of points is so large that a good decision as to where to draw a straight line through them cannot be made, the conclusion that $y$ depends linearly on $x$ should be seriously questioned. If, on the other hand, the points readily define a straight line, not only does the resulting graph supply evidence that $x$ and $y$ are indeed linearly related, but the value of $a$ in Equation 1 may be obtained by finding the line's slope. Note that this is a convenient method of getting the average value of $a$. Moreover, a value for $b$ representing an average result of all the plotted data can be read directly off the graph. This simple procedure is equivalent to the much more tedious one of determining the best values of $a$ and $b$ by a "best fit" calculation.

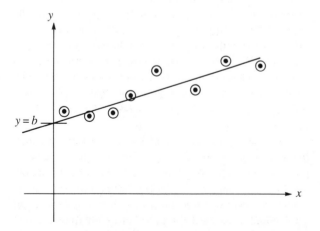

**Figure 1**  A "best fit" straight line drawn through a set of experimental points

A linear relation between $x$ and $y$ is easily recognized when the points are plotted, but other relationships are not so obvious. Thus, suppose the relationship to be investigated were

$$y = ax^2 + b \qquad (2)$$

Plotting $y$ against $x$ would give a curve, but it would be very difficult to distinguish this curve from the curve resulting from, say, $y = ax^3 + b$. In fact, the straight line is the only graph that is really obvious. However, if the validity of Equation 2 is to be tested graphically, a new variable $u \equiv x^2$ can be introduced so that Equation 2 becomes

$$y = au + b \qquad (3)$$

Then, if Equation 2 is valid, a plot of $y$ against $u$ (that is, against $x^2$) will yield a readily recognizable straight line. This procedure may be used in many cases where a new variable $u$ may be substituted for a function of $x$ to produce a linear relation such as Equation 3.

A special case of this procedure arises when $y$ depends exponentially on $x$, so that

$$y = Ae^{ax} \qquad (4)$$

where $e$ is the base of natural logarithms and is approximately equal to 2.718. This situation occurs often enough to merit special treatment. The first step is to take the natural logarithm of both sides of Equation 4:

$$\ln y = \ln Ae^{ax} = \ln A + \ln e^{ax} = \ln A + ax \qquad (5)$$

Equation 5 is just like Equation 1 except that $\ln y$ rather than $y$ is to be plotted against $x$. This plot will thus be linear if Equation 4 is valid, and the $y$ intercept ($b$ in Equation 1) will be $\ln A$. Hence $A$ can be found by taking the antilog of the intercept.

Because this situation arises often, special graph paper is printed on which the graduations along the ordinate ($y$ axis) are logarithmically rather than linearly spaced. This means that if a value of $y$ is plotted on the given ordinate scale, the actual position of the point along the $y$ axis will be proportional to the logarithm of $y$. In other words, plotting a value of $y$ on this special graph paper automatically takes the logarithm, making a separate calculation of $\ln y$ for each value of $y$ unnecessary. However, note that commercial logarithmic graph paper is set up for common (base 10) logarithms rather than natural (base $e$) logarithms. Taking the common log of both sides of Equation 4 yields

$$\log y = \log Ae^{ax} = \log A + \log e^{ax}$$
$$= \log A + ax \log e$$
$$= \log A + a(0.4343)x \qquad (6)$$

The slope of the resulting straight line is now $0.4343a$ rather than just $a$. Notice also that because the $y$ axis is graduated logarithmically, the value of $A$ may be read off from the intercept directly, the paper having also automatically done the job of taking the antilog.

Another popular special case arises when a functional relation of the form

$$y = ax^n \tag{7}$$

is to be investigated. Although this can be handled by introducing $u \equiv x^n$ as already discussed, such a procedure requires calculating $x^n$ for each value of $x$. Since $n$ may be any number, positive or negative, such calculations can get tedious unless a reasonably sophisticated calculator is available. Another very convenient method is again to take the logarithm of both sides of Equation 7. Then,

$$\log y = \log a + n \log x \tag{8}$$

and a plot of $\log y$ against $\log x$ will produce a straight line with slope $n$.* You may object that since you must look up the logarithms of all the values of both $x$ and $y$, things haven't been simplified much, but again special graph paper is available that makes this calculation unnecessary. Since $\log y$ is now to be plotted against $\log x$ rather than $x$, this graph paper has both the ordinate and the abscissa graduated logarithmically. It is therefore called full log or log-log paper, whereas paper with the ordinate graduated logarithmically and the abscissa linearly is called semilog. The logarithmic scales are called 1-cycle, 2-cycle, etc., depending on the number of powers of ten covered on the axis in question. Thus, an axis graduated logarithmically from 1 to 10 is called 1-cycle; from 1 to 100, 2-cycle;

etc. Scales of up to 5 cycles are available commercially, and in the case of full log paper there are various standard combinations of numbers of cycles along the ordinate and the abscissa. Appropriate graph paper pages are included in this book as needed; refer to Experiment 33 for an example of semilog paper and to Experiment 37 for the full log type.

In drawing graphs, scales for the coordinate axes should be chosen so that the curve extends over most of the graph sheet and so that decimal parts of units are easily determined. This can be done if each small division is made equal to one, two, five, or ten units. The same scale need not be used for both axes. The independent variable should be plotted along the $x$ axis and the dependent variable along the $y$ axis. Each axis should be labeled with the name of the quantity being plotted and the scale divisions used. The numbers should increase from left to right and from bottom to top. Each graph should have a title indicating what the curve is intended to show.

Each point should be plotted as a dot surrounded by a small circle, which shows where the point is located even if the dot is obscured by the curve drawn through it. A straight line (or smooth curve if a straight-line plot is not being sought) should then be drawn through the dots. The curve need not pass through all the dots but should be drawn so as to fit them as closely as possible, as already mentioned. In general, as many points will lie on one side of the curve as on the other. The extent to which the plotted points coincide with the curve is a measure of the accuracy of the results.

## SIGNIFICANT FIGURES

The numbers dealt with in mathematics are exact numbers. When mathematicians write 2, they mean 2.00000 . . . , and all subsequent calculations assume that the 2 means exactly two, not the tiniest fraction more or less. In physics the situation is very different. Many of the numbers dealt with come from measurements of physical quantities, and these can never be exact. For example, suppose that a distance is measured with an ordinary centimeter rule and found to be 5.23 cm. In this measurement, the 3 is an estimate, for the smallest divisions on a centimeter rule are millimeters (tenths of a centimeter). The 3 represents a guess as to where between the 5.2 and 5.3 cm divisions the end of the measured distance lies. The statement that the distance was found to be 5.23 cm does not mean that it is exactly 5.23 cm but merely that it is probably not less than 5.22 cm or more than 5.24 cm. If a high-quality micrometer had been used, the distance might have been found to be 5.2347 cm, where

the 7 represents a guess as to where the micrometer's index line fell between the .234 and the .235 divisions. Thus the micrometer yields a much more precise value of the measured length than does the centimeter rule, but it too is not exact. More precise measurement methods might give further decimal places that cannot be determined with the micrometer any more than the 4 and the 7 could be found with the centimeter rule. Thus, when the result of the centimeter rule measurement has been written as 5.23 cm, it doesn't mean that the distance is exactly 5.23 cm or that zeros can be written after the 3. Nothing can be written after the 3 because the instrument being used gives no information as to what to write there.

The 5, the 2, and the 3 in the centimeter rule measurement are called *significant figures* because they each give trustworthy information about the size of the physical quantity being measured. The centimeter rule is quite good enough as a length-measuring instrument to determine that the length in question lies between 5.2 and 5.3 cm, and the 3 in the next place represents a significant guess as to where between 5.2

---

*Notice that in this case there is no intercept in the usual sense. $\log y = \log a$ and $y = a$ when $x = 1$.

and 5.3 cm the actual length lies. The centimeter rule measurement is thus *good to three significant figures*, whereas the micrometer measurement gave five significant figures, the micrometer being a much more precise length-measuring instrument than the centimeter rule. The 3 in the centimeter rule measurement and the 7 in the micrometer measurement are less significant than the other figures but are still considered significant because they give some real information about the desired length even though there is some doubt about their actual values. Clearly, however, if there is some doubt about them, any figures that might get written to the right of them in the respective cases would be meaningless. In particular, one should be careful not to write zeros there. If the length measurement made with the centimeter rule were recorded as 5.230 cm, the zero would be a significant figure and would mean that somehow someone was able to interpolate between the 5.2 and 5.3 divisions to 1/100 rather than just 1/10 of the space between them. Indeed, the micrometer measurement shows that the figure to the right of the 3 should be 4, so that putting a zero there says something that isn't true. Always take care to distinguish between zeros that are significant and those that are not. In general, zeros that merely serve to place the decimal point are not significant. Thus, if the length measurement were to be stated in meters, the two zeros in 0.0523 m would not be significant. They merely place the decimal point appropriately in the three-significant-figure measurement. However, if in measuring the distance with the centimeter rule the end of this distance appeared to fall right opposite the .2 cm division following the 5 cm mark, it would be recorded as 5.20 cm and the zero would be significant. In general, zeros appearing to the right of figures that are already to the right of the decimal point must be regarded as significant, for if they weren't they wouldn't be there. Zeros between other figures and the decimal point should usually be regarded as serving only to place the decimal point. The example of the length of 0.0523 m is typical, there being no doubt that the zeros are not significant. There are some ambiguous cases, however. Suppose that a certain race course is found to be 1.2 km long. As written, this is a two-significant-figure measurement. The same result may be given as 1200 m. Here again the zeros are not significant but must be present in order to properly locate the decimal point. Without the knowledge that the original measurement of 1.2 km contained only two significant figures, however, there is no way to tell whether these zeros are significant or not. In such cases, the experimenter must refer to the measuring instrument to determine how many significant figures are justified.

There is usually no problem in deciding how many significant figures a given measurement should contain, but difficulties arise when these numbers are used in calculations. This is because mathematics assumes that all numbers are exact and thus automatically fills all places to the right of the last significant figure with zeros even though this is physically wrong. The calculations then often produce a great many figures that look as if they were significant but really are not, for clearly no mathematical manipulation can give a result whose precision is greater than that of the quantities put into it. Some examples may serve to show how this problem should be handled.

1. Addition and subtraction: When carrying out addition or subtraction by hand, do not carry the result beyond the first column that contains a doubtful figure. This means that all figures lying to the right of the last column in which all figures are significant should be dropped. Thus in obtaining the sum of these numbers

| 806.5 | they should be written as | 806.5 |
|---|---|---|
| 32.03 | | 32.0 |
| 0.0652 | | 0.1 |
| 125.0 | | 125.0 |
| | | 963.6 |

Note that, in dropping nonsignificant figures, the last figure retained should be unchanged if the first figure dropped is less than 5 and should be increased by 1 if the first figure dropped is 5 or greater. This is a normal convention to which this book will adhere.

If an electronic calculator is used, the numbers to be added may be entered without the bother of determining which figures to drop, in which case all figures will appear in the sum. This is like adding the numbers as given on the left in the example above. The result will be 963.5952. You must then look at the data and observe that in two of the numbers being added there is no indication of what the figure in the second decimal place should be. The result must therefore be rounded off to one decimal place by dropping the 952. Since the 9 is equal to or greater than 5, the figure in the first decimal place is raised by one to give 963.6 as before.

2. Multiplication and division: The operations of multiplication and division usually produce many more figures than can be justified as significant, so that results must be properly rounded off. The rule is to retain in the result only as many figures as the number of significant figures in the least precise quantity in the data. Suppose the area of a plate is to be measured. A centimeter rule is used to find that the plate has a length of 7.62 cm and a width of 3.81 cm. As in the earlier example with the centimeter rule, these measurements each contain three significant figures, of which the third is doubtful. If the area is now found by multiplying $3.81 \times 7.62$ either by hand or with a calculator, 29.0322 cm$^2$ will be obtained. This number appears to have six significant figures, but the two original quantities have only three each. Therefore, only three significant figures should be retained in the

result, which should be written as 29.0 cm². With some exceptions, the measurements to be made in the experimental work covered in this book will contain three or four significant figures, so that when an electronic calculator is used, the final result must be rounded off to the number of figures that can be justified by the data as being significant.

## THEORY OF ERRORS

All measurements are affected by errors; this means that measurements are always subject to some uncertainty. There are many different types of errors, such as personal, systematic, instrumental, and accidental errors. Personal errors include blunders, such as mistakes in arithmetic, in recording an observation, or in reading scale divisions. Another important kind of personal error is known as personal bias, such as trying to fit the measurements to some preconceived idea, or being prejudiced in favor of the first observation. Systematic errors are characterized by their tendency to be in one direction only, either positive or negative. For example, if a meter stick is slightly worn at one end, and measurements are taken from this end, then a constant error will occur in all these measurements. Instrumental errors are those introduced by slight imperfections in the manufacture or calibration of the instrument. The worn meter stick just mentioned or an electrical meter that has not been properly set to zero with no input are examples of instrumental errors. Note that such errors are usually also systematic. Accidental errors are deviations beyond the control of the observer. These errors are due to jarring, noise, fluctuations in temperature, variations in atmospheric pressure, and the like. Included in this category are variations in observed data due to inherently random processes such as the intersurface actions that produce the force we call friction and the radioactive decay of atomic nuclei. Since the causes just listed for accidental errors are essentially random in nature, all these causes of data variation are subject to treatment by statistical methods, as will be discussed here.

It will be assumed in these experiments that instrumental errors due to improper calibration, zeroing, etc., have been prevented by proper inspection and adjustment of the equipment and that care has been taken to eliminate systematic errors, personal errors, and personal bias. There remain accidental errors, which make themselves known by causing a spread in the values obtained when a given measurement is repeated several times. Two examples may serve to illustrate how this comes about.

Consider first the distance measurement with the centimeter rule discussed in connection with significant figures. In the measurement of 5.23 cm, the 3 was doubtful, being an interpolation between the 5.2 and 5.3 cm divisions, which are the smallest divisions on the centimeter rule. If a two-significant-figure result were adequate, the distance could have been quoted as 5.2 cm. In this case, if the measurement were repeated many times, even by different experimenters, the likelihood is that 5.2 cm would be obtained each time. No accidental error is revealed because the measuring instrument is not being pushed to the limit of its precision and random processes in the experiment (such as small variations in the length of the rule and/or the distance being measured due to temperature fluctuations) are negligible compared to the smallest scale unit in the measuring instrument (the millimeter divisions on the centimeter rule in this case). The precision here is said to be limited by the scale of the instrument.

However, if the distance measurement is repeated with an estimated interpolation made each time between the 5.2 and 5.3 cm divisions, the same estimate may not always be made. This would be especially true if each measurement were made by a different experimenter who had no knowledge of the others' results. Thus, one might guess 5.22 cm, another 5.21, another 5.24, etc. To handle this situation, a mean or average of the various measured values is calculated. As will be discussed shortly, this average is more accurate than any one of the measurements alone and can in fact be shown to improve in accuracy as the square root of the number of individual measurements made. Clearly, a way of improving the accuracy of experimental data is to measure each quantity many times, and an important matter of judgment in experimental work is to decide on how many times a given quantity is to be measured. In this regard, it must be remembered that to measure something $N$ times takes $N$ times as long as measuring it once, but the accuracy obtained by doing so is only $\sqrt{N}$ times as great. Thus, if a certain measurement takes one minute, making it ten times will take only ten minutes but will yield over three times the accuracy. However, making it a hundred times will take an hour and forty minutes, but this investment in time will only yield another threefold increase in accuracy. Clearly, a compromise based on the accuracy required, the time needed for a particular measurement, and the time available must be reached in each case.

A second example, one dominated by random processes inherent in the experiment itself, is that of the range of a spring-operated gun. The experimental setup is shown in Figure 2. The spring gun consists of a tube containing a spring-loaded plunger. A small steel ball is placed in the tube against the plunger. The plunger is pulled back a given distance, compressing the spring by a known amount, and is then released sharply, propelling the ball out of the tube. The ball

**Figure 2** *The range of a spring gun*

strikes the floor at a horizontal distance $x$ from the end of the tube (the gun's muzzle), this distance being the range in question.

If this experiment is repeated under conditions made as identical as possible to those in effect on the first try (the ball is carefully put back in the tube, the plunger is pulled back by a distance made as closely equal as possible to that used the first time, and care is taken to release the plunger in the same way), will the ball strike the floor at exactly the same point? Simple theory predicts that it will, but small variations in the distance the plunger was pulled back, in the state of the spring, and in the condition of the surfaces of the ball and the inside of the tube—all random, uncontrollable effects—will cause the measured range to vary somewhat on subsequent shots. Indeed, no one would really expect successive shots from a gun to all land in precisely the same spot even though the gun was clamped

in a fixed position and given the same charge each time. Instead, a spread of impact points would be expected, as shown in the plan view of the spring-gun experiment in Fig. 3. The extent of the spread may be reduced by using great care in the experimental technique (wiping off the ball after each shot, handling it with plastic gloves to prevent getting fingerprints on it, taking care in the measurement of how far the plunger is retracted, and releasing the plunger smartly each time), but the spread can never be reduced to zero. The size of the spread is a measure of the precision of the experiment. An estimate of this precision is very desirable in all experimental work, and the following discussion will show how the extent of the spread can be used to express such an estimate quantitatively.

According to statistical theory, the arithmetic mean or average of a number of observations will give the most probable result. This is clear from the results

**Figure 3** *The range of a spring gun. Plan view*

of the range experiment illustrated in Fig. 3. If a single number is to be quoted as the range of the spring gun, it should be the distance from the gun muzzle to the center of the distribution of impact points. In the absence of any peculiar experimental effects, we expect the distribution of points to be densest near the center, to thin out as we go away from the center, and to be symmetrical (to show as many impact points beyond the center as short of it). Hence in this normal case the *average* range $\bar{x}$ is also the *median* (the midpoint of the distribution with as many points with bigger $x$ as with smaller $x$) and the *most probable* value (the point near which there is the greatest density of points). Thus the first step in data analysis is to find the average of the distances from the gun muzzle to all the individual impact points. This is shown on the left in Table I. Note that these measurements are made with a meter stick and could therefore be given to one more significant figure by interpolation between the millimeter divisions on the stick. If this order of precision were wanted, each value of $x$ would be measured several times by different investigators, each of whom would make an interpolation, and an average value obtained for each $x$. Then every entry in the left-hand column of Table I could be quoted to two decimal places (five significant figures). However, this would take a great deal of time and effort, all of which would be wasted because the spread in the data is several centimeters, making the fifth significant figure in each measurement nonsignificant in the final result. In other words, the random effects in the experiment dominate the picture and limit the usable precision of the measuring instrument.

One obvious way of expressing the extent of the spread in a set of experimental data is to note the deviation of each measurement from the average or arithmetic mean just found. In the example of the spring-gun range experiment, these deviations (differ-

ences between each measurement and the average) are tabulated in the middle column of Table I and their average is then computed. Note that, in computing this average, no account is taken of the algebraic signs of the deviations. A deviation represents an error—a difference between a particular measurement and the average of all the measurements, this average being the closest available approximation to the true value of the quantity being measured. Which way the deviation lies makes no difference; it is still an error. The average error is a measure of the scatter of the observed values about their average. The average deviation thus found is therefore often called the average error, and for the purposes of the elementary laboratory, it may be taken as the possible error in the mean value. Consequently the result of the range measurement should be written as 133.9 ± 3.1 cm to show that the true value of the range has a high probability of lying between 130.8 cm (133.9 − 3.1 cm) and 137.0 cm (133.9 + 3.1 cm). Actually, a statistical analysis shows that if a very large number of range measurements were made, 57.5% of them would lie inside this interval. That is, 57.5% of the impact points would be between 130.8 and 137.0 cm from the gun muzzle.

Statistical theory also presents some other useful ways of stating the accuracy of an experimental result. For example, the fact that the average of a set of measurements gets more and more accurate in proportion to the square root of the number of measurements made can be reflected in the stated error by dividing the average error by $\sqrt{N}$, where $N$ is the number of measurements. The result is called the *average deviation of the mean* (A.D.). Thus,

$$\text{A.D.} = \frac{\text{a.d.}}{\sqrt{N}} \qquad (9)$$

where a.d. stands for the average deviation *from the* mean, that is, the average error already discussed. The A.D. is a measure of the deviation of the arithmetical mean from the true value and is in this context generally known as the probable error. The significance of the A.D., from probability theory, is that the chances are 50% that the true value of the quantity being observed will lie within ±A.D. of the mean. Thus, in the example of the spring gun, the mean of the measured ranges is 133.9 cm and the average deviation from the mean (the a.d.) is 3.1 cm, which says that on the average the readings differ from the mean (133.9 cm) by ±3.1 cm. The average deviation *of* the mean (the A.D.) is $3.1/\sqrt{11} = 0.9$ cm, which says that the chances that the true value of the range will lie in the interval 133.9 ± 0.9 cm are 50%, while the chances that it will lie outside this interval are also 50%.

Another (and, from the standpoint of statistical theory, most important) measure of the *dispersion* (scatter of experimental points) is the *standard deviation*. This is defined as the square root of the average

**Table I**  *The Range of a Spring Gun*

| Range, cm | Deviations, cm | Deviations squared, cm² |
|---|---|---|
| 134.2 | +0.3 | 0.09 |
| 139.5 | +5.6 | 31.36 |
| 133.0 | −0.9 | 0.81 |
| 136.6 | +2.7 | 7.29 |
| 129.4 | −4.5 | 20.25 |
| 127.8 | −6.1 | 37.21 |
| 130.6 | −3.3 | 10.89 |
| 136.5 | +2.6 | 6.76 |
| 135.3 | +1.4 | 1.96 |
| 131.9 | −2.0 | 4.00 |
| 138.1 | +4.2 | 17.64 |
| 11)1472.9 | 11)33.6 | 11)138.26 |
| $\bar{x}$ = 133.9 cm | a.d. = 3.1 cm | 12.57 |
| | | $\sigma = \sqrt{12.57} = 3.5$ cm |

of the squares of the individual deviations, or, mathematically, by

$$\sigma = \sqrt{\frac{(x_1 - \bar{x})^2 + (x_2 - \bar{x})^2 + \cdots + (x_N - \bar{x})^2}{N}}$$

(10)

where $\sigma$ is the standard deviation, $x_1, x_2, \ldots, x_N$ are the $N$ individual measurements, and $\bar{x}$ is their average. Note that the signs of the various deviations make no difference in calculating $\sigma$ since each is squared. An example of a standard deviation is given in the right-hand column in Table I. Like other measures of dispersion, the standard deviation gives information about how closely the distribution is grouped about the mean. Statistical analysis shows that for a large number of normally distributed measurements, 68.3% of them will fall within the interval $x \pm \sigma$. In the results of our hypothetical range experiment, this is $133.9 \pm 3.5$ cm, and after a large number of firings we would expect to find that about 68% of the impact points lay between 130.4 and 137.4 cm from the gun muzzle.

When a very large number of measurements of a given quantity are made and variations between the different values obtained are due to truly random effects, a *normal distribution* of these values will be found. The word *distribution* as used here means an expression of the relative frequency with which the different observed values occur. Such an expression often takes the form of a graph in which the number of observed values in a small interval centered on a particular value of $x$ is plotted against $x$. Thus, suppose in the experiment with the spring gun a very large number of range observations were made. We could, for example, count the number of such observations falling in the interval $127.0 \pm 0.5$ cm and plot this number as the ordinate of a point whose abscissa was 127 cm. Another point would be the number of observations falling between 127.5 and 128.5 cm plotted with an abscissa of 128 cm, and this process could be continued until we found somewhere beyond 140 cm that there were no more observed points to plot. A smooth curve could then be drawn through the plotted points. According to our earlier discussion, this curve should show a maximum at the mean value $\bar{x} = 133.9$ cm and should fall off symmetrically on either side. Such a curve, called the *normal curve*, was first discovered by a famous French mathematician, De Moivre, while working on certain problems in games of chance. It was also derived independently by Laplace and Gauss, who made statistical use of it and found that it accurately represents the errors of observation in scientific measurements. The curve is also known as the *normal probability* curve because of its use in the theory of probability, as the *normal curve of error*, and as the *Gaussian curve*. Here error is used to

mean a deviation from the true value. Whenever any measurements are made in which there are random fluctuations, the results predicted by the normal curve are found to be valid. Thus it has been found that this curve describes very well many distributions that arise in the fields of the physical sciences, biology, education, and the social sciences.

The mathematical representation of the normal curve is given by the equation

$$y = \frac{N}{\sigma\sqrt{2\pi}} e^{-(x - \bar{x})^2/2\sigma^2}$$

(11)

where $y$ represents the distribution function. In accordance with the foregoing discussion of how the normal curve is obtained in a physical case, the number $y\Delta x$ is the number of measurements of $x$ (out of a very large total number $N$) that fall within the very small interval $\Delta x$ centered on the value of $x$ for which $y$ was computed. Thus suppose we want to know how many measurements of $x$ will fall in the vicinity of a certain value $x_0$. Putting $x_0$ into Equation 11 for $x$, we calculate the corresponding value $y_0$ for $y$. Then $y_0\Delta x$ is the number of measurements of $x$ expected to be found in the interval $x_0 - \frac{1}{2}\Delta x$ to $x_0 + \frac{1}{2}\Delta x$. Thus $y_0$ is to be interpreted as the number of measurements *per unit interval in $x$* falling in the neighborhood of $x_0$. Note that $y_0\Delta x$ is the area of a tall, thin rectangle $\Delta x$ wide located at $x_0$ on the $x$ axis of the coordinate system in which the normal curve is plotted and extending up to the curve (a height $y_0$). Because this area represents the number of measurements of $x$ falling within $\Delta x$, the area under the complete curve should be equal to $N$, the total number of measurements made. The factor $1/\sigma\sqrt{2\pi}$ in Equation 11, called a normalizing factor, is chosen to bring this about. It follows that the area under the curve between ordinates erected at some pair of values $x = a$ and $x = b$ is the number of measurements falling between $a$ and $b$.

The curve is bell-shaped and symmetrical about the line $x = \bar{x}$. It has a maximum for this value of $x$ (bearing out the idea that the mean value of $x$ is also the most probable value) and falls quite rapidly toward the $x$ axis on both sides. For different normal distributions, the curve has the same general shape, but its steepness, height, and location along the $x$ axis will depend on the values of $N$, $\bar{x}$, and $\sigma$. The characteristic properties of the normal curve can be studied very readily by representing it in a new set of variables by means of a mathematical transformation. The first step is to divide Equation 11 through by $N$. The quantity $(y/N)\Delta x$ is then the fraction of the total number of measurements or the probability of obtaining a measurement in the narrow interval $\Delta x$ centered on the value of $x$ for which $y$ is computed. Put another way, $y/N$ is the probability *per unit interval in $x$* of getting a value of $x$ lying in that interval whenever a measurement of $x$ is made. Clearly, the area under the curve of $y/N$ is unity, corresponding to the fact that the probability of getting *some* value of $x$ is 1 or 100%. Next, we take the origin of our transformed coordinate system to be at the arithmetic mean and use the standard deviation as the unit of measurement along our new horizontal axis. This is done by choosing a new variable $t$ related to $x$ by the equation $t = (x - \bar{x})/\sigma$. When this is substituted into Equation 11 with $N$ divided out and the normalizing factor appropriately modified, we obtain in place of $y(x)$ a new function $\phi(t)$ given by

$$\phi(t) = \frac{1}{\sqrt{2\pi}} e^{-t^2/2} \tag{12}$$

This is the *standard form* of the normal curve, where $\phi(t)$ is the distribution and $t$ is the variable. For this distribution, the average value of $t$ is zero and the standard deviation is equal to one. The advantage of representing the normal curve in the standard form is that the area under the curve between any two values of $t$ may be calculated once and for all. These values are tabulated and can be obtained from tables of probability integrals. Any normal distribution may then be expressed in the standard form and the required calculations of the characteristics of the distribution can easily be made.

Some interesting properties of this curve, which is shown in Fig. 4, deserve mention. Note that theoretically it extends from $-\infty$ to $+\infty$, but practically it is so close to the axis beyond $t = \pm 3$ that the area under the curve beyond these points is negligible. The variable $t$ is measured in units of $\sigma$ along the horizontal axis, and the mean, the median, and the most probable value all coincide at the origin ($t = 0$). The total area under the curve is equal to 1. Hence the area under any portion of the curve represents the relative frequency (expressed as a fraction of unity or as a percent) with which measurements in that interval occur. Numerical values of such areas may be obtained from tables and can be changed into the actual frequencies of occurrence by multiplying by $N$.

The *percentage distribution* of area under the normal curve is given in Fig. 4, where $\sigma$ is the unit of measurement. The significance of the values given in the figure is that if the values of $x$ are normally distributed, the probability that a value chosen at random will fall within the range $\bar{x} \pm \sigma$ is 0.683. The probability that it will lie within the range $\bar{x} \pm 2\sigma$ is 0.9545. The probability that it will lie within the range $x \pm 3\sigma$ is 0.9973. Thus the probability that it will lie outside of this range is only 0.0027 or 0.27%. Hence a deviation

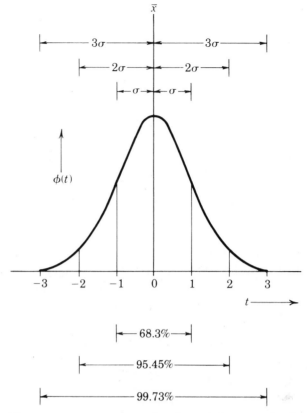

**Figure 4** *The percentage distribution of area under the normal curve*

of $3\sigma$ on both sides of the arithmetic mean includes practically the whole of a normal distribution. The probability that any particular value of $x$ will lie in the range $\bar{x} \pm$ a.d. is also easily found from Equation 12 and Fig. 4. The average deviation from the mean (a.d.) is simply the average $\bar{t}$ of $t$ expressed in units of $\sigma$ and can be calculated from Equation 12. The result is

$$\bar{t} = \sqrt{\frac{2}{\pi}} = 0.798$$

$$\text{a.d.} = \sigma\bar{t} = \sigma\sqrt{\frac{2}{\pi}} = 0.798\sigma \tag{13}$$

The area under the curve of Fig. 4 between $-0.798$ and $+0.798$ is then found to be 0.575 or 57.5% of the total area. This means that, as noted earlier, 57.5% of a large number of measurements will fall within the interval $\bar{x} \pm$ a.d. Similarly, the area under the curve between $-1$ and $+1$ is found to be 0.683 of the total area so that, also as noted earlier, 68.3% of a large number of measurements will fall within the interval $\bar{x} \pm \sigma$.

## PERCENT ERROR

The error in a measured quantity is often conveniently expressed as a percent of the quantity itself. Since the true value is usually not known, one of the errors previously discussed must be used in calculating this percent. For this purpose, the probable error is usually chosen because it is the same as the A.D. and therefore the easiest to calculate from the tabulated data. This percent error, also called the percent deviation of the mean, or the percent A.D., is equal to the A.D. divided by the arithmetic mean of the measured values and multiplied by 100 to give the result in terms of a percent as desired. That is

$$\% \text{ A.D.} = \frac{\text{A.D.}}{M} \times 100\%$$

where $M$ is the arithmetic mean. This is the quantity usually considered in judging the accuracy of a series of measurements. In the example of the range of the spring gun, the percent A.D. is

$$\frac{0.9}{133.9} \times 100\% = 0.67\%$$

If the true or accepted value of a quantity is known, the actual error can be calculated as the difference between the result obtained from the experiment (the mean value $M$ of the measurements) and the true value $M_t$. The relative error is then the ratio of the error to the true value, and the percent error is this ratio times 100%. Thus,

$$\% \text{ error} = \frac{M - M_t}{M_t} \times 100\%$$

For example, suppose the velocity of sound in dry air at 0°C is measured and found to be 333.1 m/s, while the accepted value is 331.4 m/s. The error is 1.7 m/s. The relative error is 1.7/331.4 or 0.005. The percent error is

$$\frac{1.7}{331.4} \times 100\% = 0.5\%$$

There is no definite value for the allowable percent error to be expected in the following experiments. In many cases it is reasonable to expect results within 1%, while in others the error may be 5% or more, depending on the apparatus used. However, all measurements should be made with the greatest care, so as to reduce the error as much as possible.

## CALCULATING WITH ERRORS

Whenever an experimental result is used in a calculation, account must be taken of the fact that an error is associated with it. Suppose we have two results $x_1$ and $x_2$ with respective standard deviations $\sigma_1$ and $\sigma_2$. We quote these results as $x_1 \pm \sigma_1$ and $x_2 \pm \sigma_2$ on the basis that the 68% chance that the true values lie within these ranges is good enough.* If the theory of the experiment requires these two quantities to be added, the sum is $x_1 + x_2 \pm \sigma_1 \pm \sigma_2$. It would, of course, be very nice if the error in one measurement was in the opposite direction from and therefore cancelled the error in the other, but this happy event can hardly be counted on. It is much safer to assume the worst, that is, that the errors are in the same direction so that the sum becomes $x_1 + x_2 \pm (\sigma_1 + \sigma_2)$. We conclude that to be on the safe side we should add the errors in the individual quantities to obtain the error in the sum. Statistics show, however, that this approach is unduly pessimistic and that, in fact, when standard deviations are being used, the standard deviation $\sigma_s$ in the sum is the square root of the sum of the squares of the individual standard deviations. Thus,

$$\sigma_s = \sqrt{\sigma_1^2 + \sigma_2^2}$$

The procedure is identical in the case of a subtraction, but care should be taken to note that the percent error can increase tremendously when two quantities of about the same value are to be subtracted one from the other. In such a case, we get $x_1 - x_2 \pm \sqrt{\sigma_1^2 + \sigma_2^2}$, and if $x_1$ and $x_2$ are nearly equal, their difference may be smaller than the error. This means that the errors associated with $x_1$ and $x_2$ are large enough so that we cannot tell whether the quantities are equal or slightly unequal and hence whether or not a difference actually exists. Consider, for example, two automobiles driving down the highway with one slowly passing the other. The problem is to measure the passing speed, that is, the difference between the speeds of the individual automobiles. The best way to do this (since we are not interested in the individual speeds) is to measure the relative speed directly, but suppose experimental difficulties made this impossible so that the only data obtainable were the speedometer readings in the two cars. We find that one reads 61 miles per hour and the other 62 miles per hour. Can we conclude that one car is passing the other at the rate of 1 mph? Not really, for automobile speedometers are good to only two significant figures with the last one in doubt. There is therefore an error of about 1 mph in each speedometer, so that their readings should be reported as $61 \pm 1$ and $62 \pm 1$ mph respectively. This being the case, the difference must be given as $1 \pm \sqrt{1^2 + 1^2} = 1 \pm 1.4$ mph. From this result alone we cannot tell whether the cars are proceeding side by side or even, if they are not, which is passing the other. Certainly there seems to be a bias in favor of one car's doing the passing, but a reasonably certain conclusion simply cannot be drawn on the basis of the

---

* For the experiments covered in this manual, the a.d., which is easier to find than $\sigma$, will usually be good enough.

available data. Thus, if we really want to know which car is passing the other and how fast, a method must be developed for measuring their relative velocity directly.* Although this does not seem to be a difficult task in our example of the cars, the progress of research in physics has often been marked by breakthroughs due to someone's inventing a method for direct measurement of a quantity that had been formerly obtainable only as the difference between two other quantities.

If the two quantities $x_1 \pm \sigma_1$ and $x_2 \pm \sigma_2$ are to be multiplied, we have

$$(x_1 \pm \sigma_1)(x_2 \pm \sigma_2) = x_1 x_2 \pm \sigma_1 x_2 \pm \sigma_2 x_1 + \sigma_1 \sigma_2$$

On the assumption that the errors are reasonably small, $\sigma_1 \sigma_2$ can be dropped on the basis that the product of two small quantities is negligibly small. Assuming the worst case of errors that are in the same direction and therefore add, we get

$$(x_1 \pm \sigma_1)(x_2 \pm \sigma_2) \approx x_1 x_2 \pm (\sigma_1 x_2 + \sigma_2 x_1)$$

This result is inconvenient to handle in computation, but note that the percent error (the percent standard error in this case) is

$$\% \text{ error} = \frac{\sigma_1 x_2 + \sigma_2 x_1}{x_1 x_2} \times 100\%$$

$$= \left(\frac{\sigma_1}{x_1} + \frac{\sigma_2}{x_2}\right) \times 100\% = \frac{\sigma_1}{x_1} \times 100\% + \frac{\sigma_2}{x_2} \times 100\%$$

In other words, the percent error in the product is the sum of the percent errors in each factor. It is easy to show that the same holds true for division.

Raising an experimental quantity to an integer power is treated just like multiplication, for the integer power merely tells how many times the quantity is to be multiplied by itself. Thus in calculating the area $x^2$ of a square of measured side $x$, the exponent 2 is simply an instruction to multiply $x$ by $x$ and has no error associated with it. If the error in $x$ is $\sigma$, we have

$$(x \pm \sigma)^2 = x^2 \pm 2\sigma x + \sigma^2 \approx x^2 \pm 2\sigma x$$

and the percent error is

$$\pm \frac{2\sigma x}{x^2} \times 100\% = 2\frac{\sigma}{x} \times 100\%$$

or twice the percent error in the measurement of $x$. The situation may also be regarded as the product of $x \pm \sigma$ with itself, with error $\pm \sigma x \pm \sigma x$, but note that in this case the two errors $\sigma x$ *must* be considered to add since they are not separate but identical and hence neces-

sarily in the same direction. In general, it can be shown that if a measured quantity is raised to any power $n$, integer or fractional, the percent error in the result will be $n$ times the percent error in the quantity. Notice that this is not $n$ times the error but $n$ times the *percent* error. Consequently, great care must be taken in the measurement of a quantity that is to be raised to a power greater than one, for the percent error in the result will be greater than that in the measurement by a factor of that power. A quantity raised to a high power may be thought of as "dominating" the result in that the error in this quantity will be much more important than the error in other quantities that may be multiplied or added in without being raised to such a power.

Conversely, the percent error in quantities raised to a power less than one is less important because only the fraction of that error represented by the less-than-unity value of $n$ appears in the result. For example, in the experiment on the simple pendulum, in which the theory says that the period is proportional to the square root of the pendulum's length, only half the percent error in the length measurement will show up in the calculated period. But in an investigation of the Stefan-Boltzmann law, which states that the power radiated by a hot surface is proportional to the fourth power of the surface temperature, the temperature measurements will be extremely critical.

Finally, the case of the exponential (a mathematical number such as the base $e$ of natural logarithms, with which no error is associated, being raised to a power given by a measured quantity) must be considered. Suppose that the theory under investigation calls for $e$ to be raised to the power $x$ where $x$ is a measurement containing standard error $\sigma$. We have

$$e^{x \pm \sigma} = e^x e^{\pm \sigma} = e^x \left(1 \pm \sigma + \frac{\sigma^2}{2} \pm \cdots\right)$$

where we have used the property $e^{a+b} = e^a e^b$ of exponentials and have expanded $e^{\pm \sigma}$ in its power series. Since $\sigma$ is a reasonably small quantity, terms in $\sigma^2$, $\sigma^3$, etc. can be neglected to give

$$e^{x \pm \sigma} \approx e^x(1 \pm \sigma) = e^x \pm \sigma e^x$$

The percent error is then $\pm \sigma \times 100\%$, which means that here *the error in the measured quantity is numerically equal to the relative error in the result and hence to the percent error after multiplication by 100%.* Thus, in any experiment in which $e$ is to be raised to a measured quantity, particular care must be exercised in making the measurement in order to keep the error as small as possible. There is, in fact, a famous case (that of the thermionic emission of electrons from a heated metal surface) in which the theory, which predicts that the current will be proportional to some power of the temperature multiplied by an exponential

---

* We could also substitute speedometers of much higher precision, but this is an inelegant approach, as most of the precision will be wasted in getting the desired result.

having the temperature in the exponent, has never been really satisfactorily checked. There is no question about the exponential dependence, but it so dominates the result that even very careful experimental data has not been able to verify by what, if any, power of the temperature the exponential should be multiplied. Such are the challenges, the frustrations, and the fascination of experimental physics. It is hoped that the student will in some measure taste all three while performing the work described in this manual.

# Measurement of Length, Mass, Volume, and Density

# 1

Physics is a quantitive experimental science and as such is largely a science of measurement. Many measuring instruments of great accuracy have been developed to meet the requirements of the physics laboratory. The measurement of length is of fundamental importance in scientific work; hence it is fitting to begin experimental work with this type of measurement.

Mass is another fundamental quantity which must often be measured, and mass determinations will also be done in this experiment. Finally, if the dimensions of a regular geometrical object have been measured, its volume may be calculated; and if its mass is also known, its density may be found. Density is one of the important characteristics of materials. The engineer uses tables of densities to determine the weights of bridges and other structures that cannot be weighed. From the dimensions the volume can be computed and then from tables of densities the total weight can be calculated.

In this experiment the dimensions of various objects will be measured by means of a meter ruler, vernier caliper, and micrometer caliper, and the probable error in these measurements will be determined. There will also be an opportunity to compare the English and metric units of length. The mass of each object will be found with a triple-beam balance, and the density of the material of which each object is made will be obtained.

## THEORY

It is important to be able to use the proper instruments when measuring various physical quantities. The simplest way to measure length is to use an ordinary meter stick or an English ruler. For ordinary measurements, a length is determined by a rough comparison with the scale—the result is given at a glance and in round numbers. For a more precise determination, the scale must be very accurately made and must be read to a fraction of its smallest scale divison.

As discussed in the Introduction, when an instrument is used to the limit of its precision, certain errors occur that cannot be eliminated; these are called accidental errors. Thus, when a series of measurements of a physical quantity is made, the individual results usually differ among themselves because of the accidental errors involved. The best value of the quantity measured, that is, the most probable value, is the arithmetic mean (the average of the values obtained).

The precision of measurement can usually be increased by using more complex and more accurate equipment and by taking precautions to eliminate errors as much as possible. However, you should always try to make measurements with the greatest accuracy attainable with the given apparatus.

**The Vernier Caliper** When a measurement is made with an ordinary meter stick, it is necessary to estimate the tenths of a millimeter, the millimeter being the smallest division on the scale. A vernier helps in reading accurately the fractional part of a scale division. It is a small auxiliary scale that slides along the main scale. The graduations of the vernier are different than those of the main scale, but they are simply related.

The vernier caliper, which is illustrated in Fig. 1.1, consists of a fixed part and a movable jaw. The fixed part includes a stem, on which is engraved the main scale, and a fixed jaw attached to the stem. The movable jaw is free to slide on the fixed stem and has a vernier scale engraved on it. The fixed scale is divided into centimeters and millimeters. The vernier is divided so that ten divisions on it cover the same interval as nine divisions on the main scale. Hence, the length of each vernier division is $\frac{9}{10}$ the length of a main-scale division. When the jaws are closed, the first line at the left end of the vernier, called the zero line or the index, coincides with the zero line on the main scale. However, the first vernier division is 0.1 mm away from the first main-scale division, the second vernier division is 0.2 mm away from the second main-scale division, and so on. If the jaws are slightly opened, it is easy to tell what fraction of a main-scale division the vernier index has moved by noting which vernier division coincides with a main-scale division.

A measurement is made with the vernier caliper by closing the jaws on the object to be measured and then reading the position where the zero line of the

**Figure 1.1** *The vernier caliper*

vernier falls on the main scale. This setting of the zero line of the vernier on the main scale is an incomplete measurement, however, because no attempt was made to estimate to a fraction of a main-scale division. The fractional part of a main-scale division is obtained by noting which line on the vernier coincides with a line on the main scale. Several possible vernier readings are illustrated in Fig. 1.2.

Frequently a vernier does not read zero when the jaws are completely closed. In such cases a zero correction must be applied to every reading; this correction may be either positive or negative. If the errors in the direction of increasing main-scale readings are called positive, then the zero correction is always made by subtracting the zero reading from the final reading.

Vernier scales are attached to many different kinds of instruments, and not all of them divide a main-scale division into ten equal parts. In any given case, the number of vernier divisions equals the number of equal parts into which a main-scale division is to be divided. The vernier works because the total number of vernier divisions spans a lesser number (usually one less) of main-scale divisions.

**The Micrometer Caliper** The micrometer caliper (see Fig. 1.3) is an instrument used for the accurate measurement of short lengths. Essentially, it consists of a carefully machined screw mounted in a strong frame. The object to be measured is placed between the end of the screw and the projecting end of the frame, called the anvil. The screw is then advanced until the object is gripped gently between the two jaws of the instrument. Most micrometers are provided with a ratchet arranged to slip on the screw as soon as a light and constant force is exerted on the object. By using the ratchet, it is possible to tighten up the screw by the same amount each time and also to avoid using too great a force. If this arrangement is absent, great care should be taken not to force the screw, for the instrument may be easily damaged.

The micrometer caliper used in this experiment consists of a screw with a pitch of 0.5 mm, a longitudinal scale engraved along a barrel containing the screw, and a circular scale engraved around a thimble which rotates with the screw and moves along the scale on the barrel. The longitudinal scale is divided into millimeters. The circular scale has 50 divisions. Since the pitch of the screw is 0.5 mm, which is the

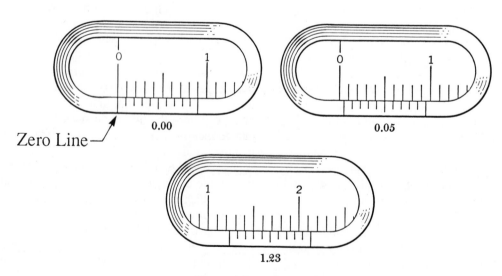

Zero Line

0.00

0.05

1.23

**Figure 1.2** *Vernier readings*

**Figure 1.3** *The micrometer caliper*

distance advanced by the screw in turning through one revolution, it is clear that rotating the thimble through one scale division will cause the screw to move a distance of $\frac{1}{50}$ of 0.5 mm, or 0.01 mm. Hence, readings may be taken directly to one hundredth of a millimeter, and by estimating tenths of a thimble-scale division, they may be taken to one thousandth of a millimeter.

The micrometer is read by noting the position of the edge of the thimble on the longitudinal scale and the position of the axial line of the barrel on the circular scale, and adding the two readings. The reading of the main scale gives the measurement to the nearest whole main-scale division; the fractional part of a main-scale division is read on the circular scale. Since two revolutions of the screw are required to make it advance a distance of 1 mm, it is necessary always to be careful to note whether the reading on the circular scale refers to the first half or the second half of a millimeter. (In Fig. 1.3 the reading is 7.75 mm.)

The micrometer should be checked for a zero error, for it may not read zero when the jaws are completely closed. In such cases a zero correction has to

be applied to every reading and may be either positive or negative. The value of the zero reading is obtained by rotating the screw until it comes in contact with the anvil and then noting the reading on the circular scale.

**The Triple-Beam Balance** The masses of various objects will be determined in this experiment with a triple-beam balance, so called because it has three beams, each provided with a sliding weight, as shown in Fig. 1.4. These weights have different sizes, the largest sliding along a scale with a notch every 100 g, the next on a scale with a notch every 10 g and the smallest on a scale with 1-g main divisions and 0.1-g subdivisions. The unknown mass is placed on the tray, the 100-g and 10-g weights set in the appropriate notches (determined by trial), and the smallest slider adjusted for balance. The triple-beam balance can thus be read to 0.1 g, or, if one is willing to interpolate between the subdivisions, to about .05 g. However, great care is required to make the second decimal place really significant. In Fig. 1.4, the balance is set to 371.4 g.

In this and all further work in physics, be very

**Figure 1.4** *The triple-beam balance*

careful to distinguish between mass and weight, which are not the same despite the very common confusion between the two. Mass is the property of material bodies that makes them hard to accelerate. Even if a body is taken out into deep space far from the earth, the sun, and the other planets, it will not become massless—it will still have inertia; that is, a force will still be required to change its velocity. Weight, on the other hand, is simply the force of gravity with which the earth attracts a body. Since this force depends on the distance between the earth and the body, the body will indeed become weightless if taken far away (meaning many thousands of miles). The confusion arises because apparently the property of material bodies called mass is responsible for their mutual gravitational attraction, the two being in fact proportional. Thus the weight of a body (that is, the force with which it is attracted to the earth) is proportional to its mass; if its weight is measured, its mass is also known. Scales and balances measure weight.

The triple-beam balance compares the weight of the unknown object with that of the fixed masses on the beams. However, because mass and weight are proportional, the scales on the beams can be calibrated in grams (a unit of mass) rather than in newtons (a unit of force), even though the weight, that is, the force of gravity, is what is actually being measured. There is nothing wrong with this—a measuring instrument's scale can always be graduated in units proportional to those it is actually measuring—but a physicist is nevertheless careful not to forget the distinction between mass and weight. For a further discussion of this matter, see the section on units in the Introduction.

In the present experiment the masses of several objects will be determined and their volumes calculated from the measurements of their dimensions. The density of the material of which each is made, defined as the material's mass per unit volume, is then calculated from

$$D = \frac{M}{V}$$

where $M$ is the mass in grams, $V$ is the volume in cubic centimeters, and $D$ is the density, whose units are thus seen to be grams per cubic centimeter.

For the metal cylinders and the wire (which may be considered a long, thin cylinder) the volume is given by

$$V = \pi r^2 L = \frac{\pi d^2 L}{4}$$

where $r$ is the cylinder's radius, $d$ is the diameter ($2r$), and $L$ is the cylinder's length. For the irregular body the volume will be determined by measuring the volume of liquid that it displaces.

## APPARATUS

1. Triple-beam balance
2. English ruler
3. Metric ruler (meter stick)
4. Vernier caliper
5. Micrometer caliper
6. Three cylinders of different metals

    a. aluminum
    b. brass
    c. iron
7. Length of copper wire
8. An irregular solid
9. A 250-mL graduated cylinder

## PROCEDURE

**1.** Measure the length of the piece of copper wire with the metric ruler. Read the position of both ends of the wire, estimating to the nearest tenth of a millimeter. Record both readings in centimeters and read to 0.01 cm. Make four independent measurements, that is, use a different part of the ruler for each measurement, and record all the readings.

**2.** Measure the length of the wire with the English ruler. Read the position of both ends of the wire, estimating to the nearest tenth of the smallest scale division. Record both readings, expressed in inches, to the nearest 0.01 in. Make four independent measurements, that is, use a different part of the ruler for each measurement, and record all the readings.

**3.** Determine the zero reading of the vernier caliper. This is the reading when the jaws of the instrument are in contact with each other. Be sure that the jaws are clean so that no grit or other foreign matter gets between them and prevents true contact. Record the zero reading to 0.01 cm. Make four independent determinations of this reading, that is, open and close the jaws before each setting, and record your results.

**4.** Measure the length and diameter of each cylinder with the vernier caliper. These measurements are made by closing the jaws of the caliper on the length or diameter of the cylinder being measured and reading the position where the zero line of the vernier falls on the main scale. The fractional part of a main-scale division is obtained by noting which line on the vernier coincides with a line of the main scale. Record the reading in centimeters and read to 0.01

cm. Make four independent measurements, that is, open and close the jaws before each setting. This is most conveniently done by measuring the diameter of each cylinder in turn once, then measuring the length of each in turn once, and repeating the complete sequence four times. Record all readings.

**5.** Determine the zero reading of the micrometer caliper, that is, the reading when the surfaces of the anvil and the screw end are in contact. Be sure these surfaces are very clean, since even a small speck of dust can give a false reading. The screw end may be brought almost into contact with the anvil by turning the thimble directly, but actual contact must always be made by turning the ratchet slowly until it clicks several times. If the ratchet is absent, great care must be taken not to force the screw; merely let the screw end approach the anvil very slowly and stop turning the screw as soon as the two surfaces touch. Record the value of this reading in centimeters and read to 0.0001 cm, estimating to one tenth of the smallest scale division. Make four independent determinations of the zero reading, that is, open and close the instrument before each setting, and record the readings.

**6.** Measure the diameter of the copper wire with the micrometer caliper. The measurement is made by placing the wire between the screw end and the anvil and advancing the screw until the wire is gripped be-

tween the anvil and screw-end surfaces. Again the ratchet should be used. If the instrument has no ratchet, care should be taken not to force the screw. Record the reading in centimeters and read to 0.0001 cm, estimating to one tenth of the smallest scale division. Make six independent measurements of the diameter of the wire, that is, open and close the caliper before each setting, and make the measurements at different points along the wire's length. Also, spin the wire between measurements so that different diameters are measured each time. Record the six readings so obtained.

**7.** Determine the mass of each cylinder, the irregular solid, and the copper wire using the triple-beam balance.

**8.** Determine the volume of the irregular solid by the displacement method. Partly fill the graduated cylinder with water and read the water level. Filling the cylinder to an exact graduation division will prove convenient in the subsequent calculations. Record the water level. Suspend the solid by means of a thread and lower it into the cylinder until it is completely submerged. Read and record the water level now observed. When the first reading is subtracted from the second, the volume of the irregular solid is obtained in cubic centimeters.

# DATA

### Measurement of Wire Length Using a Metric Ruler

| Ruler readings | | Lengths | Deviations |
|---|---|---|---|
| **Left end** | **Right end** | | |
| | | | |
| | | | |
| | | | |
| | | | |
| Average values | | | |

### Measurement of Wire Length Using an English Ruler

| Ruler readings | | Lengths | Deviations |
|---|---|---|---|
| **Left end** | **Right end** | | |
| | | | |
| | | | |
| | | | |
| | | | |
| Average values | | | |

Number of centimeters in 1 inch

   Calculated value  _____

   Accepted value  _____

Difference between calculated and
   accepted values  _____

Percent error  _____

### Length and Diameter of Metal Cylinders Using a Vernier Caliper

|  | Vernier caliper readings | | | | Average |
|---|---|---|---|---|---|
|  | 1 | 2 | 3 | 4 |  |
| Zero reading |  |  |  |  |  |
| Length, aluminum cylinder |  |  |  |  |  |
| Length, brass cylinder |  |  |  |  |  |
| Length, iron cylinder |  |  |  |  |  |
| Diameter, aluminum cylinder |  |  |  |  |  |
| Diameter, brass cylinder |  |  |  |  |  |
| Diameter, iron cylinder |  |  |  |  |  |

### Diameter of Copper Wire Using a Micrometer Caliper

|  | Micrometer readings | | | | | | Average |
|---|---|---|---|---|---|---|---|
|  | 1 | 2 | 3 | 4 | 5 | 6 |  |
| Zero reading |  |  |  |  | ✕ | ✕ |  |
| Reading with wire |  |  |  |  |  |  |  |
| Diameter of wire |  |  |  |  |  |  |  |

### Determination of Density

| Object used | Mass, g | Length, cm | Diameter, cm | Volume, cc (cm³) | Density computed | Density (from Table IV) | Percent error |
|---|---|---|---|---|---|---|---|
| Aluminum cylinder |  |  |  |  |  |  |  |
| Brass cylinder |  |  |  |  |  |  |  |
| Iron cylinder |  |  |  |  |  |  |  |
| Copper wire |  |  |  |  |  |  |  |
| Irregular solid |  | ✕ | ✕ |  |  |  |  |

## CALCULATIONS _____

**1.** From the data of Procedure 1, compute the length of the copper wire for each set of measurements. Calculate the average value of the length by finding the arithmetic mean of the lengths obtained, to the proper number of significant figures.

**2.** From the data of Procedure 2, compute the length of the copper wire for each set of measurements. Calculate the average value of the length by finding the arithmetic mean of the lengths obtained, to the proper number of significant figures.

**3.** Compute the number of centimeters in one inch from your measurements in Procedures 1 and 2. Use the calculated average values of the wire's length in making the computations. Compare your result with the accepted value by finding the difference between them and the percent error.

**4.** From the data of Procedure 3, compute the average value of the zero reading of the vernier caliper. From the data of Procedure 4, compute the average value of the setting of the vernier caliper for the length and diameter of each cylinder. Making sure to include its algebraic sign properly, use the average zero reading to obtain the measured length and diameter of each cylinder, and insert these values in the table of data for the density determinations.

**5.** From the data of Procedure 5, compute the average value of the zero reading of the micrometer caliper. From the data of Procedure 6, compute the average value of the setting of the micrometer caliper for the diameter of the copper wire. Calculate the diameter of the copper wire by inserting the zero correction with the proper algebraic sign and enter your result in the table of data for the density determination.

**6.** From the data of Procedures 1 and 2, calculate the deviations of the lengths obtained from the average value of the length. Compute the average deviation from the mean (the a.d.), the average deviation of the mean (the A.D.), and the percent deviation of the mean (the percent A.D.) for both sets of measurements. (See the Theory of Errors in the Introduction.) The a.d. should be entered in your tables as the average of your deviations in Procedures 1 and 2, respectively.

**7.** Calculate the error to be associated with your result for the number of centimeters in 1 in. found in Procedure 3 and note whether the accepted value falls within the range of this error.

**8.** Compute the volume of each object measured.

**9.** Calculate the density of each substance used.

**10.** Compare your results with those given in Table IV at the end of the book and find the percent error. *Note:* If the instructor has designated the substance of which the irregular solid is made as an unknown, try to identify it by comparing your computed value with the list in Table IV.

## QUESTIONS _____

**1.** Why are several observations taken for each measurement?

**2.** (a) What is the smallest part of a centimeter than can be read or estimated with a meter stick? (b) What is the smallest part of a centimeter that can be read or estimated with your vernier caliper? (c) Which readings are more reliable? Why?

**3.** What is the smallest part of a centimeter that can be read or estimated with your micrometer caliper? This represents the sensitivity of the micrometer, for the sensitivity of a measuring instrument is the value of the smallest quantity that can be read or estimated with it.

**4.** State the number of significant figures in the data of Procedures 1, 4, and 6.

**5.** (a) What is the significance of the average deviation from the mean (the a.d.)? (b) What is the significance of the average deviation of the mean (the A.D.)?

**6.** In measuring the length and diameter of a cylinder, which dimension should be measured more carefully? Why? *Hint:* Remember that the length and diameter are being measured in order to determine the volume.

**7.** Why was the micrometer used instead of the vernier caliper in determining the diameter of the copper wire?

**8.** (a) What is the volume in cubic millimeters of the largest cylinder you measured? (b) What is its volume in liters? (c) What is its mass in kilograms?

**9.** A thin circular sheet of copper has a diameter of 30.0 cm and a thickness of 1 mm. Find the weight of the sheet in grams.

**10.** A certain graduated cylinder has an inside diameter of 4.00 cm. (a) The graduations on its side are labeled "cc" (cubic centimeters). How far apart are these graduations? (b) The cylinder is partially filled with water, and a solid sphere of radius 1.20 cm is then totally submerged therein. Through what distance does the water level rise when the sphere is so immersed?

# Addition of Vectors. Equilibrium of a Particle 2

When a system of forces, all of which pass through the same point, acts on a body, it may be replaced by a single force called the resultant. The purpose of this experiment is to show that the magnitude and direction of the resultant of several forces acting on a particle may be determined by drawing the proper vector diagram and that the particle is in equilibrium when the resultant force is zero.

## THEORY

A scalar is a physical quantity that possesses magnitude only; examples of scalar quantities are temperature, mass, and density. A vector is a quantity that possesses both magnitude and direction; examples of vector quantities are velocity, acceleration, and force. A vector may be represented by drawing a straight line in the direction of the vector, the length of the line being made proportional to the vector's magnitude. The sense of the vector (for example, whether it is pointing toward the right or toward the left) is indicated by an arrowhead placed at the end of the line.

Vectors may be added either graphically or analytically. The sum or resultant of two or more vectors is the single vector that produces the same effect. For example, if two or more forces act at a certain point, their resultant is that force which, if applied at that point, has the same effect as the two separate forces acting together. Thus, in Fig 2.1 the resultant of two forces **A** and **B** is shown. The equilibrant, defined as the force equal and opposite to the resultant, is also shown. Note that if the equilibrant is regarded as a third force (in addition to **A** and **B**), the sum of **A**, **B**, and the equilibrant is zero.

The operation of adding vectors graphically consists of constructing a figure in which a straight line is drawn from some point of origin to represent the first vector, the length of the line being proportional to the magnitude of the vector and the direction of the line being the same as the direction of the vector. From the arrowhead (nose) end of this line and at the proper angle with respect to the first vector, another line is drawn to represent the second vector, and so on with the remaining vectors. The resultant is the vector drawn from the origin or tail of the first vector to the nose of the last vector (see Fig. 2.2). If a closed polygon is formed, that is, if the nose of the last vector falls upon the tail of the first, then the resultant is zero. If the vectors represent forces, they are in equilibrium.

If no more than two vectors are to be added, the parallelogram method may be used. This consists of constructing both vectors with their tails at the origin and completing the parallelogram, as shown in Fig. 2.3. The diagonal of the parallelogram is then the resultant. Note, however, that this method is really identical to the previously described method of graphical addition by constructing the vectors nose-to-tail. Thus, the side of the parallelogram opposite **A** is parallel to and has the same length as **A**. Hence it is really a reconstruction of **A** with its tail on the nose of **B**. The resultant **R** is then drawn from the tail of **B**

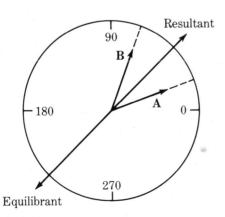

**Figure 2.1** *The resultant and equilibrant for two forces*

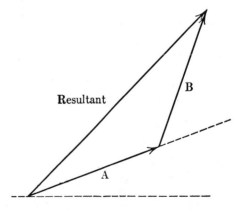

**Figure 2.2** *Graphical addition of vectors*

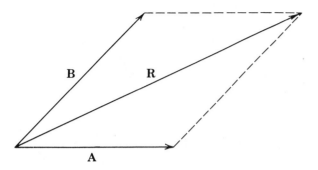

**Figure 2.3** *Parallelogram method of adding vectors*

to the nose of **A** as before. Similarly, the side of the parallelogram opposite **B** is equal to **B** and is therefore a reconstruction of **B** with its tail on the nose of **A**. Here the resultant may be considered as being drawn from the tail of **A** to the nose of **B**. The parallelogram method thus demonstrates the commutative property of vector addition: in vector addition, as in algebraic addition, the order in which the quantities are added doesn't matter. This property may be stated mathematically as

$$\mathbf{A} + \mathbf{B} = \mathbf{B} + \mathbf{A}$$

and may seem obvious, but bear in mind that mathematical processes do exist for which it does not hold.

Vectors may also be added analytically, and this is in fact the preferred method since it does not require the making of precise drawings and does not involve the inaccuracies inherent in the measurements made on them. The method is illustrated in Fig. 2.4, in which the vectors of Fig. 2.2 have been placed on the usual cartesian (x–y) coordinate system. In Fig. 2.4, the vectors **A** and **B** have been resolved into their x and y components. For example, the x component $A_x$ of the vector **A** is the projection of **A** on the x axis and is equal to $A \cos \theta_A$, where A is the magnitude of **A**. Similarly, the y component of **A** is the projection of **A** on the y axis and is given by $A \sin \theta_A$.

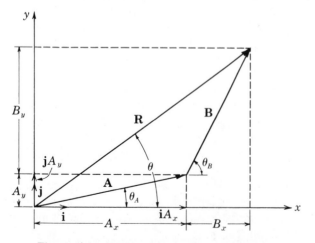

**Figure 2.4** *Addition of vectors by components*

It is easy to show how a vector is represented by its components. Fig. 2.4 introduces the two *unit* vectors **i** and **j**, which are defined as vectors of unit magnitude and with directions along the x and y axes respectively. Then $\mathbf{i}A_x$ is a vector along the x axis having magnitude $A_x$, and $\mathbf{j}A_y$ is a vector along the y axis having magnitude $A_y$. Fig. 2.4 shows that, by the parallelogram method of vector addition,

$$\mathbf{A} = \mathbf{i}A_x + \mathbf{j}A_y \qquad (2.1)$$

By the same arguments, vector **B** is given by

$$\mathbf{B} = \mathbf{i}B_x + \mathbf{j}B_y \qquad (2.2)$$

where $B_x$ and $B_y$, the x and y components of **B**, have the values $B \cos \theta_B$ and $B \sin \theta_B$, respectively.

Finally, Fig. 2.4 shows that the x component $R_x$ of the resultant **R** is just the sum of the two x components $A_x$ and $B_x$, while $R_y$ is the sum of $A_y$ and $B_y$. Thus, the analytical method consists of finding the x and y components of all the vectors to be added, adding the x components algebraically (paying due attention to sign) to get the resultant's x component, and adding the y components to get the resultant's y component. In mathematical terms,

$$R_x = A_x + B_x \qquad (2.3)$$

and

$$R_y = A_y + B_y \qquad (2.4)$$

With the x and y components of **R** known, the magnitude and direction of this vector are easily found if desired. Inspection of Fig. 2.4 shows that

$$R^2 = R_x^2 + R_y^2 \qquad (2.5)$$

and

$$\tan \theta = \frac{R_y}{R_x} \qquad (2.6)$$

However, since a vector is completely specified by its components, it is often left in this form, the magnitude and direction being calculated only if they are needed.

The conclusion that the components of a resultant are each the algebraic sum of the respective components of the vectors to be added may be arrived at analytically as well as by inspection of Fig. 2.4. Thus, in the addition of vectors **A** and **B**, these vectors may be given by Equations 2.1 and 2.2 so that

$$\mathbf{R} = \mathbf{A} + \mathbf{B} = \mathbf{i}A_x + \mathbf{j}A_y + \mathbf{i}B_x + \mathbf{j}B_y \quad (2.7)$$

Writing **R** as $\mathbf{i}R_x + \mathbf{j}R_y$ and grouping terms appropriately in Equation 2.7 leads to

$$\mathbf{i}R_x + \mathbf{j}R_y = \mathbf{i}(A_x + B_x) + \mathbf{j}(A_y + B_y) \quad (2.8)$$

This can be true only if $R_x = A_x + B_x$ and, separately, $R_y = A_y + B_y$ in agreement with Equations 2.3 and 2.4.

The apparatus used in this experiment (see Fig. 2.5) consists of a horizontal force table graduated in degrees and provided with pulleys that may be set at any desired angles. A string passing over each pulley supports a weight holder upon which weights may be placed. A pin holds a small ring to which the strings are attached and which acts as the particle. When a test for equilibrium is to be made, the pin is removed; if the forces are in equilibrium, the particle will not be displaced.

**Figure 2.5** *Force table*

## APPARATUS

1. Force table
2. Four pulleys
3. Four weight hangers
4. Set of slotted weights, including:
   four 100-g weights
   four 50-g weights

    two 20-g weights
    two 10-g weights
5. Set of small weights (5-, 2-, 2-, 1-g)
6. Protractor
7. Metric ruler

## PROCEDURE

**1.** Mount a pulley on the 20° mark on the force table and suspend a mass of 100 g over it. Mount a second pulley on the 120° mark and suspend a mass of 200 g over it. Draw a vector diagram to scale, using a scale of 0.2 N/cm, and determine graphically the direction and magnitude of the resultant by using the parallelogram method. Note that your weights are labeled according to their masses in grams but that you are dealing with forces measured in newtons.

**2.** Check the result of Procedure 1 by setting up the equilibrant on the force table. This will be a force equal in magnitude to the resultant, but pulling in the opposite direction. Set up a third pulley 180° from the calculated direction of the resultant and suspend masses over it giving a weight equal to the magnitude of the resultant. Cautiously remove the center pin to see if the ring remains in equilibrium. Before removing the pin, make sure that all the strings are pointing exactly at its center; otherwise the angles will not be correct.

**3.** Mount the first two pulleys as in Procedure 1, with the same weights as before. Mount a third pulley on the 220° mark and suspend a mass of 150 g over it. Draw a vector diagram to scale and determine graphically the direction and magnitude of the resultant. This may be done by adding the third vector to the sum of the first two, which was obtained in Procedure 1. Now set up the equilibrant on the force table and test it as in Procedure 2.

**4.** Clamp a pulley on the 30° mark on the force table and suspend a mass of 200 g over it. By means of a vector diagram drawn to scale, find the magnitude of the components along the 0° and the 90° directions. Set up these forces on the force table as they have been determined. These two forces are equivalent to the original force. Now replace the initial force by an equal force pulling in a direction 180° away from the original direction. Test the system for equilibrium.

## CALCULATIONS _____

**1.** Calculate the resultant in Procedure 1 by solving for the third side of the force triangle algebraically. Find the magnitude of the resultant vector by using the law of cosines and the direction from the law of sines. Compare your result with that obtained in Procedure 1. How do these predicted results compare with your experimental finding in Procedure 2?

**2.** Calculate the resultant in Procedure 1 by using the analytical method of adding vectors. Compare this result with those already obtained in Procedures 1 and 2 and in Calculation 1.

**3.** Calculate the resultant in Procedure 3 by the analytical method. Note that you have already found the components of two of the force vectors in Calculation 2 and need only find and add the components of the third. Compare your result with that obtained in Procedure 3.

## QUESTIONS _____

**1.** State how this experiment has demonstrated the vector addition of forces.

**2.** In Procedure 3, could all four pulleys be placed in the same quadrant or in two adjacent quadrants and still be in equilibrium? Explain.

**3.** State the condition for the equilibrium of a particle.

**4.** The forces used in this experiment are the weights of known masses, that is, the forces exerted on these masses by gravity. Bearing this in mind, explain the function of the pulleys.

**5.** The analytical method of adding vectors expressed in terms of their components may be applied to vectors in three dimensions, for which graphical work is inconvenient. Find the magnitude of the resultant of the vectors $\mathbf{A} = \mathbf{i}12 - \mathbf{j}37 + \mathbf{k}58$ and $\mathbf{B} = \mathbf{i}5 + \mathbf{j}30 - \mathbf{k}42$, where $\mathbf{i}$, $\mathbf{j}$, and $\mathbf{k}$ are unit vectors along the $x$, $y$, and $z$ axes, respectively.

**6.** The $x$ and $y$ components of a certain force are measured and found to be $68 \pm 3$ and $42 \pm 2$ N, respectively. Calculate the direction and magnitude of this force, expressing your result to the proper number of significant figures and showing the error in both magnitude and angle. Note that the error in the latter is obtained from the error in its tangent. Devise a procedure for doing this and show your work.

# Equilibrium of a Rigid Body 3

When a rigid body is acted upon by a system of forces that do not all pass through the same point, a change may be produced in the angular (rotational) velocity of the body as well as in its linear (translational) velocity. Under certain conditions the body will be in equilibrium, that is, there will be no tendency for either its translational or rotational motion to change. If it is at rest, it will remain at rest. This experiment presents a study of the conditions for the equilibrium of a rigid body under the action of several forces.

## THEORY

In the preceding experiment, the equilibrium of a particle was investigated, and it was found that the particle was in equilibrium (had no tendency to change its translational velocity) when the resultant force acting on it was zero. In that case, the particle was a body so small that it could be considered a point and therefore the question of where on the body the various forces were applied did not arise. Moreover, rotation was not considered, since rotation of a very small body is of little consequence.

More generally, bodies of appreciable extent must be considered. Then the points on a body at which forces are applied become important and rotation of the body as a unit is a new motion that must be considered in addition to its translational motion, that is, its displacement as a whole from one location to another. The bodies in question are assumed to be rigid, which means that their parts do not change their distances from one another. Internal forces (between atoms and molecules) hold a rigid body together. These forces maintain all parts in fixed relative positions and are assumed strong enough to do this no matter what external forces may be applied. The body then remains rigid and unbroken, and the internal forces need be considered no further.

A force's tendency to produce rotation of an extended rigid body is called the torque or the moment of the force. It is equal to the product of the force's magnitude and the perpendicular distance from the axis of rotation to the force's line of action. This definition is illustrated in Fig. 3.1. Here a force $\mathbf{F}$ is shown applied to a disk wheel at point $P$ a distance $r$ from the wheel's axis. The dashed line is the line of action of force $\mathbf{F}$ and is simply a line constructed through $P$ along the direction of $\mathbf{F}$. It therefore includes the arrow that represents $\mathbf{F}$ and is really no more than an extension of that arrow in both directions. The perpendicular distance from the axis of rotation to the line of action of $\mathbf{F}$ is shown as $l$ in Fig. 3.1, and the magnitude of the torque is $Fl$. The distance $l$ is called the lever arm or moment arm of the force $\mathbf{F}$ about the wheel's axis.

Since torque has a direction associated with it, it is a vector, and in general this vector lies along the axis of rotation, that is, perpendicular to the plane determined by the line of action of the force and the lever arm. In Fig. 3.1 this plane is the plane of the page, and the torque vector may be imagined as being perpendicular to the page and lying along the wheel's axis, which is shown in Fig. 3.1 as a point representing the end view of the axis. This will be the situation in all cases encountered in elementary mechanics, so that all torque vectors will be parallel to each other and only their sense need be considered when they are added. The usual convention that torques tending to produce counterclockwise rotation are taken to be positive whereas those tending to produce clockwise rotation are taken to be negative will be adhered to throughout this book.

An important problem arises when the lever arm of the force of gravity on (the weight of) a rigid body is to be found about some given axis. The force of gravity acts on every little piece of the body and is

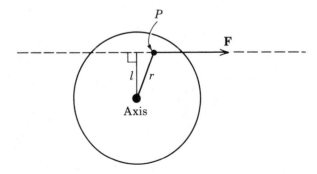

**Figure 3.1** *The torque or moment of a force*

therefore distributed over the extent of the body rather than being applied at one particular point. However, it is found that the body can be balanced, meaning that it can be supported at a single point about which (if the point is properly chosen) it will show no tendency to rotate. But if the body has no tendency to rotate, the total torque on it must be zero. Moreover, if the body is just being supported, the upward supporting force must be exactly equal to its weight. Fig. 3.2 now shows that if the entire weight $W$ of the body is considered to act at the balance point, then the lever arms of the weight and the supporting force will be the same about any arbitrarily chosen axis, and their respective torques will be equal and opposite. The net torque will be zero and there will be no tendency for the body to rotate. Since, in fact, the body is observed not to rotate when supported at its balance point, the actual situation is equivalent to the hypothetical one in which the entire weight of the body is considered to be applied at the balance point. In fact, correct results will be obtained if the weight is taken to act at the balance point, and this point is therefore called the *center of gravity*. Thus, the moment or lever arm of the weight of a rigid body about any given point may be calculated by assuming the body's entire weight to be concentrated at its center of gravity.

The foregoing shows that for a rigid body to be in equilibrium, it is not enough for the resultant force to be zero. That merely guarantees that the body's translational velocity will not change. If the body's rotational velocity is not to change either, the net torque due to all the applied forces must also be zero. There are thus two conditions for the equilibrium of a rigid body: (1) the vector sum of all the forces acting on the body must be zero and (2) the algebraic sum of all the torques about any axis must be zero. The first condition means that the sum of the forces in any

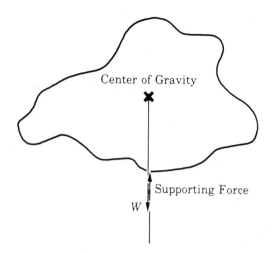

**Figure 3.2** *Center of gravity*

direction must be equal to the sum of the forces in the opposite direction. The second condition means that the sum of the torques producing clockwise rotation around any point must be equal to the sum of the torques producing counterclockwise rotation around the same point. The present experiment illustrates these principles using a meter stick as the rigid body. The arrangement is shown in Fig. 3.3. The meter stick is supported at an axis of rotation near its center and various forces are applied to it by weights and the supporting stand. The torque due to each force is calculated and the unknown force or distance computed by application of the two conditions of equilibrium. The weight and location of the center of gravity of the meter stick are also found and the torque due to this weight is calculated as if the weight were concentrated at this point.

**Figure 3.3** *A meter stick as a rigid body in equilibrium*

## APPARATUS _____

1. Meter stick
2. Three meter stick knife-edge clamps
3. Stand for supporting the meter stick with a clamp
4. Set of hooked weights
5. Triple-beam balance
6. Body of unknown mass (about 100 g)

## PROCEDURE _____

**1.** Weigh the meter stick and record the weight. Note that weight is a force and should be expressed in newtons even though your scales may read in grams.

**2.** Weigh the three meter stick clamps together and compute their average weight.

**3.** Find the center of gravity of the meter stick to the nearest 0.5 mm by balancing it in one of the clamps.

**4.** Put another clamp near one end of the stick and hang a 100-g weight from it. Slide the stick through the supporting clamp until the position of balance is found. Record the position of the axis of rotation—that is, the point of support—and the position of the weight.

**5.** Put another clamp near the other end of the stick and suspend a 200-g weight from it. Leaving the other weight in place, find the new point of balance by again sliding the meter stick through the supporting clamp. Record the position of the 200-g weight and of the new point of balance.

**6.** Remove the weights and clamps. Clamp the meter stick in the support clamp at its center of gravity. Put another clamp near one end of the stick and suspend the body of unknown weight from it. Slide another clamp, with a 200-g weight attached, along the other end of the stick until the position of equilibrium is found. Record the positions of both weights.

**7.** Weigh the body of unknown mass used.

## DATA _____

Weight of the meter stick _____
Weight of the three meter stick clamps _____
Average weight of the meter stick clamps _____
Center of gravity of the meter stick _____
Position of the 100-g weight _____
Position of the axis of rotation _____
Position of the 200-g weight _____
New point of balance _____

Position of the unknown weight _____
New position of the 200-g weight _____
Actual weight of the unknown mass _____
Weight of the meter stick computed by the method of moments _____
Weight of the unknown mass computed by the method of moments _____

## CALCULATIONS _____

**1.** Compute the weight of the meter stick from the data of Procedure 4 by the method of moments. Compare your result with that obtained by direct weighing of the stick. In particular, note whether the two measurements agree within the errors associated with each.

**2.** Using the point of support as the axis in Procedure 5, compute the moment of force of each of the weights and also of the meter stick, assuming its weight to be concentrated at its center of gravity. Add all these moments together, paying attention to their algebraic signs. Compare this net torque with zero, noting in particular whether zero lies within the error associated with your result.

**3.** Compute the weight of the body used in Procedure 6 by the method of moments. Compare the measured weight of this body with the computed weight, noting in particular whether the two weights agree within the experimental errors involved.

## QUESTIONS

**1.** State the two conditions for the equilibrium of a rigid body and tell how this experiment demonstrates their validity.

**2.** In Calculation 2, why was the supporting force exerted on the meter stick by the stand not considered?

**3.** Why was the meter stick clamped at its center of gravity in Procedure 6? Was there any advantage in doing this?

**4.** Using the data of Procedure 5, assume the axis of rotation to be at one end of the stick and compute the moments of all the forces about this axis. The total upward force at the point of support is equal to the sum of all the downward forces. Add all the positive moments together and all the negative moments together and then find their algebraic sum. What conclusion do you draw from the result?

**5.** A meter stick is pivoted at its 50-cm mark but does not balance because of nonuniformities in its material that cause its center of gravity to be displaced from its geometrical center. However, when weights of 150 and 200 g are placed at the 10-cm and 75-cm marks, respectively, balance is obtained. The weights are then interchanged and balance is again obtained by shifting the pivot point to the 43-cm mark. Find the mass of the meter stick and the location of its center of gravity.

# Uniformly Accelerated Motion. The Atwood Machine 4

In accordance with Newton's first law of motion, when the resultant of all the forces acting on a body is zero, if the body is at rest it will remain at rest, and if it is in motion, it will continue to move with constant speed in a straight line. Newton's second law of motion describes what happens if the resultant is different from zero. This law states that if an unbalanced force is acting on a body, it will produce an acceleration in the direction of the force, the acceleration being directly proportional to the force and inversely proportional to the mass of the body. If the acceleration is constant, the body is said to be moving with uniformly accelerated motion. The purpose of this experiment is to measure the acceleration of a given mass produced by a given force and to compare it with that calculated from Newton's second law of motion.

## THEORY

The Atwood machine consists of two weights connected by a light, flexible string that passes over a light pulley, as shown in Fig. 4.1. The pulley should be as nearly frictionless as possible. The machine is used in measuring the acceleration produced by an arbitrarily chosen force acting upon a given mass. Once the mass and the force have been chosen, the acceleration produced is determined by Newton's second law of motion

$$F = ma \qquad (4.1)$$

where $F$ is the net force in newtons acting on a body, $m$ is the mass of the body in kilograms, and $a$ is the acceleration in meters per second per second.

In the Atwood machine, the total mass that is being accelerated is the sum of the two masses. The driving force, which is expressed in newtons, is the difference in the weights on the two ends of the string. The important feature of this device is that a large mass (the sum) is being accelerated by a small force (the weight difference) so as to produce accelerations that are considerably smaller than the acceleration of gravity and therefore easier to measure. Moreover, the total mass being accelerated may be kept constant while the force is varied by transferring masses from one side of the moving system to the other. Before measurements can be made, however, the force due to friction in the pulley bearing must be determined. This is done by transferring masses from the ascending side to the descending side until the mass on the descending side moves downward with constant velocity when given a very slight push. The difference between the weights at the two string ends is then just compensating for the opposing frictional force and may be taken as equal to that force.

For convenience, the distance through which the descending mass falls should be the same in all of the observations. The starting point is taken as the position of the moving system in which one of the masses rests on the floor. As this mass ascends, the other will descend an equal distance, and the stopping point is taken as the instant at which this mass strikes the floor. The distance traversed should be about

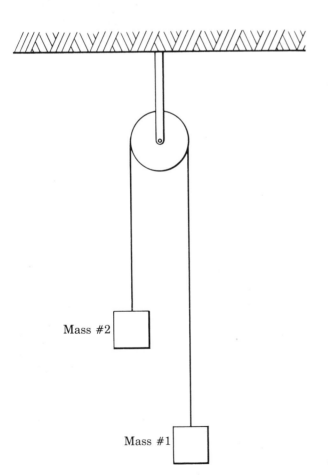

**Figure 4.1** *The Atwood machine*

Mass #2

Mass #1

**Figure 4.2** *Demonstration Atwood machine with electric timing device*

1.5 m. The time required for the mass to move through this distance is measured with a stopwatch.

In uniformly accelerated motion, the velocity is increased by the same amount in each succeeding second. The distance traveled is equal to the average velocity multiplied by the time. Since the system starts from rest and experiences a uniform acceleration, the final velocity will be twice the average velocity. From the time taken to acquire this final velocity, the corresponding acceleration can be computed. The equations of motion involved in uniformly accelerated motion follow. The distance is given by

$$s = \overline{v}t \qquad (4.2)$$

where $s$ is the distance in meters, $\overline{v}$ is the average velocity in meters per second, and $t$ is the time in seconds. The average velocity is given by

$$\overline{v} = \frac{v_1 + v_2}{2} \qquad (4.3)$$

where $\overline{v}$ is the average velocity; $v_1$ is the initial velocity, which is zero in this case; and $v_2$ is the final velocity. The final velocity is given by

$$v_2 = v_1 + at \qquad (4.4)$$

where $v_2$ is the final velocity; $v_1$ is the initial velocity, which is zero in this case; $a$ is the acceleration in meters per second per second; and $t$ is the time in seconds.

## APPARATUS

1. Pulley clamped to a vertical rod
2. Two weight holders
3. Set of slotted weights
4. Set of small weights, including
    one 10-g weight
    one 5-g weight
    five 2-g weights
    one 1-g weight
5. String (strong fishing line)
6. Stopwatch or stop clock
7. 2-m stick

## PROCEDURE

**1.** Using a total mass of about 2 kg, determine the force of friction in the machine by transferring masses from the ascending side to the descending side until the mass on the descending side moves downward with constant velocity when given a very slight push. Record the mass on the descending side and the mass on the ascending side. Be sure to include the mass of each weight hanger. Arrange the weights so as to have five 2-g weights on the ascending side when the machine is in the balanced condition. CAUTION: Always stand clear of the suspended weights because the string may break.

**2.** Transfer two more grams from the ascending side to the descending side, and thus determine the acceleration produced by a net force of 0.004 kg × 9.8 m/s² = 0.0392 N. Begin the observation when the ascending mass is on the floor, starting the stopwatch at the instant when you let the weights go, and stopping it at the instant when the descending mass strikes the floor. Make and record four independent observations.

**3.** Measure the distance traversed.

**4.** Repeat Procedure 2, using accelerating forces of 0.008 × 9.8, 0.012 × 9.8, 0.016 × 9.8, and 0.02 × 9.8 N, by transferring two additional grams each time. Make four independent observations for each accelerating force.

# DATA _____

Distance traversed _____  Ascending mass, for constant velocity _____

Descending mass, for constant velocity _____  Force of friction _____

| Descending mass | Ascending mass | Net force | Time 1 | 2 | 3 | 4 | Average |
|---|---|---|---|---|---|---|---|
| | | | | | | | |
| | | | | | | | |
| | | | | | | | |
| | | | | | | | |
| | | | | | | | |

| Net force | Average velocity | Final velocity | Acceleration experimental value | Acceleration from $F = ma$ | Percent error |
|---|---|---|---|---|---|
| | | | | | |
| | | | | | |
| | | | | | |
| | | | | | |
| | | | | | |

# CALCULATIONS _____

**1.** Calculate the force due to friction by subtracting the value of the ascending mass from that of the descending mass as found in Procedure 1 and multiplying this difference by the acceleration of gravity. Record your result, being careful to use proper units. Note that all forces should be expressed in newtons.

**2.** Compute the average time taken with each accelerating force. From the known distance and the time taken in each case, compute the average velocity corresponding to each accelerating force.

**3.** Calculate the final velocity for each set of observations.

**4.** Calculate the acceleration produced by each accelerating force from the values of the final velocity and the time.

**5.** Compute the theoretical value for the acceleration from Newton's second law of motion. Assuming this to be the correct value, compute the percent error of the observed value for each case.

## QUESTIONS _____

**1.** (a) State what this experiment tested. (b) State the relation between force and acceleration observed in this experiment.

**2.** What is the difference between uniform motion (constant velocity) and uniformly accelerated motion?

**3.** What is the advantage of transferring masses from one side to the other, instead of adding masses to one side?

**4.** If you gave the system an initial velocity different from zero, how would this affect your results?

**5.** Why should the pulley be as light as possible as well as nearly frictionless?

**6.** Compute the a.d. for each of your sets of time measurements, which will give you an error to be associated with each experimental value of the acceleration. Does the accepted value determined from Newton's second law fall within your limits of error in each case? If not, try to explain why not.

**7.** Using Equations 4.2, 4.3, and 4.4, derive an expression for $s$ as a function of $t$ for this case of uniformly accelerated motion.

# The Ballistic Pendulum. Projectile Motion 5

The principle of conservation of momentum follows directly from Newton's laws of motion. According to this principle, if there are no external forces acting on a system containing several bodies, then the momentum of the system remains constant. In this experiment, the principle is applied to the case of a collision, using a ballistic pendulum. A ball is fired by a gun into the pendulum's bob. The initial velocity of the ball is determined in terms of the masses of the ball and the bob and the height to which the bob rises after impact. This velocity can also be obtained by firing the ball horizontally and allowing it to fall freely toward the earth. The velocity is then determined in terms of the range and the vertical distance of fall. It is the object of this experiment to study the law of conservation of momentum and the elements of projectile motion; more precisely, the purpose is to determine the initial velocity of a projectile (1) by measurements of its range and vertical distance of fall and (2) by means of a ballistic pendulum. If the theory is correct, the two results should agree.

## THEORY

The momentum of a body is defined as the product of the mass of the body and its velocity. Newton's second law of motion states that the net force acting on a body is proportional to the time rate of change of momentum. Hence, if the sum of the external forces acting on a body is zero, the linear momentum of the body is constant. This is essentially a statement of the principle of conservation of momentum. Applied to a system of bodies, the principle states that if no external forces act on a system containing two or more bodies, then the momentum of the system *as a whole* does not change.

In a collision between two bodies, each one exerts a force on the other. These forces are equal and opposite, and if no other forces are brought into play, the total momentum of the two bodies is not changed by the impact. Hence, the total momentum of the system after collision is equal to the total momentum of the system before collision. During the collision the bodies become deformed and a certain amount of energy is used to change their shape. If the bodies are perfectly elastic, they will recover completely from the distortion and will return all of the energy that was expended in distorting them. In this case, the total kinetic energy of the system also remains constant. If the bodies are not perfectly elastic, they will remain permanently distorted, and the energy used up in producing the distortion is not recovered.

Inelastic impact can be illustrated by a device known as the ballistic pendulum, which is sometimes employed to determine the speed of a bullet. If a bullet is fired into a pendulum bob and remains imbedded in it, the momentum of the bob and bullet just after the collision is equal to the momentum of the bullet just before the collision. This follows from the law of conservation of momentum. The velocity of the pendulum before collision is zero, so that the total momentum of the system is then just that of the bullet alone. After the collision, the pendulum and the bullet move with the same velocity. Thus, conservation of the system's momentum requires

$$mv = (M + m)V \qquad (5.1)$$

where $m$ is the mass of the bullet in kilograms, $v$ is the velocity in meters per second of the bullet just before the collision, $M$ is the mass of the pendulum bob, and $V$ is the common velocity of the bob and bullet just after the collision.

As a result of the collision, the pendulum with the imbedded bullet swings about its point of support, and the center of gravity of the system rises through a vertical distance $h$. From a measurement of this distance, it is possible to calculate the velocity $V$. The kinetic energy of the system just after the collision must be equal to the increase in potential energy of the system as the pendulum reaches its highest point. This follows from the law of conservation of energy; here we assume that the loss of energy due to friction at the point of support is negligible. The energy equation gives

$$\frac{1}{2}(M + m)V^2 = (M + m)gh \qquad (5.2)$$

where $M$ is the mass of the pendulum bob, $m$ is the mass of the bullet, $V$ is their common velocity just after the collision, $g$ is the known value of the acceleration of gravity, and $h$ is the vertical distance through which the center of gravity of the system rises. The left-hand side of Equation 5.2 represents the kinetic energy of the system just after the impact; the right-hand side represents the change in the potential energy of the system. Solving Equation 5.2 for $V$ yields

$$V = \sqrt{2gh} \qquad (5.3)$$

By substituting this value of $V$ and the values of the masses $M$ and $m$ in Equation 5.1, the velocity of the bullet before the collision can be calculated.

The velocity of the bullet can also be determined from measurements of the range and vertical distance of fall when the bullet is fired horizontally and allowed to fall to the floor without striking the pendulum bob.

The motion of a projectile is a special case of a freely falling body in which the initial velocity may be in any direction with respect to the vertical. The projectile's path is a curve produced by a combination of the uniform velocity of projection and the velocity due to the acceleration of gravity. This type of motion may be studied very effectively by considering it as made up of two independent motions, one of constant speed in the horizontal direction, and the other of constant acceleration in the vertical direction. Here the problem is simplified by neglecting air resistance and also by firing the projectile horizontally so that there is no vertical component of the initial velocity $v$.

Because the initial velocity is in the horizontal or $x$ direction, in which there is no acceleration, the horizontal distance covered is given by

$$x = vt \qquad (5.4)$$

where $t$ is the time of flight in seconds. In the $y$ direction, however, the initial velocity is zero, so that the motion in this direction is that of a freely falling body starting from rest. Therefore, the distance traveled vertically is given by

$$y = \frac{1}{2}gt^2 \qquad (5.5)$$

where $y$ is the vertical distance moved and $g$ is the known acceleration of gravity. Combining Equations 5.4 and 5.5 to eliminate the time, in which we have no interest, and solving for $v$ yields

$$v = x\sqrt{\frac{g}{2y}} \qquad (5.6)$$

The apparatus used in this experiment (see Fig. 5.1) is a combination of a ballistic pendulum and a spring gun. The pendulum consists of a massive cylindrical bob, hollowed out to receive the projectile and suspended by a strong, light rod pivoted at its upper end. The projectile is a brass ball that is fired into the pendulum bob and is held there by a spring in such a position that its center of gravity lies on the axis of the suspension rod. A brass index is attached to the pendulum bob to indicate the height of the center of gravity of the loaded pendulum. When the projectile is fired into the bob, the pendulum swings upward and is caught at its highest point by a pawl that engages a tooth on a curved rack.

**Figure 5.1** *The Blackwood ballistic pendulum*

## APPARATUS

1. Blackwood ballistic pendulum
2. Triple-beam balance
3. Metric steel scale
4. $11\frac{1}{2} \times 1\frac{11}{16} \times \frac{1}{2}$-in. plate
5. 2-m stick
6. 1-m stick
7. Plumb bob
8. Carbon paper
9. White paper and masking tape
10. Level

## PROCEDURE

**1.** In the first part of the experiment, the initial velocity of the projectile is obtained from measurements of the range and fall. The apparatus should be set near one edge of a level table. The pendulum is not used in this part and should be swung up onto the rack so that it will not interfere with the free flight of

the ball. In some cases, the rack itself will interfere and must be removed. Your instructor will supply the proper tool for doing this. The pendulum must then be held up out of the way when the gun is fired. Check that the apparatus is level; if it is not, place sheets of paper under the feet as required.

**2.** Get the gun ready for firing by placing the ball on the end of the firing rod and pushing it back, compressing the spring until the trigger is engaged. The ball is fired horizontally so that it strikes a target placed on the floor. Fire the ball several times and determine approximately where it strikes the floor. Place a sheet of white paper on the floor so that the ball will hit it near its center; secure the paper with masking tape. Cover this sheet with a sheet of carbon paper, carbon side down, so that a record will be made of the exact spot where the ball strikes. Fire the ball ten more times. *Note:* It is imperative that the apparatus not move on the bench between shots. If you have trouble keeping it in position, ask your instructor to provide a clamp to fix it to the bench. *Do not overtighten the clamp.*

**3.** Measure the range for each shot; this is the *horizontal* distance from the point of projection to the point of contact with the floor. To do this, place the metal plate, long edge down, on the apparatus base so that it extends forward from the gun's firing pin. Then place the meter stick edge down on the top edge of the plate and hold it there with one end butted against the end of the firing pin. With the gun fired (that is, not cocked) this is very closely the position of the center of the ball at the instant it leaves the gun. Hang the plumb bob from the meter stick at some convenient position from which the plumb line can hang down beyond the table edge (see Fig. 5.2). Place a second sheet of white paper on the floor under the plumb bob, secure it in position with masking tape, and mark the spot indicated by the plumb bob's point. Make sure the bob hangs freely and has stopped swinging when you make the mark. Note and record the reading on the meter stick at the point where you have hung the plumb line; this is the *horizontal* distance from the tip of the firing pin to the mark on the floor. You can now use the 2-m stick to

**Figure 5.2** *Setup for range measurements*

measure the distance from your mark to each of the ten carbon spots made by the ball when you fired it in Procedure 2. Record all your data as called for in the data table, noting that the range of each shot is the pertinent distance along the floor measured with the 2-m stick *plus* the distance from the plumb-bob mark to the firing pin read from the 1-m stick as described above.

**4.** Measure the ball's vertical fall, that is, the vertical distance of the point of projection above the floor. For this measurement, use the meter stick as a straightedge. Lay it on top of the metal plate as shown in Fig. 5.2 but broadside rather than edge down. In this position its upper surface should just slide under the ball when the latter is on the firing pin so that this surface is at the level of the bottom of the ball at the instant of firing. Because the meter stick projects beyond the edge of the table, you can use the 2-m stick to measure the vertical distance from the floor to its upper surface. This should then also be the height of the bottom of the ball above the floor at the time of firing. Note that it is the distance to the *bottom* of the ball that you want, as it is the *bottom* of the ball (not its center) that strikes the floor.

**5.** Return the pendulum to its central (straight down) position and allow it to swing freely. Get the gun ready for firing, and, when the pendulum is at rest, pull the trigger, thereby firing the ball into the pendulum bob. This will cause the pendulum with the ball inside it to swing up along the rack where it will be caught at its highest point. Record the notch on the curved scale reached by the pawl when it catches the pendulum. To remove the ball from the pendulum, push it out with your finger or a rubber-tipped pencil while holding up the spring catch.

**6.** Repeat Procedure 5 four more times, recording the position of the pendulum on the rack each time.

**7.** From the data of Procedures 5 and 6, compute the average value of the position of the pendulum on the rack. Set the pendulum with the pawl engaged in the notch that corresponds most closely to the average reading. Measure the vertical distance $h_1$ from the base of the apparatus to the index point attached to the pendulum. This index point is at the vertical position of the center of gravity of the pendulum and ball. Use the metric steel scale and read the measurement to 0.1 mm.

**8.** With the pendulum hanging in its lowest position, measure the vertical distance $h_2$ from the base of the apparatus to the index point.

**9.** Loosen the thumbscrew holding the axis of rotation of the pendulum and carefully remove the pendulum from its support. Weigh the pendulum and the ball separately and record the values obtained.

# DATA

Horizontal distance from firing pin
   to plumb bob mark                    _____

| Shots | Distance from plumb bob | Horizontal range $x$ | Deviations |
|---|---|---|---|
| 1 | | | |
| 2 | | | |
| 3 | | | |
| 4 | | | |
| 5 | | | |
| 6 | | | |
| 7 | | | |
| 8 | | | |
| 9 | | | |
| 10 | | | |
| Averages | | | |

Vertical distance of fall $y$ _____

Highest position of ballistic
  pendulum:
    Reading of curved scale
    Average reading _____

Vertical distance $h_1$ or the average
  height of the center of gravity
  of the pendulum in its highest
  position _____

Vertical distance $h_2$ or the height
  of the center of gravity of the
  pendulum in its lowest position _____

Vertical distance $h$ through which
  the center of gravity of the
  pendulum was raised _____

Velocity of projection from range
  and fall measurements _____

Trials

1. _____ 2. _____ 3. _____ 4. _____ 5. _____

Mass of the pendulum bob _____

Mass of the ball _____

Velocity of the pendulum and ball
  just after the collision _____

Velocity of the ball before the
  collision _____

Difference between this value of $v$
  and that found from the range
  measurement _____

Percent difference _____

## CALCULATIONS _____

**1.** From the data of Procedure 3, compute the average range of the projectile.

**2.** Calculate the deviation of each of your range measurements from your average and then calculate the average deviation (a.d.).

**3.** Compute the velocity of projection from Equation 5.6. Attach the proper error to your result by using the a.d. found in Calculation 2 as the error in $x$ and attaching no errors to your measurement of $y$ or the known value of $g$.

**4.** From the data of Procedures 7 and 8, calculate the vertical distance $h$ through which the center of gravity of the loaded ballistic pendulum was raised as a result of the collision.

**5.** Compute the value of $V$, the common velocity of the pendulum bob and ball just after the collision, by using Equation 5.3.

**6.** Calculate the velocity of the ball before the collision by substituting the value of $V$ found in Calculation 5 and the measured values of the masses of the pendulum bob and of the ball in Equation 5.1.

**7.** Compare the values of $v$ obtained by your two methods by finding the difference between them and expressing it as a percent. Note also whether your value of $v$ from Calculation 6 falls within the range of error found in Calculation 3.

# QUESTIONS _____

**1.** Using the data of your experiment, calculate the kinetic energy of the ball just before impact from the value of the velocity of the ball obtained with the ballistic pendulum and the mass of the ball.

**2.** Calculate the kinetic energy of the pendulum bob and ball just after impact from the value of their common velocity and of their masses.

**3.** (a) Using the results of Questions 1 and 2, calculate the fractional loss of energy during this inelastic impact. Express it in percent. (b) What became of the energy lost?

**4.** (a) Compute the ratio of the mass of the pendulum bob to the total mass of the bob and the ball. Express it in percent. (b) How does this ratio compare with the fraction of energy lost during the impact?

**5.** Compare the momentum and the kinetic energy of an automobile weighing 2000 lb and moving with a velocity of 60 mi/h with the momentum and the kinetic energy of a projectile weighing 70 lb and moving with a velocity of 2500 ft/s.

**6.** A bullet of mass 10 g is fired horizontally into a block of wood of mass 2 kg and suspended like a ballistic pendulum. The bullet sticks in the block and the impact causes the block to swing so that its center of gravity rises 0.1 m. Find the velocity of the bullet just before the impact.

**7.** Using your range data from Procedures 1–3, calculate the average deviation of the mean (A.D.) and the standard deviation ($\sigma$). Refer to the section titled "Theory of Errors" in the Introduction for definitions and explanations of these quantities. Then find the error to be attached to your value of the spring gun's muzzle velocity $v$ when the A.D. is used as the error in the range and also when $\sigma$ is so used. In each case, see if your value of $v$ from the ballistic pendulum measurement falls within the error range obtained.

**8.** In your calculations of $v$ from your range measurements, no attempt was made to attach errors to the values you used for the acceleration of gravity $g$ or the distance of vertical fall $y$. Justify your ignoring the errors that must exist for these quantities. Also justify the fact that you did not calculate an error for the value of $v$ you obtained from your measurements with the ballistic pendulum.

**9.** (a) Derive Equation 5.6 from Equations 5.4 and 5.5. (b) Derive an equation giving the gun's muzzle velocity $v$ in terms of $h$ and the masses of the ball and pendulum by combining Equations 5.1 and 5.3.

# Uniform Circular Motion 6

Whenever a body moves in a circular path, a force directed toward the center of the circle must act on the body to keep it moving in this path. This force is called centripetal force. The reaction, which is equal and opposite, is the pull of the body on the restraining medium and is called centrifugal force. The purpose of this experiment is to study uniform circular motion and to compare the observed value of the centripetal force with the calculated value.

## THEORY

If a body moves with constant speed in a circle, it is said to be moving with uniform circular motion. Even though the speed is constant, the velocity is continuously changing because the direction of the motion is continuously changing. Thus such a body has an acceleration. It can be shown that the direction of the acceleration is always toward the center of the circle and that its magnitude is given by

$$a = \frac{v^2}{r}$$

where $v$ is the speed of the body in meters per second and $r$ is the radius of the path in meters.

A force is necessary to produce this acceleration, and this force is called centripetal force because it is always directed toward the center of rotation. By Newton's second law of motion, the magnitude of this force is given by the relation

$$F = ma \quad \text{or} \quad F = m\frac{v^2}{r}$$

where $F$ is the force in newtons, $m$ is the mass in kilograms, and $v$ and $r$ are the same as before. But by Newton's third law of motion, an equal and opposite force is exerted by the body on the restraining medium because of the inertia of the body. This reaction is called centrifugal force.

The centripetal force can also be expressed in terms of the angular speed, since

$$v = r\omega \quad \text{and} \quad \omega = 2\pi f$$

where $v$ is the linear speed and $r$ the radius of the path as before, $\omega$ is the angular speed in radians per second, and $f$ is the number of revolutions per second. Thus

$$F = mr\omega^2 \quad \text{or} \quad F = 4\pi^2 f^2 rm$$

where $F$ is the centripetal force in newtons.

In the apparatus used for this experiment (see Fig. 6.1) a stretched spring provides the centripetal force necessary to keep a mass rotating in a circle of a particular measurable radius. The rotator is adjusted to a rotational speed such that the back of the mass just touches the sensing probe of a sensitive indicator. The indicator's pointer then rises to stand oppo-

**Figure 6.1** *Centripetal force apparatus*

site an index ridge at the end of the shaft on which the apparatus turns. As this index and the pointer are near the center of rotation, they are easily visible regardless of the rotational speed. This speed is kept constant at the value just described and is measured by timing some convenient number of turns. The force required to stretch the spring by the same amount is subsequently measured by hanging weights from it.

## APPARATUS

1. Centripetal force apparatus
2. Motor with coupler and means for counting turns
3. Stopwatch or stop clock
4. Supporting rods, hook, right-angle clamp, and bench clamp
5. Weight hanger and weights
6. Vernier caliper

## PROCEDURE

**1.** When you come into the laboratory, examine the rotating assembly carefully in order to understand how it works. Do not change the spring tension adjustment; it has already been set by your instructor. Note and record the value of the rotating mass. This value is either marked on the mass itself or will be given to you.

**2.** Set up the bench clamp, a vertical rod, right-angle clamp, horizontal rod, and hooked collar so as to suspend the apparatus vertically. Hook the weight hanger to the eye on the rotating mass and add weights carefully until this mass is pulled down just enough to touch the sensitive indicator and bring the pointer to the index. Record the mass required to do this. Don't forget to include the mass of the weight hanger.

**3.** With the weights still on the hanger so that the rotating mass is in the position established in Procedure 2, use the vernier caliper to measure the mass's radius of rotation. The measurement is made from the point on the shaft above the index (the axis of rotation) to the line engraved around the mass, which marks the location of its center of gravity in that direction.

**4.** Unhook the apparatus from the weight hanger and supporting rods and mount it on the motor shaft by means of the coupling provided. Mount the motor unit securely on the bench using the bench clamp and check that all mounting clamps and screws are tight.

**5.** Make sure the speed control is at its lowest position and turn on the motor. As you gradually increase the speed, note that until the rotating mass has moved out to the radius at which it engages the indicator, the apparatus is highly unbalanced and will shake quite badly. Increase the speed until you see the pointer rise, but be careful not to increase it too much. CAUTION: It is dangerous to let the apparatus rotate at excessive speeds. Best results will be obtained if you get the pointer to rise just above the index and then slow the rotation very carefully, making very small speed changes and watching the results. Your aim is to keep the motor speed at the value for which the pointer stands exactly opposite the index, but this adjustment is very critical, and you may want to practice a while before attempting to take data. Holding a sheet of white paper behind the apparatus as you watch the pointer and index will help you see them more clearly.

**6.** When you have the speed set at the correct value so that the pointer is opposite the index, measure this speed by timing a counted number of turns. If you are using a revolution counter, note its reading and then engage it and start the stop clock simultaneously. Disengage it and simultaneously stop the clock after about a minute and record the number of turns and the elapsed time. The setup shown in Fig. 6.1 uses a variable speed motor with an 18 : 1 reduction gear on one end and a pointer mounted on the slowly turning shaft thus provided. Revolutions of this shaft are easily counted visually. In performing this experiment it is a good plan to have one observer pay strict attention to the proper adjustment of the speed while the other counts turns and operates the stop clock.

**7.** Repeat Procedures 5 and 6 three more times so as to have a total of four measurements of the rotation speed. Record the number of revolutions counted and the elapsed time in each case.

## DATA

Value of the rotating mass ⎯⎯⎯⎯

Mass hung on the spring ⎯⎯⎯⎯

Total mass necessary to stretch the spring ⎯⎯⎯⎯

Radius of rotation ⎯⎯⎯⎯

Centripetal force calculated from the theory ⎯⎯⎯⎯

Measured value of the centripetal
    force         _____

Difference between calculated and
    measured value    _____

Percent discrepancy          _____

| Trial | Number of turns | Time interval | Rotation speed | Deviation |
|-------|-----------------|---------------|----------------|-----------|
| 1 | | | | |
| 2 | | | | |
| 3 | | | | |
| 4 | | | | |
| Averages | | | | |

## CALCULATIONS _____

**1.** From the data of Procedures 5–7, calculate the speed of rotation for each set of readings. Note that dividing the number of turns by the elapsed time gives you revolutions per second, that if you used a reduction gear you must multiply by the gear ratio to get the rotational speed of the apparatus, and that you must then multiply by $2\pi$ to get your result in radians per second.

**2.** Compute the average of your four speed measurements, the deviation of each from the average, and the average deviation (a.d.).

**3.** Calculate the centripetal force from the theory using the average speed of rotation obtained in Calculation 2.

**4.** Find the directly measured value of the centripetal force by adding the value of the rotating mass given you in Procedure 1 to the mass found in Procedure 2 and multiplying by the acceleration of gravity to get the force stretching the spring in newtons. Record this result in the space provided in the data sheet.

**5.** Compare the calculated value of the centripetal force with the value measured directly by subtracting one from the other and computing the percent discrepancy.

**6.** Using the a.d. found in Calculation 2 as the error in your rotational speed measurement, find the resulting error in your calculated value of the centripetal force and see whether the directly measured value falls within the limits of error so established.

## QUESTIONS _____

**1.** State what the experiment checks.

**2.** (a) How does the centripetal force vary with the speed of rotation for a constant radius of the path? (b) How does it vary with the radius of the path for a constant speed of rotation?

**3.** Distinguish between centripetal force and centrifugal force. Explain in what direction each force is acting and on what it is acting.

**4.** Calculate at what speed the earth would have to rotate in order that objects at the equator would have no weight. Assume the radius of the earth to be 6400 km. What would be the linear speed of a point on the equator? What would be the length of a day (time from sunrise to sunset) under these conditions?

**5.** Engines for propeller-driven aircraft are limited in their maximum rotational speed by the fact that the tip speed of the propeller must not approach the speed of sound in air (Mach 1). Taking 6 ft as a typical diameter for a propeller of a light airplane and 1100 ft/s as the speed of sound, find the upper limit on the rpm (revolutions per minute) of the propeller shaft.

# Friction 7

Whenever a body slides along another body, a resisting force is called into play that is known as the force of friction. This is a very important force and serves many useful purposes—a person could not walk without it, nor could a car propel itself along a highway without the friction between the tires and the road surface. On the other hand, friction is very wasteful; it reduces the efficiency of machines because work must be done to overcome it, and this energy is wasted as heat. The purpose of this experiment is to study the laws of friction and to determine the coefficient of friction between two surfaces.

## THEORY

Friction is the resisting force encountered when one surface slides over another; this force acts along the tangent to the surfaces in contact. The force necessary to overcome friction depends on the nature of the materials in contact, their roughness or smoothness, and on the normal force, but not on the area of contact, within wide limits, or the speed of the motion. It is found experimentally that the force of friction is directly proportional to the normal force. The constant of proportionality is called the coefficient of friction.

When the contacting surfaces are actually sliding one over the other, the force of friction is given by

$$F_r = \mu_k N \qquad (7.1)$$

where $F_r$ is the force of friction and is directed parallel to the surfaces and opposite to the direction of motion, $N$ is the normal force, and $\mu_k$ the coefficient of friction. The subscript k stands for *kinetic*, meaning that $\mu_k$ is the coefficient that applies when the surfaces are moving one with respect to the other. It is therefore more precisely called the coefficient of *kinetic* or *sliding* friction. Note carefully that $F_r$ is always directed *opposite* to the direction of motion. This means that if you reverse the direction of sliding, the frictional force reverses too. In short, it is always against you. Friction is therefore called a *nonconservative force* because energy must be used to overcome it no matter which way you go. This is in contrast to a so-called *conservative force* such as gravity, which is against you on the way up but with you on the way down. Thus energy expended in lifting an object may be regained when the object redescends, but the energy used to overcome friction is dissipated—i.e., lost or made unavailable—as heat. As you will see in your later study of physics, the distinction between conservative and non-conservative forces is a most important one, fundamental to our concepts of heat and energy.

A method of checking the proportionality of $F_r$ and $N$ and of determining the proportionality constant $\mu_k$ is to have one of the surfaces in the form of a plane placed horizontally with a pulley fastened at one end. The other surface is the bottom face of a block that rests on the plane and to which is attached a cord that passes over the pulley and carries weights. These are varied until the block moves at constant speed after having been started with a slight push. Since there is thus no acceleration, the net force on the block is zero, which means that the frictional force is equal to the tension in the cord. This tension, in turn, is equal to the total weight attached to the cord's end. The normal force between the two surfaces is equal to the weight of the block and can be increased by placing weights on top of the block. Thus, corresponding values of $F_r$ and $N$ can be found, and plotting them will show whether $F_r$ and $N$ are indeed proportional. The slope of this graph gives $\mu_k$.

When a body lies at rest on a surface and an attempt is made to push it, the pushing force is opposed by a frictional force. As long as the pushing force is not strong enough to start the body moving, the body remains in equilibrium, which means that the frictional force automatically adjusts itself to be equal to the pushing force and thus to just balance it. There is, however, a threshold value of the pushing force beyond which larger values will cause the body to break away and slide. We conclude that in the static case (body at rest) the frictional force automatically adjusts itself to keep the body at rest up to a certain maximum. But if static equilibrium demands a frictional force larger than this maximum, static equilibrium conditions will cease to exist because this force is not available, and the body will start to move. This situation may be expressed in equation form as

$$F_r \le \mu_s N \qquad \text{or} \qquad F_{r_{max}} = \mu_s N \qquad (7.2)$$

where $F_r$ is the frictional force in the static case, $F_{r_{max}}$ is the maximum value this force can assume, and $\mu_s$ is the coefficient of *static* friction. It is found that $\mu_s$ is slightly larger than $\mu_k$, which means that a somewhat larger force is needed to break a body away and start it sliding than is needed to keep it sliding at constant speed once it is in motion. This is why a slight push is necessary to get the block started for the measurement of $\mu_k$.

One way of investigating the case of static friction is to observe the so-called *limiting angle of repose,* defined as the maximum angle to which an inclined plane may be tipped before a block placed on the plane just starts to slide. The arrangement is illustrated in Fig. 7.1. The block has weight $W$ whose component $W \cos \theta$ (where $\theta$ is the plane angle) is perpendicular to the plane and is thus equal to the normal force $N$. The component $W \sin \theta$ is parallel to the plane and constitutes the force urging the block to slide down the plane. It is opposed by the frictional force $F_r$, and as long as the block remains at rest, $F_r$ must be equal to $W \sin \theta$. If the plane is tipped up until at some value $\theta_{max}$ the block just starts to slide, then $F_{r_{max}} = W \sin \theta_{max}$. But $F_{r_{max}} = \mu_s N = \mu_s W \cos \theta_{max}$. Hence, $W \sin \theta_{max} = \mu_s W \cos \theta_{max}$ or

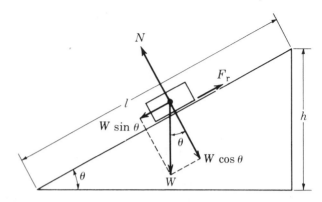

**Figure 7.1** *The inclined plane*

$$\mu_s = \frac{\sin \theta_{max}}{\cos \theta_{max}} = \tan \theta_{max} \qquad (7.3)$$

Thus, if the plane is gradually tipped up until the block just breaks away and the plane angle is then measured, the coefficient of static friction is equal to the tangent of this angle, which is called the *limiting angle of repose*. It is interesting to note that $W$ cancelled out in the derivation of Equation 7.3, so that the weight of the block doesn't matter.

## APPARATUS

1. Board with pulley at one end
2. Rods, bench clamp, and right-angle clamp for supporting the inclined plane
3. Wood block with a cord attached to it
4. Glass block

5. Set of known weights
6. Weight hanger
7. Triple-beam balance
8. Protractor
9. Lintless dust cloth or paper wipers

## PROCEDURE

**1.** Weigh the wood block and record the weight in newtons.

**2.** Place the board in a horizontal position on the laboratory table with its pulley projecting beyond the table's edge. Be sure that the surfaces of both the board and the wood block are clean, dry, and free of any dust or grit. Wipe them off if necessary with a clean, dry, lintless cloth or paper wiper. *After this has been done, do not touch these surfaces with your hands.* Handle the block with the cloth or a wiper and set it down *only* on the clean board. Begin the experiment by setting the block on the board with its largest surface in contact with the board's surface. Run the cord attached to the block over the pulley and attach it to the weight hanger. Place some weights on the hanger and slowly increase the load until it is just sufficient to keep the block sliding slowly with constant speed after it has been started with a very small push. Record this load. Don't forget to include the weight of the hanger.

**3.** Repeat Procedure 2 placing masses of 200,

400, 600, 800, and 1000 g successively on top of the wood block. Record the load needed to produce constant speed in each case.

**4.** Turn the wood block on its side and repeat Procedure 2 with a mass of 400 g on top of the block. Record the load needed.

**5.** Again turn the wood block with the largest surface in contact with the plane and place 400 g on top of the block. Gradually increase the load on the hanger until the block just starts to move, without any initial push. Be careful to place the weights on the hanger gently so as not to jerk the cord. Notice whether this time the block moves with uniform speed or whether it is being accelerated. Record the load needed under these conditions. Repeat this procedure twice to obtain three independent measurements of the required load.

**6.** Set up the board as an inclined plane. Place the wood block on the plane with its largest surface in contact, and gradually tip the plane up until the block just breaks away and starts to slide down. Be very careful to tip the plane slowly and smoothly so as to get a precise value of the angle with the horizontal at which the block just breaks away. This is the limiting angle of repose $\theta_{max}$. Measure it by means of a protractor and record the result obtained in three separate trials. These trials should be independent, meaning that in each case the plane should be returned to the horizontal, the block placed on it, and the plane carefully tipped up until the limiting angle of repose is reached.

**7.** Repeat Procedure 6 using the glass block. Record the limiting angle of repose obtained in three independent trials.

## DATA _____

Weight of wood block _____

| Position of block | Mass placed on the block | Total normal force | Force to keep block moving uniformly |
|---|---|---|---|
| Flat | 0 g | | |
| Flat | 200 g | | |
| Flat | 400 g | | |
| Flat | 600 g | | |
| Flat | 800 g | | |
| Flat | 1000 g | | |
| On side | 400 g | | |

Coefficient of kinetic friction $\mu_k$ from graph _____

Coefficient of kinetic friction $\mu_k$ from Procedure 4 _____

Total normal force for Procedure 5 _____

| Trial | Force to start block moving | $\mu_s$ | Deviation |
|---|---|---|---|
| 1 | | | |
| 2 | | | |
| 3 | | | |
| Averages | | | |

| Block used | Trial 1 | | | Trial 2 | | | Trial 3 | | | Averages | |
|---|---|---|---|---|---|---|---|---|---|---|---|
| | $\theta_{max}$ | $\mu_s$ | Deviation | $\theta_{max}$ | $\mu_s$ | Deviation | $\theta_{max}$ | $\mu_s$ | Deviation | $\mu_s$ | Deviation |
| Wood | | | | | | | | | | | |
| Glass | | | | | | | | | | | |

Difference between your two values of $\mu_s$ _____

Percent error in agreement of the two values of $\mu_s$ _____

## CALCULATIONS

**1.** From the data of Procedures 2 and 3, plot a curve using the values of the total normal force as abscissas and the values of the force of friction as ordinates. See if your curve is a straight line and obtain the coefficient of kinetic friction $\mu_k$ for wood on wood by finding its slope.

**2.** Use the data of Procedure 4 to calculate the coefficient of kinetic friction for the case of the wood block sliding on its side. Record your result and see how it compares with the value of $\mu_k$ obtained from your graph.

**3.** From the data of Procedure 5, compute the coefficient of static friction $\mu_s$ for wood on wood from each of your three trials. Calculate an average value of $\mu_s$, the deviation for each trial, and the a.d. Record your result on the data sheet.

**4.** From the data of Procedure 6, calculate $\mu_s$ for wood on wood from each of your three trials. Calculate an average value of $\mu_s$, the deviation for each trial, and the a.d. Record your result on the data sheet. Compare this value of $\mu_s$ with that obtained in Procedure 5 by finding the difference between the two values and the percent error in their agreement. See whether your two values agree within their limits of error.

**5.** From the data of Procedure 7, calculate $\mu_s$ for glass on wood for each of your three trials and obtain the average value, the deviation for each trial, and the a.d. Record your result on the data sheet.

## QUESTIONS _____

**1.** Explain in your own words why it is necessary that the block move at constant velocity in Procedures 2–4.

**2.** (a) How does the coefficient of friction depend upon the normal force between the surfaces in contact? (b) How does it depend upon the area of the surfaces in contact?

**3.** How does the coefficient of static friction compare with the coefficient of kinetic friction for the same surfaces, areas, and normal forces? In this connection explain what happened in Procedure 5.

**4.** Calculate the force needed to pull a mass of 20 kg at a uniform slow speed up a plane inclined at an angle of 30° with the horizontal if the coefficient of kinetic friction is 0.20.

**5.** Using an average value of the limiting angle of repose obtained from your data in Procedure 6 and the value of $\mu_k$ from your graph, find the acceleration of the block after it just breaks away in Procedure 6.

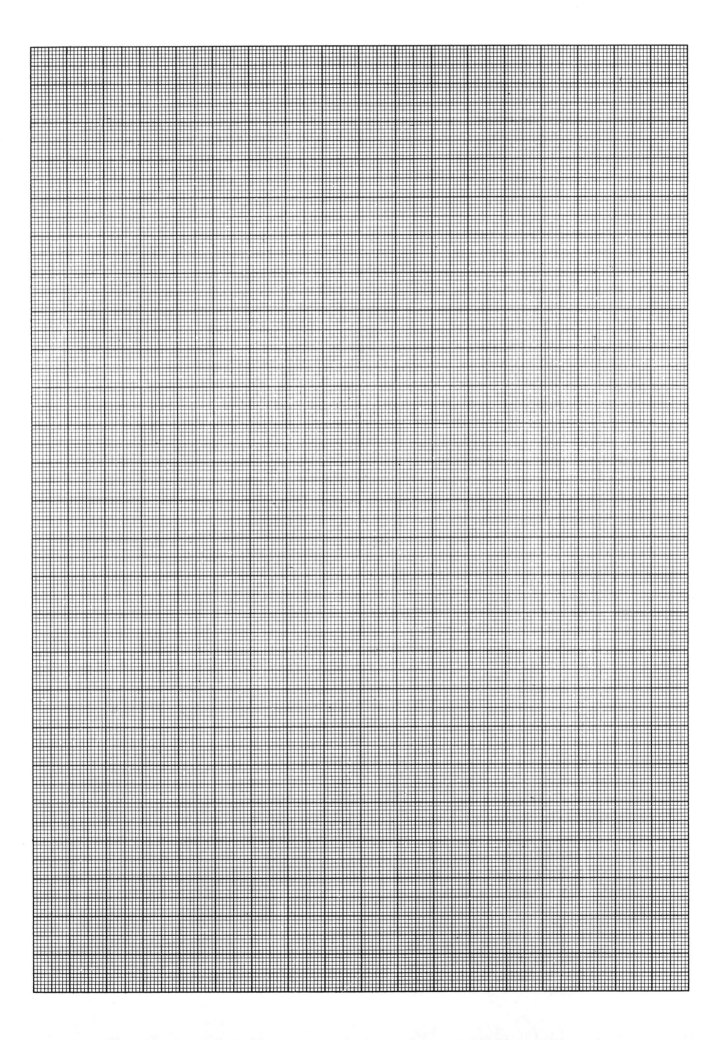

# Simple Machines and the Principle of Work 8

A simple machine is a device by which a force applied at one point is changed into another force applied at a different point for some useful advantage. The pulley system, the jack, the lever, and the inclined plane are well-known examples of such machines. The purpose of this experiment is to study an inclined plane, a wheel and axle, and a pulley system and to determine their characteristics.

## THEORY

A simple machine is a device by means of which a force applied at one point (called the input) gives rise to another force acting at some other point (the output). The purpose of the machine is to make the new force different in magnitude and/or direction from the applied force in such a way as to make some particular piece of work possible or at least easier. We must recall that in physics the word *work* means a force acting through a distance and is given by the product of that force and the distance through which it acts. The units of work are thus foot-pounds or newton-meters (joules) or dyne-centimeters (ergs). Clearly, a given amount of work may be accomplished by a small force moving through a large distance or a large force moving through a small distance; in either case, the product of force and distance will be the same. A simple machine may be thought of as a device for changing the "mix" of force and distance. Ideally (that is, in the absence of losses due to friction), the work put into a simple machine will be the same as that received from it, the purpose of the machine being to increase the force at the expense of distance or vice versa. The machine may also change the direction of the force in the interests of convenience.

A common example is that of the 2000-lb car that must be raised 6 in. off the road in order to change a tire. The work required is 1000 ft·lb, but the average person cannot do this directly. To lift the car at all requires a 2000-lb effort, which is beyond most people's capacity. The fact that the car need be raised only 6 in. is no help—it might as well be 6 mi! 

However, by the use of a jack, the required 2000-lb force is easily developed, the car gets raised the necessary 6 in., and the 1000 ft·lb of work gets done. The operator accomplishes this by applying a force of, say, 50 lb at the jack handle, the jack multiplying this force by 40. But the operator has to pump the handle up and down many times. By the time the car has been raised 6 in., the handle will have been pushed down through a total distance of 6 × 40 = 240 in. or 20 ft. The work of 20 × 50 = 1000 ft·lb at the input is the same as that of ($\frac{1}{2}$) × 2000 = 1000 ft·lb at the output, but the proportion of force and distance involved has been drastically altered. The operator's hand has had to move through a much larger distance than the car has been lifted, but this is usually considered a small price to pay for the ability to do the job at all.

The ratio of the force delivered by a simple machine at its output to the force applied at its input is called its *mechanical advantage*. In the above example of an ideal machine in which there are no frictional losses, this ratio is the same as the ratio of the distance moved at the input to the distance moved at the output. In any practical machine, however, frictional losses do exist and must be supplied by additional work at the input. Thus, in the jack, frictional forces in its mechanism might require that a force of 60 lb rather than 50 be exerted on the handle in order to develop the 2000-lb lifting force at the output. The *actual mechanical advantage* (A.M.A.) is then only 2000/60, or 33.3 instead of 40. The handle must nevertheless be pumped through the same total of 20 ft to raise the car 6 in. as before. Thus, under the ideal conditions of no friction, the mechanical advantage the jack would have remains the ratio of the total distance moved by the handle to the distance the car is raised: 20 ÷ $\frac{1}{2}$, or 40. This is called the *ideal mechanical advantage* (I.M.A.) and is given for any simple machine by the ratio of the distance moved at the input to the distance moved at the output. In equation form

$$\text{I.M.A.} = \frac{s}{h} \qquad (8.1)$$

where $s$ is the distance through which the force applied at the machine's input moves and $h$ is the distance through which the force delivered by the machine moves. The actual mechanical advantage is given by

$$\text{A.M.A.} = \frac{P}{F} \qquad (8.2)$$

where $P$ is the output force exerted by the machine (such as the weight of the car being lifted) and $F$ is the applied force (such as the force applied to the jack handle). Since in most cases the primary purpose of the simple machine is to multiply force, the actual mechanical advantage will be the machine's most important characteristic.

Ideally, the same amount of work is delivered at the output of a simple machine as is put in at the input, but in all actual cases extra work must be put in to make up for frictional losses in the machine. We are therefore interested in the machine's efficiency, which is a measure of how much of the work put in is actually delivered at the output. Specifically, the efficiency of a machine is defined as the ratio of the work delivered at the machine's output to the work done at the input. Thus,

$$\text{Efficiency} = \frac{\text{output work}}{\text{input work}} = \frac{hP}{sF} \qquad (8.3)$$

Note that for the ideal machine the efficiency is 1 or 100% since all of the work put in is delivered at the output. For practical machines, the efficiency will be a fraction less that 1 and may be expressed as a percentage by multiplying by 100. Note also that since the I.M.A. is $s/h$ and the A.M.A. is $P/F$, the efficiency can be expressed as

$$\text{Efficiency} = \frac{\text{A.M.A.}}{\text{I.M.A.}} \qquad (8.4)$$

In the example of the jack, in which frictional losses required the input force to be 60 lb rather than 50 lb, the above considerations show that the efficiency is $2000(\frac{1}{2})/60(20) = 33.3/40 = .83$ or 83%.

In this experiment an inclined plane, a wheel and axle, and a pulley system will be studied. In all of these cases a known load is raised against the gravitational force, and the applied force is measured. The ideal mechanical advantage is easily derived in each case as follows.

**Inclined Plane**  When a block rests on an inclined plane as shown in Fig. 8.1, its weight W, which acts vertically downward, may be resolved into two components, one perpendicular to the plane and one parallel to it. The component of the weight perpendicular to the plane is $W \cos \theta$, and the component parallel to the plane is $W \sin \theta$, where $\theta$ is the angle made by the plane with the horizontal. The plane also exerts a force on the block. If the contacting surfaces are frictionless, this force (designated by $N$ in Fig. 8.1) is

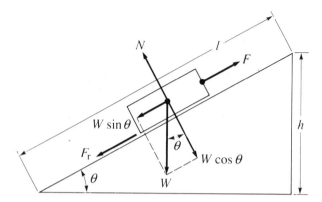

**Figure 8.1**  *The inclined plane as a simple machine*

perpendicular to the plane and is equal to the component of the block's weight perpendicular to the plane. In the absence of friction, a force $W \sin \theta$ directed up the plane will then keep the block in equilibrium, either at rest or moving up or down the plane with uniform velocity. Thus, the plane enables us to raise the block to height $h$ with a force $W \sin \theta$ instead of a force $W$, which would be required to lift the block directly. The plane allows the required force to be reduced by a factor of the sine of the plane angle. However, the distance through which the block must be moved to get it to the top of the plane is $l$, so that the work done is $(W \sin \theta)l$. Since, as Fig. 8.1 shows, $l \sin \theta = h$, this work is the same as $Wh$, the work required to lift the block directly. Notice that for the inclined plane

$$\text{I.M.A.} = \frac{l}{h} = \frac{l}{l \sin \theta} = \frac{1}{\sin \theta}$$

Thus, the ideal mechanical advantage of an inclined plane is the reciprocal of the sine of the plane angle (the cosecant of the plane angle).

If there is friction between the block and the plane, the force $F_r$ of friction will be directed down the plane when the block is being pulled upward by the applied force $F$ shown in Fig. 8.1. Hence the value of $F$ needed to keep the block moving up the plane at constant speed will be $W \sin \theta + F_r$, where $F_r$ is the force of kinetic or sliding friction and is given by

$$F_r = \mu_k N = \mu_k W \cos \theta$$

Then

$$F = W \sin \theta + \mu_k W \cos \theta$$

and

$$\text{A.M.A.} = \frac{W}{W \sin \theta + \mu_k W \cos \theta}$$

$$= \frac{1}{\sin \theta + \mu_k \cos \theta} \qquad (8.5)$$

Moreover, the input work is $Fl = Wl(\sin \theta + \mu_k \cos \theta)$, so that the efficiency is

$$\text{Efficiency} = \frac{Wh}{Fl} = \frac{Wh}{Wl(\sin \theta + \mu_k \cos \theta)}$$

$$= \frac{\sin \theta}{\sin \theta + \mu_k \cos \theta}$$

$$= \frac{1}{1 + \mu_k \cot \theta} \qquad (8.6)$$

Thus, the efficiency is reduced below 100% by the addition of the $\mu_k$ term in the denominator. Only in the case of $\mu_k = 0$ (no friction) do we get a 100% efficient inclined plane.

**Wheel and Axle**  Figure 8.2 is an end view of a wheel and axle. The load $W$ is supported by a cord wrapped around an axle whose radius is $r$, while the operator applies force $F$ to another cord wrapped around a wheel of radius $R$ fixed on the axle. The load $W$ exerts a clockwise torque $Wr$ about the axis of the system, and the operator exerts a counterclock-wise torque $FR$ about this axis by pulling with force $F$ on the cord wrapped around the wheel. If there is no friction in the bearings, equilibrium will exist with $W$ at rest or ascending or descending at constant speed if the two torques are equal. Thus, $W = (R/r)F$ and the ideal mechanical advantage is the ratio $R/r$ of the wheel radius to the axle radius. That this is the I.M.A. as defined in Equation 8.1 may be seen from the fact that when the wheel and axle turns through one revolution, a length of the load-supporting cord equal to one axle circumference gets wrapped around the axle, so that the load is raised by a distance $2\pi r$. Similarly, the operator must draw a length $2\pi R$ of cord off of the wheel. Therefore, the input force $F$ moves through this distance. Hence,

$$\text{I.M.A.} = \frac{s}{h} = \frac{2\pi R}{2\pi r} = \frac{R}{r}$$

If there is friction in the bearings, the operator must exert an additional torque beyond $FR$ to overcome this friction and must therefore pull with a force greater than $F$. The A.M.A. and the efficiency are thus reduced accordingly.

**Pulley System**  A typical pulley system is shown in Fig. 8.3. In the absence of friction in the pulleys, the tension in the cord is the force $F$ with which the operator pulls on the end. But in effect, three strands of the cord pull on the load $W$, two via the pulley that is

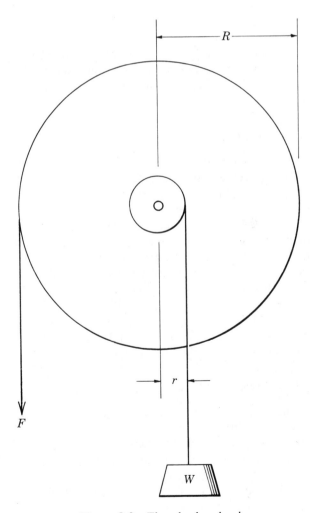

**Figure 8.2**  *The wheel and axle*

**Figure 8.3**  *Pulley system*

attached to $W$ and one that is attached to $W$ directly. Thus, the upward pull on $W$ is $3F$. At equilibrium ($W$ at rest or moving up or down at constant speed), $3F$ must be equal to $W$, and the ideal mechanical advantage is 3. If there is friction in the pulleys, a force larger than $W/3$ must be exerted on the cord by the operator and the actual mechanical advantage becomes less than 3. However, in either case, lifting the weight $W$ through 1 ft requires that each strand attached to it be shortened by 1 ft, which means that 3 ft of cord must be pulled out of the system. Thus, force $F$ must move through a distance three times that

through which the load is raised, confirming that I.M.A. = 3. An interesting point to note here is that in any pulley system consisting of a set of pulleys attached to a fixed support and another set attached to and moving with a load, the ideal mechanical advantage is equal to the number of strands of the cord attached to the load either directly or via the load-end pulleys. A pulley system using just one cord is, of course, assumed. Note that the uppermost pulley in Fig. 8.3 contributes nothing to the mechanical advantage. It merely gives us the convenience of pulling down on the cord at $F$ rather than having to pull up.

## APPARATUS

1. Board with pulley at one end
2. Rods, bench clamp, and right-angle clamp for supporting the inclined plane
3. Wood block with a cord attached to it
4. Set of known weights
5. Weight hanger
6. Triple-beam balance
7. Meter stick
8. Protractor
9. Wheel and axle
10. Pulley system
11. Spring balance
12. Lintless dust cloth or paper wipers

## PROCEDURE

### Inclined Plane

**1.** Weigh the wood block and record the weight.
**2.** Measure the length of the board and record this value.
**3.** Adjust the board as an inclined plane and set it at an angle of 10° to the horizontal with the pulley at the upper end. Be careful about measuring the angle so that precise values of the sine and cosine may be obtained.
**4.** Clean the surface of the plane and the bottom of the block with the lintless dust cloth or paper wipers. *After this has been done, do not touch these surfaces with your hands.*
**5.** Lay the block on the plane, run the cord over the pulley, and attach the weight hanger to the cord's end. Place some weights on the hanger and slowly increase this load until it is just sufficient to keep the block sliding slowly up the plane with constant speed after it has been started with a very small push. Record the load so determined.
**6.** Measure the height $h$ of the inclined plane. Be careful in doing this not to include the thickness of the board. Thus, if you measure $h$ from the laboratory bench surface, measure it to the edge where the *bottom* surface of the board meets the board's upper end. Record the value of $h$.
**7.** Set the plane at an angle of 20° and repeat Procedures 3, 5, and 6. Again use care in measuring the plane angle and the height.

**8.** Set the plane at an angle of 30° and repeat Procedures 3, 5, and 6, once more using care to obtain precise measurements of your new values of $\theta$ and $h$.

### Wheel and Axle

**9.** Wind one string several times around the axle of the wheel-and-axle assembly. Wind the other string several times in the reverse direction around the wheel.
**10.** Hang a 500-g mass on the string coming from the axle. This is the load to be lifted. Attach the weight hanger to the string coming from the wheel and place weights on it until the load moves upward at a small, constant speed. Record the total weight needed to do this.
**11.** Measure the diameter of the wheel up to the surface on which the string is wound. Similarly, measure the diameter of the axle. Record these values.

### Pulley System

**12.** With a spring balance, measure the force necessary to lift a mass of 1 kg with the pulley system set up in the laboratory. Observe the force while the mass is being lifted at a slow, uniform speed and record the value of this force.
**13.** Count the number of supporting strands in the pulley system and record it.

## DATA _____

### Inclined Plane

Weight of wood block _____    Length of board used as plane _____

|  | Angle of the inclined plane | | |
|---|---|---|---|
|  | 10° | 20° | 30° |
| Force needed to pull block up plane |  |  |  |
| Height $h$ of the inclined plane |  |  |  |
| Sine of the plane angle $\theta$ |  |  |  |
| Cosine of the plane angle $\theta$ |  |  |  |

### Wheel and Axle

Load used in the wheel-and-axle experiment _____    Diameter of the wheel _____
Force needed to lift this load _____    Diameter of the axle _____

### Pulley System

Load on the pulley system _____    Number of supporting strands _____
Force needed to lift this load _____

### Mechanical Advantage and Efficiency

| Machine | Actual mechanical advantage | Ideal mechanical advantage | Efficiency | Coefficient of kinetic friction |
|---|---|---|---|---|
| Inclined plane at 10° |  |  |  |  |
| Inclined plane at 20° |  |  |  |  |
| Inclined plane at 30° |  |  |  |  |
| Wheel and axle |  |  |  |  |
| Pulley system |  |  |  |  |

## CALCULATIONS _____

**1.** From the data of Procedures 2, 3, and 6, calculate the sine of the plane angle by dividing the height of the plane by its length. Compare this result with the value given in a table of trigonometric functions or by your calculator for the sine of 10°. Your two values should agree to three significant figures. Record your calculated value in the space provided on the data sheet.

**2.** Using the formula $\sin^2\theta + \cos^2\theta = 1$, calculate the value of cos 10° from your value of sin 10°. Check your result with the tabulated value of cos 10° and enter it on the data sheet.

**3.** Repeat Calculations 1 and 2 for the 20° and 30° plane angles using the data obtained in Procedures 7 and 8.

**4.** From the data of Procedures 1, 3, and 5, determine the actual mechanical advantage of the inclined plane set at an angle of 10°. The ideal mechanical advantage is the reciprocal of the sine of the plane angle already found. Record this I.M.A. and calculate and record the 10° inclined plane's efficiency.

**5.** Use either Equation 8.5 or 8.6 to compute the coefficient of kinetic friction between the block and the plane.

**6.** Repeat Calculations 4 and 5 for the inclined plane set at 20° using the data from Procedure 7.

**7.** Repeat Calculations 4 and 5 for the inclined plane set at 30° using the data from Procedure 8.

**8.** From the data of Procedures 9–11, determine the actual mechanical advantage, the ideal mechanical advantage, and the efficiency of the wheel and axle used.

**9.** From the data of Procedures 12 and 13, determine the actual mechanical advantage, the ideal mechanical advantage, and the efficiency of the pulley system used.

## QUESTIONS

**1.** Should the coefficients of kinetic friction obtained for the three angular settings of the inclined plane be the same? Are they?

**2.** The instructions for Calculation 1 suggest that your calculated value of the sine of the plane angle is better than the value obtained by looking up the sine of this angle as measured by the protractor. Why should this be true?

**3.** (a) State how the mechanical advantage of the inclined plane varies with the inclination of the plane. (b) Does this apply to both the ideal and the actual mechanical advantage?

**4.** (a) How does the efficiency of the inclined plane vary with the inclination of the plane? (b) Explain the reason for this.

**5.** A body weighing 400 lb is pulled up an inclined plane 15 ft long and 5 ft high. The force required to pull the body up the plane at a slow uniform speed is 200 lb. Compute the ideal mechanical advantage, the actual mechanical advantage, and the efficiency of the plane under these conditions. Also compute the coefficient of kinetic friction between the body and the plane.

**6.** Calculate the amount of work done in drawing a mass of 20 kg at a uniform slow speed up a plane 2 m long, making an angle of 30° with the horizontal, if the coefficient of kinetic friction is 0.20.

**7.** How does the uppermost pulley in the system of Fig. 8.3 affect the A.M.A. of this machine? How does it affect the efficiency? Discuss briefly why you think (or do not think) it is worth having.

**8.** The block and tackle of Fig. 8.3 is used to lift a stone weighing 300 lb. The force required is 125 lb. Calculate the actual mechanical advantage and the efficiency of this simple machine.

# Young's Modulus 9

All bodies are deformed in some way by the application of a force. An elastic body is a body that changes in size or shape upon the application of a distorting force, but returns to its original condition when that force is removed. But there is a limit to the magnitude of the force that may be applied if the body is to return to its original condition. This is called the elastic limit. A greater force will cause a permanent distortion or even a break. In this experiment, a steel wire is stretched by the application of a force, and the resulting changes in length are measured. From these measurements, Young's modulus for a steel wire is calculated.

## THEORY

When a wire is stretched by a mass hanging from it, the wire is said to experience a tensile strain, which is defined as the elongation per unit length. The force acting on the wire is a measure of the tensile stress, which is defined as the force per unit area over which it is acting.

Hooke's law states that as long as a body is not strained beyond its elastic limit, the stress is proportional to the strain. Thus, the ratio of stress to strain is a constant and is called a modulus of elasticity. For the case of a wire stretched by a force, this constant is called Young's modulus. Thus,

$$Y = \frac{\text{force per unit area}}{\text{elongation per unit length}} = \frac{F/A}{e/L} = \frac{FL}{eA}$$

where $Y$ is Young's modulus in newtons per square meter, $F$ is the force in newtons, $L$ is the length of the wire in meters, $e$ is the elongation in meters, and $A$ is the area of cross section of the wire in square meters.

The arrangement of the apparatus used in this experiment is shown in Fig. 9.1. The elongations produced in this experiment are very minute, hence a method called an "optical lever" is used to magnify them. This consists of a vertical mirror mounted on a horizontal T-shaped base having pointed legs at the three extremities. The wire whose elongation is to be measured is clamped at its upper end and has a small cylinder attached to it near the lower end. This cylinder is free to move up and down through a hole in a shelf supporting the mirror and its base. The two front feet of the base rest on the shelf and form an axis of rotation for the mirror. The rear foot rests on the cylinder and moves up and down with it due to the elongation of the wire. A laser is arranged to shine a narrow light ray on the mirror, which reflects it to a vertical strip of paper attached to a wall or screen two or three meters away (see Fig. 9.1). The spot where the ray strikes the paper when the wire is unstretched is marked. As the wire is stretched by applying a load, the angular movement of the mirror rotates the reflected ray so that the spot moves up the paper strip. The elongation can be readily computed

**Figure 9.1** *Young's modulus apparatus*

from the relation

$$e = \frac{Sd}{2D} \qquad (9.1)$$

where $e$ is the elongation in meters, $D$ is the distance in meters from the mirror to the paper strip, $d$ is the perpendicular distance from the moving foot of the optical lever to the line joining the other two feet,

and $S$ is the distance along the paper strip from the mark made before the load was applied to the position of the spot after elongation. This relation is derived by using the principle of optics that the reflection of a beam of light falling on a plane mirror is turned through an angle twice as large as the angle through which the mirror is rotated, the incident beam remaining fixed.

## APPARATUS

1. Young's modulus apparatus
2. Laser with support stand
3. Mirror mounted on T-base (optical lever)
4. Roll of white calculator paper and masking tape
5. Slotted weights (ten 1-kg weights)
6. Micrometer caliper
7. Vernier caliper
8. 2-m stick

## PROCEDURE

1. Measure the length of the steel wire used, that is, the distance from the lower end of the upper chuck to the top of the cylinder at the lower end of the wire.
2. Measure the diameter of the wire in five places, using the micrometer caliper.
3. Measure the distance of the moving foot of the optical lever from the axis of rotation. This may be done by pressing the pointed legs on a piece of white paper and constructing the required distance. Use the vernier caliper for this measurement.
4. Carefully place the optical lever with its two front feet in the groove provided on the rigid support and its rear foot on the cylinder fixed to the wire.
5. Apply a load of 1 kg to the wire to take up any slack.
6. Mount the laser so that its beam strikes the mirror and is reflected to a wall or screen about 2 m away. The position of the laser, the cylinder on which the optical lever's moving foot rests, and the orientation of the mirror should be adjusted so that the feet of the optical lever are approximately in a horizontal plane and the path of the ray from the mirror to the wall or screen is nearly horizontal. CAUTION: Although the low-power laser used here cannot hurt you even if its beam falls directly on your skin, *do not look into the beam at any time.*
7. Tape a strip of white calculator paper to the wall or screen, positioning it vertically with the laser light spot falling near its center. Mark the position of

the spot on the paper.
8. With the 2-m stick, carefully measure the distance from the mirror to the mark just made. A convenient way of doing this is to make the distance exactly 2 m by positioning the apparatus so that the stick just fits between the wall or screen at the position of the mark and the mirror.
9. Add a second kilogram to the load on the wire and mark the new position of the laser light spot on the paper.
10. Add 1 kg at a time up to a total of 10 kg, marking the spot position after each added load.
11. Remove the load, 1 kg at a time, marking the spot position for each load. *Note:* The values of the load to be used apply only to a steel wire. If a different kind of wire is used, the proper values will be suggested by the instructor. Label each of your marks with the pertinent load value and indicate whether the load was increasing or decreasing.
12. Turn off the laser and with the vernier caliper measure the distance from the zero mark made with the initial 1-kg load to each of the marks made as the load was increased. These are your values of $S$ for an increasing load. Measure the distance from the zero mark made when you had only the 1-kg load left at the end of the run to each of the marks made as the load was decreased. These are your values of $S$ for a decreasing load. Record all these measurements in the data table.

## DATA

Length of wire used _____    Diameter of wire _____
Micrometer readings:

Trials

1. _____  2. _____  3. _____  4. _____  5. _____

Average reading     _____     Zero reading     _____

Cross-sectional area of the wire     _____     Young's modulus (computed value)     _____

Distance from front legs to the rear leg of the optical lever     _____

Distance from mirror to scale     _____     Young's modulus (from Table V at the end of the book)     _____

Percent error     _____

| Total load, kg | Value of S (increasing load) | Value of S (decreasing load) | Average value of S | Elongation e |
|---|---|---|---|---|
| 2 | | | | |
| 3 | | | | |
| 4 | | | | |
| 5 | | | | |
| 6 | | | | |
| 7 | | | | |
| 8 | | | | |
| 9 | | | | |
| 10 | | | | |

## CALCULATIONS

**1.** Compute the average of the readings for the diameter of the wire. This will be the value of the diameter when properly corrected for the zero reading of the micrometer.

**2.** Calculate the cross-sectional area of the wire in square meters.

**3.** Compute the elongation corresponding to each load using the average value of $S$ obtained in each case. Note that as the 1-kg load corresponds to your zero mark, the elongation calculated with the total load of 2 kg is that produced by a load of 1 kg; the elongation calculated with the total load of 3 kg is that produced by 2 kg; etc.

**4.** Plot a curve on the graph paper using loads in newtons as abscissas and the resulting elongations in meters as ordinates. Note that the load producing a given elongation is that due to the number of kilograms on the weight hanger minus the 1 kg used for the zero reading. Thus, the maximum load appearing as an abscissa in this graph is 9 kg $\times$ 9.8 = 88.2 N.

**5.** Compute Young's modulus for the wire from the slope of your graph. Compare your result with the value given in Table V at the end of the book and find the percent error.

## QUESTIONS _____

**1.** State what the curve shows and hence what the experiment checked.

**2.** Why was not the whole length of the wire measured?

**3.** Define elasticity, elastic limit, stress, and strain.

**4.** (a) State Hooke's law. (b) Does your curve agree with it? (c) How would the curve look if the elastic limit were exceeded?

**5.** Derive Equation 9.1 from a consideration of the apparatus geometry and the principle of optics stated in the Theory section.

# Rotational Motion 10

The motion of the flywheel of a steam engine, an airplane propeller, and any rotating wheel are examples of a very important type of motion called rotational motion. If a rigid body is acted upon by a system of torques, the body will be in equilibrium, as far as rotational motion is concerned, if the sum of the torques about any axis is zero. This means that if the body is at rest, it will remain at rest; if it is rotating about a fixed axis, it will continue to rotate about the same axis with uniform angular speed. However, if an unbalanced torque is acting on the body, it will produce an angular acceleration in the direction of the torque, the acceleration being proportional to the torque and inversely proportional to the moment of inertia of the body about its axis of rotation. The purpose of this experiment is to study rotational motion, to observe the effect of a constant torque upon a body that is free to rotate, and to determine the resulting angular acceleration and the moment of inertia of that body.

## THEORY

Rotation refers to motion about some axis in space. In describing this type of motion, the angular displacement, the angular velocity, and the angular acceleration must be given. The unit of angular measurement usually used is the radian. This is the angle subtended at the center of a circle by an arc equal in length to the radius. The angular velocity is the time rate of change of angular displacement. It is equal to the angle through which the body rotates divided by the time during which the rotation takes place. The angular acceleration is the time rate of change of angular velocity.

The equations of motion involved in uniformly accelerated angular motion are discussed below. The angular displacement is given by

$$\theta = \overline{\omega}t$$

where $\theta$ is the angular distance in radians, $\overline{\omega}$ is the average angular velocity in radians per second, and $t$ is the time in seconds. The average angular velocity is given by

$$\overline{\omega} = \frac{\omega_1 + \omega_2}{2}$$

where $\overline{\omega}$ is the average angular velocity, $\omega_1$ is the initial angular velocity, and $\omega_2$ is the final angular velocity. The final angular velocity is given by

$$\omega_2 = \omega_1 + \alpha t$$

where $\alpha$ is the angular acceleration in radians per second per second, $t$ is the time in seconds, and $\omega_1$ and $\omega_2$ have the meanings already assigned.

The corresponding equations involved in uniformly accelerated linear motion are

$$s = \overline{v}t$$

$$\overline{v} = \frac{v_1 + v_2}{2}$$

and

$$v_2 = v_1 + at$$

where $s$ is the distance, $\overline{v}$ is the average velocity, $v_1$ is the initial velocity, $v_2$ is the final velocity, $a$ is the acceleration, and $t$ is the time.

Using the radian as the unit of angular measurement gives rise to some very simple relations between the linear quantities and the angular quantities. Suppose that $s$ represents the distance a particle moves around the circumference of a circle of radius $r$; let $v$ be the linear speed and $a$ the linear or tangential acceleration of the particle. Then the angular distance $\theta$, the angular velocity $\omega$, and the angular acceleration $\alpha$ are given by the relations

$$s = r\theta \tag{10.1}$$

$$v = r\omega \tag{10.2}$$

$$a = r\alpha \tag{10.3}$$

When a torque is applied to a body free to rotate about a fixed axis, then the body will acquire an angular acceleration given by the relation

$$\tau = I\alpha \tag{10.4}$$

where $\tau$ is the sum in meter-newtons of all the torques about the fixed axis of rotation, $I$ is the moment of inertia in kilogram-meters$^2$ of the body about the same axis, and $\alpha$ is the angular acceleration in radians per second per second.

The moment of inertia is the inertia of a body as regards rotation about an axis. The moment of inertia of a body with respect to an axis is the sum of the products obtained by multiplying the mass of every particle of the body by the square of its distance from the axis. For a simple geometrical solid it can be readily determined by the use of integral calculus. For example, the moment of inertia of a uniform cylinder, or disk, with respect to its longitudinal axis is given by

$$I = \frac{1}{2}MR^2 \qquad (10.5)$$

where $M$ is the mass of the cylinder or disk; $R$ is its radius in meters; and $I$ is the moment of inertia in kilogram-meters$^2$.

The apparatus used in this experiment is shown in Fig. 10.1. It consists of a disk mounted on a shaft which is carried in a very good bearing in the base plate so that the disk is free to rotate with a minimum of friction. On its underside (and thus not visible in the figure), the disk carries a drum with three diameters around any one of which a string may be wound so as to exert a torque on the rotating assembly. The string passes over a pulley and has a weight attached to the end. When this weight is released, it descends with a constant linear acceleration, while the rotating assembly turns with a constant angular acceleration. This situation is illustrated in a simplified form in Fig. 10.2. Note that in the actual apparatus the disk is mounted in a horizontal plane so that a second disk,

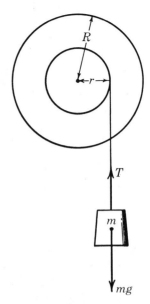

**Figure 10.2**   *Forces acting on a falling mass*

a ring, and a bar may be loaded onto it to vary the moment of inertia of the complete assembly. A pulley is thus required to change the string direction from horizontal to vertical. This pulley is chosen to be very light and small so as to contribute a negligible moment of inertia, and its contribution to the friction in the apparatus will be included with the friction in the bearing supporting the disk. As we will neglect the effect of the pulley, Fig. 10.2 is drawn as if the disk of radius $R$ were in a vertical plane so that the string, which is wound around a drum of radius $r$, can be shown going straight down to the falling mass $m$. We see that there are two forces acting on this

**Figure 10.1**   *Apparatus for the study of rotational motion*

mass: the downward force of gravity, that is, the mass's weight $mg$, and the upward pull of the string, namely its tension $T$. The resultant force acting on $m$ is thus

$$F = mg - T$$

But by Newton's second law of motion $F = ma$. Therefore,

$$mg - T = ma \qquad (10.6)$$

where $a$ is the acceleration of the falling mass in meters per second per second.

From Fig. 10.2 we see that the torque acting on the rotating assembly is

$$\tau_a = Tr \qquad (10.7)$$

However, a second torque $\tau_f$ due to friction in the bearings is also acting, so that the resultant torque $\tau$ is the difference $\tau_a - \tau_f$. Note that the frictional torque is subtracted because friction always operates in the direction opposing the motion.

After an appropriate rearrangement of terms, substitution in Equation 10.4 then yields

$$\tau_a = I\alpha + \tau_f \qquad (10.8)$$

Equation 10.8 shows that the angular acceleration of the rotating assembly depends on the applied torque, the frictional torque, and the moment of inertia. Assuming the last two remain constant during the experiment, we note that a plot of $\tau_a$ against $\alpha$ should be a straight line with slope $I$ and intercept $\tau_f$. In this experiment, three different drum radii are available, so that $\tau_a$ can be varied by changing both the drum radius $r$ and the mass $m$. The resulting measured values of $\alpha$ will allow an experimentally determined plot of Equation 10.8 to be made. This plot will serve as both a check on the theory (as it should be a straight line) and a means for finding both the moment of inertia $I$ of the rotating assembly and the frictional torque. The experiment also allows for the setting up of various rotating assemblies, all of whose moments of inertia can be calculated from their dimensions, so that a further check on the theory can be made by comparing the calculated and measured values of $I$ in each case.

## APPARATUS

1. Rotational motion apparatus, consisting of a main disk and drum, base plate with bearing and shaft, auxiliary disk, steel ring, and rectangular plate
2. Pulley on rod with mounting clamp
3. Weight hanger with 50- and 100-g weights
4. Stopwatch or stop clock
5. Meter stick
6. Vernier calipers
7. Triple-beam balance
8. Level
9. String

## PROCEDURE

**1.** Set the base plate of the apparatus on the bench and mount the pulley on one end as shown in Fig. 10.1. Position the base plate so that when the string hangs down over the pulley, it will clear the edge of the bench. Place the shaft in the bearing and level the apparatus using the level provided.

**2.** Observe the three-diameter drum attached to the main disk and measure each diameter with the vernier calipers. Measure the main disk diameter with the meter stick and weigh this disk on the triple-beam balance. Record all these measurements.

**3.** Wind the string on the largest-diameter drum in a single layer. Choose a length of string so that when the weight hanger is attached to the free end, the main disk is placed on the shaft, and the string is passed over the pulley, the weight hanger will strike the floor on the way down with two or three turns of the string still left on the drum.

**4.** For the first set of measurements use a 50-g weight. Assemble the apparatus as shown in Fig. 10.1 with the weight at its highest convenient position. Measure the distance the weight has to fall, i.e., the distance from the bottom of the weight hanger to the floor. Choose the weight's highest position to give some convenient value for this measurement. One meter is an obvious choice if the bench is high enough. Always start the weight from this position.

**5.** Find the time required for the 50-g weight to reach the floor. With the string wound around the largest drum and the weight at the top of its path as described in Procedures 3 and 4, release the disk and start the stopwatch at the same instant; make sure you give no impulse to the disk as you release it so that it truly starts from rest. Stop the watch at the instant the weight strikes the floor.

**6.** Repeat Procedure 5 twice to obtain three independent observations of the time, and record these results.

**7.** Repeat Procedures 4–6 using the 100-g weight.

**8.** Repeat Procedures 3–7 with the string wound on the intermediate-diameter drum.

**9.** Repeat Procedures 3–7 with the string wound on the smallest-diameter drum.

**10.** Measure the diameter of the auxiliary disk with the meter stick and weigh it on the triple-beam balance. Record these measurements and then place this disk on the shaft on top of the main disk.

**11.** Repeat Procedures 3–9 with the auxiliary disk on top of the main disk so that the rotating assembly consists of the two disks together.

**12.** Weigh the steel ring on the triple-beam balance. Then use the meter stick to measure its inside and outside diameters. Record these measurements, remove the auxiliary disk from the apparatus, and replace it with the steel ring. Notice that the ring has two locating pins which fit into holes in the upper surface of the main disk so that the ring will be properly centered.

**13.** Repeat Procedures 3–9 with the steel ring in place on the main disk so that the rotating assembly is the combination of the disk and the ring.

**14.** Measure the length and width of the rectangular plate with the meter stick and then weigh the plate on the triple-beam balance. Record these measurements and then substitute the plate for the ring on top of the main disk.

**15.** Repeat Procedures 3–9 with the rectangular plate in place on the main disk so that the rotating assembly is the combination of the disk and the plate.

## DATA

Mass of the main disk and drum   _____    Distance weight falls   _____

Diameter of the main disk   _____    Radius of the main disk   _____

Diameter of the largest drum   _____    Radius of the largest drum   _____

Diameter of the intermediate drum   _____    Radius of the intermediate drum   _____

Diameter of the smallest drum   _____    Radius of the smallest drum   _____

Moment of inertia of the main disk and drum from the slope of the line on the graph   _____

Frictional torque from the $y$ intercept on the graph   _____

Moment of inertia of the main disk and drum calculated from the formula $\frac{1}{2}MR^2$   _____

Percent difference between the measured and calculated values of the moment of inertia
    of the main disk   _____

Mass of the auxiliary disk   _____    Distance weight falls   _____

Diameter of the auxiliary disk   _____    Radius of the auxiliary disk   _____

Moment of inertia of the main and auxiliary disks from the slope of the line on the
    graph   _____

Moment of inertia of the auxiliary disk alone   _____

Frictional torque from the $y$ intercept on the graph   _____

Moment of inertia of the auxiliary disk calculated from the formula $\frac{1}{2}MR^2$   _____

Percent difference between the measured and calculated values of the moment of
    inertia of the auxiliary disk   _____

Mass of the steel ring   _____    Distance weight falls   _____

Inner diameter of the steel ring   _____    Inner radius of the steel ring   _____

Outer diameter of the steel ring   _____    Outer radius of the steel ring   _____

Moment of inertia of the main disk and steel ring from the slope of the line on the
    graph   _____

Moment of inertia of the steel ring alone   _____

Frictional torque from the $y$ intercept on the graph   _____

Moment of inertia of the steel ring calculated from the formula $\frac{1}{2}M(R_1^2 + R_2^2)$   _____

Percent difference between the measured and calculated values of the moment of
    inertia of the steel ring   _____

Mass of the rectangular plate _____    Distance weight falls _____

Length of the rectangular plate _____    Width of the rectangular plate _____

Moment of inertia of the main disk and rectangular plate from the slope of the line on
the graph _____

Moment of inertia of the rectangular plate alone _____

Frictional torque from the $y$ intercept on the graph _____

Moment of inertia of the rectangular plate calculated from the formula $\frac{1}{12}M(a^2 + b^2)$ _____

Percent difference between the measured and calculated values of the moment of inertia
of the rectangular plate _____

## Main Disk Alone

|  | Large-diameter drum | | Medium-diameter drum | | Small-diameter drum | |
|---|---|---|---|---|---|---|
|  | 50-g mass | 100-g mass | 50-g mass | 100-g mass | 50-g mass | 100-g mass |
| Time of fall of mass:   1. |  |  |  |  |  |  |
| 2. |  |  |  |  |  |  |
| 3. |  |  |  |  |  |  |
| Average |  |  |  |  |  |  |
| Average velocity of falling mass |  |  |  |  |  |  |
| Final velocity of falling mass |  |  |  |  |  |  |
| Acceleration of falling mass |  |  |  |  |  |  |
| Angular acceleration |  |  |  |  |  |  |
| Tension in the string |  |  |  |  |  |  |
| Torque acting on the rotating assembly |  |  |  |  |  |  |

## Main Disk and Auxiliary Disk

|  | Large-diameter drum | | Medium-diameter drum | | Small-diameter drum | |
|---|---|---|---|---|---|---|
|  | 50-g mass | 100-g mass | 50-g mass | 100-g mass | 50-g mass | 100-g mass |
| Time of fall of mass:   1. |  |  |  |  |  |  |
| 2. |  |  |  |  |  |  |
| 3. |  |  |  |  |  |  |
| Average |  |  |  |  |  |  |
| Average velocity of falling mass |  |  |  |  |  |  |
| Final velocity of falling mass |  |  |  |  |  |  |
| Acceleration of falling mass |  |  |  |  |  |  |
| Angular acceleration |  |  |  |  |  |  |
| Tension in the string |  |  |  |  |  |  |
| Torque acting on the rotating assembly |  |  |  |  |  |  |

## Main Disk and Steel Ring

|  | Large-diameter drum | | Medium-diameter drum | | Small-diameter drum | |
|---|---|---|---|---|---|---|
|  | 50-g mass | 100-g mass | 50-g mass | 100-g mass | 50-g mass | 100-g mass |
| Time of fall of mass:      1. |  |  |  |  |  |  |
| 2. |  |  |  |  |  |  |
| 3. |  |  |  |  |  |  |
| Average |  |  |  |  |  |  |
| Average velocity of falling mass |  |  |  |  |  |  |
| Final velocity of falling mass |  |  |  |  |  |  |
| Acceleration of falling mass |  |  |  |  |  |  |
| Angular acceleration |  |  |  |  |  |  |
| Tension in the string |  |  |  |  |  |  |
| Torque acting on the rotating assembly |  |  |  |  |  |  |

## Main Disk and Rectangular Plate

|  | Large-diameter drum | | Medium-diameter drum | | Small-diameter drum | |
|---|---|---|---|---|---|---|
|  | 50-g mass | 100-g mass | 50-g mass | 100-g mass | 50-g mass | 100-g mass |
| Time of fall of mass:      1. |  |  |  |  |  |  |
| 2. |  |  |  |  |  |  |
| 3. |  |  |  |  |  |  |
| Average |  |  |  |  |  |  |
| Average velocity of falling mass |  |  |  |  |  |  |
| Final velocity of falling mass |  |  |  |  |  |  |
| Acceleration of falling mass |  |  |  |  |  |  |
| Angular acceleration |  |  |  |  |  |  |
| Tension in the string |  |  |  |  |  |  |
| Torque acting on the rotating assembly |  |  |  |  |  |  |

## CALCULATIONS _____

**1.** From the data of Procedures 4, 5, and 6, calculate the average time of descent.

**2.** Similary, calculate the average time of descent for each set of data in Procedures 7, 8, and 9.

**3.** From the data of Procedure 2, compute the radius of each drum and the radius of the main disk.

**4.** Using the average time of fall, compute the average velocity of the falling mass for each set of observations.

**5.** From the value of the average velocity, calculate the final velocity of the mass as it strikes the floor, for each set of observations. Notice that since the system starts from rest and experiences a uniform acceleration, the final velocity will be twice the average velocity.

**6.** From the average time of fall and the final velocity, calculate the acceleration of the falling mass for each set of observations.

**7.** The value of the acceleration found in Calculation 6 is also the tangential acceleration of a point on the circumference of the axle. Using Equation 10.3, find the angular acceleration $\alpha$ for each set of observations.

**8.** Using Equation 10.6, calculate the tension in the string for each set of observations from the values of the acceleration found in Calculation 6.

**9.** Calculate the applied torque for each set of observations by using Equation 10.7.

**10.** Plot the values of the angular acceleration $\alpha$ along the $x$ axis and the corresponding values of the applied torque $\tau_\alpha$ along the $y$ axis. Draw the best straight line through the six points.

**11.** Determine the value of the moment of inertia $I$ of the main disk and drum by computing the slope of the straight line drawn in Calculation 10.

**12.** Determine the value of the frictional torque $\tau_f$ by reading the value of the $y$ intercept of the straight line drawn in Calculation 10.

**13.** Calculate the moment of inertia $I$ of the main disk from Equation 10.5 where $M$ is the disk's mass as measured in Procedure 2 and $R$ its radius as found in Calculation 3.

**14.** Find the percent difference between the measured and calculated moments of inertia found in Calculations 11 and 13.

**15.** Repeat Calculations 1–12 for the main disk and auxiliary disk using the data obtained in Procedures 10 and 11.

**16.** Find the moment of inertia of the auxiliary disk by subtracting the moment of inertia of the complete rotating assembly from that of the main disk alone found in Calculation 11.

**17.** Calculate the moment of inertia of the auxiliary disk from the formula $I = \frac{1}{2}MR^2$ using the mass of this disk measured when you repeated Procedure 2 and the radius obtained when you repeated Calculation 3.

**18.** Find the percent difference between the measured and calculated moments of inertia found in Calculations 16 and 17.

**19.** Repeat Calculations 1–12 for the main disk and steel ring using the data of Procedures 12 and 13. Notice that in repeating Calculation 3 for the steel ring, both an inside and an outside radius must be found.

**20.** Repeat Calculations 16–18 to find a measured and calculated value for the moment of inertia of the steel ring and the percent difference between these results. The formula for the ring's moment of inertia is

$$I = \frac{1}{2}M(R_1^2 + R_2^2) \tag{10.9}$$

where $M$ is the mass of the ring measured when you repeated Procedure 2 and $R_1$ and $R_2$ are the two radii obtained when you repeated Calculation 3.

**21.** Repeat Calculations 1–12 for the main disk and rectangular plate using the data of Procedures 14 and 15. Notice that in repeating Calculation 3 for the rectangular plate there is no radius to measure. The length and width were found in Procedure 14.

**22.** Repeat Calculations 16–18 to find a measured and calculated value for the moment of inertia of the rectangular plate and the percent difference between these results. The formula for the plate's moment of inertia is

$$I = \frac{1}{12} M (a^2 + b^2) \tag{10.10}$$

where $M$ is the mass of the plate obtained when you repeated Procedure 2 and $a$ and $b$ are the length and width measured in Procedure 14.

## QUESTIONS _____

**1.** The radius of gyration of a rotating body is the distance from the axis of rotation to the point at which the entire mass of the body may be considered concentrated without altering the moment of inertia. It is defined by the expression

$$I = MK^2 \tag{10.11}$$

where $I$ is the body's moment of inertia, $M$ its mass, and $K$ the radius of gyration in question. Use your measured values of the moments of inertia of the main disk, auxiliary disk, steel ring, and rectangular plate to find the radius of gyration of each of these objects.

**2.** Using Equations 10.5, 10.9, 10.10, and 10.11, derive a theoretical expression for the radius of gyration of each of your rotating objects. Put the appropriate values for the masses and dimensions of these objects into these expressions to get numerical values of $K$ and compare them with those found in Question 1.

**3.** Why is the radius of gyration of any wheel always smaller than the wheel's actual radius?

**4.** Consider the disk and drum of Fig. 10.2. Let the disk's mass be $M$ and assume that the drum radius $r$ is small enough so that its moment of inertia may be neglected. Also neglect the friction in the bearings, i.e., take $\tau_f$ to be zero. Using Equations 10.3–10.7, show that the downward acceleration of the falling mass is

$$a = \frac{g}{1 + \dfrac{1}{2}\dfrac{M}{m}\left(\dfrac{R}{r}\right)^2}$$

**5.** By referring to your graphs determine if it is possible for the $y$ intercept to be negative. What would this mean physically?

**6.** In your work with the main disk, you probably found that the value of the moment of inertia calculated from Equation 10.5 was larger than that obtained experimentally, i.e., from the slope of your graph. Explain why this should be so.

**7.** How do the values of the frictional torque $\tau_f$ obtained from your four graphs compare? How should they compare?

**8.** Derive Equation 10.9. If you can, also derive Equations 10.5 and 10.10. *Hint:* The derivation of these last two requires the use of calculus, but Equation 10.9 may be derived without calculus if Equation 10.5 is assumed, for a ring may be considered to be a disk of radius $R_2$, with a hole of radius $R_1$ cut out of it.

**9.** A uniform disk 0.3 m in diameter and having a mass of 2 kg is free to rotate about its horizontal axis on frictionless bearings. An object with a mass of 0.05 kg is attached to a string wound around the rim of the disk. The object is released from rest and descends with constant linear acceleration. Calculate: (a) the moment of inertia of the disk; (b) the linear acceleration of the descending object; (c) the angular acceleration of the disk; and (d) the tension in the string.

# Conservation of Angular Momentum 11

The principle of conservation of angular momentum is the rotational counterpart of the principle of conservation of linear momentum. According to this principle, the angular or rotational momentum of a rotating system remains constant provided no external torques act on it. In this experiment, the system will consist of a rotating bar with a catcher mounted on it into which a ball is fired. The system is designed this way so that the collision forces between the ball and the catcher are internal to the system. Thus, except for the friction in the bearing on which the bar is pivoted (which must be taken into account), no external torques act on the system and its angular momentum should be conserved. The purpose of this experiment is to investigate the law of conservation of angular momentum in this arrangement.

## THEORY

Just as the quantity $mv$ is the linear momentum of a moving object of mass $m$ and velocity $v$, so $I\omega$ is the angular momentum of a rotating body having moment of inertia $I$ and angular velocity $\omega$. And just as it follows from Newton's second law of motion that the linear momentum of a moving system remains constant in the absence of forces external to the system, so also do we find that the angular momentum of a rotating system is constant in the absence of external torques.

In this experiment, the rotating system consists of a heavy bar mounted on a very good bearing (so that it is free to rotate about its center with very little friction), a catcher cup mounted on one end and balanced by a counterweight mounted on the other, and a metal ball which gets fired into the catcher by a spring gun, as shown in Fig 11.1. After the ball has been fired but before it enters the catcher, it has a linear momentum $mv$, where $m$ is its mass and $v$ its velocity. However, any object moving in a straight line with a linear momentum can be considered to have an angular momentum about some selected pivot point, as illustrated in Fig 11.2. This angular momentum is found by multiplying the linear momentum $mv$ by the lever or moment arm $l$ defined as the perpendicular distance from the pivot point in question to the line of action of the linear momentum vector. As a consequence of this definition, angular momentum is sometimes called the *moment of momentum*. Like linear momentum, it is a vector; its direction lies along the supposed axis of rotation, which is a line through the pivot point perpendicular to the plane defined by that point and the line of action of $mv$. This axis is a line perpendicular to the page in Fig 11.2. The sense of the angular momentum vector along this axis is given by the right-hand rule: if the axis is gripped by the right hand so that the fingers curl in the direction of rotation, the thumb points along the axis in the di-

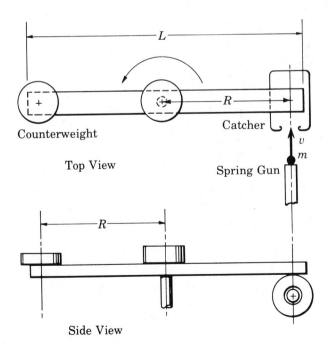

**Figure 11.1** *Angular momentum apparatus*

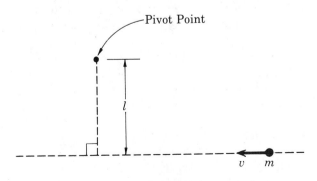

**Figure 11.2** *Moment of momentum*

rection of the vector. In the situation shown in Fig. 11.2, this is into the paper.

After the ball is in the catcher, the system (consisting of the rotating bar, the catcher with the ball in it, and the counterweight) rotates with angular velocity $\omega$. The moment of inertia of the system can be thought of as made up of three parts. The first is the moment of inertia of the bar alone and is given by

$$I_b = \frac{1}{12}ML^2$$

where $I_b$ is the moment of inertia of the bar in kilogram-meters$^2$, $M$ is the mass of the bar in kilograms, and $L$ is the bar's overall length in meters. The second is the moment of inertia of the catcher with the ball in it rotating in a circle of radius $R$, where $R$ is the distance from the pivot to the center of mass of the catcher assembly (see Fig. 11.1). This moment of inertia is given by

$$I_{cb} = m_{cb}R^2$$

where $m_{cb}$ is the total mass of the catcher, its mounting nut, and the ball. Similarly, the third part of the total moment of inertia is that of the counterweight. It moves in a circle of radius $R$ just as the catcher does and has mass $m_{cw}$ which is close to but not necessarily exactly the same as $m_{cb}$. Its moment of inertia is thus given by

$$I_{cw} = m_{cw}R^2$$

so that the total moment of inertia of the system is

$$I = I_b + I_{cb} + I_{cw}$$

$$= \frac{1}{12}ML^2 + (m_{cb} + m_{cw})R^2 \qquad (11.1)$$

The angular momentum of the system right after the ball has been caught in the catcher is then $I\omega$, where $\omega$ is the angular velocity of the bar right after the ball has been caught. The bar was at rest just before this, but the ball in flight from the spring gun had angular momentum $mvl$, where $l$ is the perpendicular distance from the ball's line of flight to the bar's pivot point (the supporting bearing shaft). In Fig. 11.1, where the axis of the catcher cup is set perpendicular to the bar, $l = R$, but the catcher axis can be set at some smaller angle $\theta$, in which case $l$ will be equal to $R \sin \theta$ and the effective angular momentum reduced accordingly. Whatever the value of $l$, the angular momentum before the collision will be $mvl$ and afterwards $I\omega$, so that

$$mvl = \left[\frac{1}{12}ML^2 + (m_{cb} + m_{cw})R^2\right]\omega \quad (11.2)$$

All the quantities in Equation 11.2 can be measured, hence its validity and thus the principle of conservation of angular momentum can be checked.

Most of these quantities are easily found by direct measurement; but two, the speed $v$ given the ball by the spring gun and the angular velocity $\omega$ of the rotating system just after the ball has been caught, have to be determined by special procedures. The initial velocity of the ball is found from measurements of its range and vertical distance of fall when it is fired horizontally and allowed to fall onto the laboratory bench without striking the catcher. The method is the same as that discussed in Experiment 5. The ball is fired horizontally, hence its horizontal velocity component is equal to $v$. This component remains constant throughout the motion, so that the horizontal distance covered in time $t$ is

$$x = vt \qquad (11.3)$$

where $x$ is in meters, $t$ is in seconds, and $v$ is in meters per second.

The initial velocity in the vertical direction is zero, hence the vertical motion is that of a freely falling body dropped from rest. Therefore, the distance traveled in the vertical direction in time $t$ is

$$y = \frac{1}{2}gt^2 \qquad (11.4)$$

where $y$ is the distance of fall, $g$ is the known value of the acceleration of gravity, and $t$ is the same time interval that appeared in Equation 11.3. If the ball, when placed on the gun's firing rod, is at a height $y$ above the bench surface, then the corresponding value of $t$ is the ball's time of flight and can be substituted in Equation 11.3 to give the range $x$. Combining Equations 11.3 and 11.4 in this way yields

$$v = x\sqrt{\frac{g}{2y}} \qquad (11.5)$$

The length $x$, which is just the distance along the bench surface from a point directly under the end of the firing rod to the impact point, is easily measured. Equation 11.5 can then be used to find $v$.

The value needed for $\omega$ is the one in effect immediately after the ball has been caught. If there were no friction in the bearing supporting the rotating system, rotation would continue at this angular velocity indefinitely. However, although the bearing is a very good one, it does exert a small frictional torque that will eventually bring the system to a stop. If the system rotates through a total angle $\phi$ before stopping and takes $t_r$ seconds to do this, then the average value $\overline{\omega}$ of $\omega$ is $\phi/t_r$ and can be measured by simply timing the rotational motion. But to a very good approximation the decelerating torque due to friction in the bearing can be considered constant, so that the rules of constant negative angular acceleration apply. One of these is that the average angular velocity is half the sum of the initial and final velocities. Since the final angular velocity is zero, $\bar{\omega} = \frac{1}{2}\omega$ or $\omega = 2\bar{\omega}$. Thus, the desired value of $\omega$ is determined by dividing the

total angle turned through by the time required for this motion and doubling the result.

## APPARATUS

1. Rotating bar with catcher and counterweight mounted on force table
2. Spring gun from Blackwood ballistic pendulum
3. Mounting rod and clamps for mounting the spring gun
4. Meter stick
5. 45° right triangle

6. Stopwatch or stop clock
7. Triple-beam balance
8. Plumb bob
9. Level
10. Carbon paper
11. Sheet of white paper and masking tape

## PROCEDURE

**1.** Weigh the bar, the catcher and its retaining nut, the counterweight, and the ball separately. Notice that the catcher can be removed from the bar by unscrewing the retaining nut, and the bar can be removed from the bearing shaft in a similar manner.

**2.** Measure the bar's total length $L$ with the meter stick. Also measure the distance from the midpoint of its central hole to the midpoint of either the catcher mounting hole or the counterweight mounting hole. This is the radius $R$ in which the catcher and counterweight revolve.

**3.** Set the spring gun up for the range measurements. To do this, mount it using the mounting rod and bench clamps so that it has a clear range over a level laboratory bench. In this part of the experiment, the rotating system is not used and should be set aside, out of the way of the ball's trajectory. Put the ball on the end of the gun's firing rod and adjust the gun mounting so that the bottom of the ball is exactly 0.5 m above the bench surface. Note that this measurement is made to the *bottom* of the ball rather than its center because it is the bottom of the ball that hits the bench.

**4.** Get the gun ready for firing by placing the ball on the end of the firing rod and pushing it back, compressing the spring until the trigger is engaged. The ball is fired horizontally along the bench so that it will strike the bench surface some distance from the gun. Fire the ball and determine approximately where it lands. Place a sheet of white paper there so that the ball will hit near its center. Secure the paper in position with masking tape. Put a sheet of carbon paper on top of the white paper with the carbon side down so that when the ball strikes, a mark will be left on the white paper at the impact point.

**5.** Fire the ball five times to obtain five range measurements. These measurements are made from a point on the bench surface directly below the end of the firing rod to each impact point. To determine the point directly below the rod end, hang the plumb bob from this end and mark the indicated point on the

bench surface. Note that you want the point directly below the center of the ball as it leaves the gun, and so you should hang the plumb bob from the very end of the firing rod with the rod in its released (fired rather than cocked) position.

**6.** You are now ready to start the conservation of angular momentum trials. Mount the rotating bar on its bearing support, put the counterweight in place on one end, and mount the catcher cup on the other. Before tightening down the cup, adjust it so that its axis is perpendicular to that of the bar. This is done accurately by placing one short edge of the 45° right triangle against the flat provided on the side of the cup and lining the perpendicular edge up with the edge of the bar. Tighten the retaining nut when the cup is properly aligned.

**7.** Place the rotating assembly in front of the gun, as shown in Fig. 11.3. The gun position will have to be changed so that the gun is aimed directly into the catcher cup. The force table supporting the rotating assembly should be placed so that when the gun's firing rod points directly into the catcher cup, the firing rod axis and the rotating bar are at right angles to each other and the indicator on the bar behind the catcher points to zero degrees on the force table scale. Note that aiming the gun at the catcher is facilitated if the two are quite close, although enough space should be left so that the ball is well clear of the firing rod before entering the catcher. Even with enough clearance in this respect, however, the end of the firing rod may be close enough to the catcher after firing to interfere with the free rotation of the bar. Be on the lookout for this, and if you see that the back of the catcher will hit the gun as it comes around, release the gun mounting clamp during the first revolution and turn the gun out of the way. The need for doing this can be avoided by beginning with the gun far enough back from the catcher, but this requires extra care in aiming and probably a few trial runs before an adjustment is found that lets the ball enter the catcher properly on every shot. Finally, level the table by placing the level on the bar and

**Figure 11.3** *Angular momentum apparatus and spring gun*

adjusting the screws in the table feet until the bar is seen to be level regardless of its orientation.

**8.** When the apparatus is working correctly, you can start taking data. Load the gun and set the rotating bar at rest in the proper position. Fire the gun, start the stopwatch at the instant the ball enters the catcher, and stop the stopwatch at the instant the bar just stops turning. While the bar is in motion, count the number of turns, and note the reading of the indicating pointer on the force table scale after rotation stops. Record both the number of turns and the pointer reading; the latter gives the number of degrees through which the bar turned in the uncompleted final revolution.

**9.** Repeat Procedure 8 twice to obtain a total of three sets of data on the angle turned through and the time required.

**10.** Reset the catcher so that its axis is at 45° to the bar. Use the 45° right triangle as in Procedure 6, but this time align the hypotenuse with the bar edge. The rotating system will then have to be moved and the gun position changed so that the gun is once more properly aimed into the catcher. Repeat Procedures 7–9 with the new catcher position.

## DATA

| | | | |
|---|---|---|---|
| Mass of the bar | _____ | Length $L$ of the bar | _____ |
| Mass of the catcher and its retaining nut | _____ | Radius $R$ of the counterweight and catcher orbits | _____ |
| Mass of the counterweight | _____ | Moment of inertia $I$ of the rotating system | _____ |
| Mass of the ball | _____ | | |

Range of the ball   1. _____   2. _____   3. _____   4. _____   5. _____

Deviations   1. _____   2. _____   3. _____   4. _____   5. _____

Average range   _____          Average deviation   _____

Velocity of propagation from range measurements   _____

| Catcher angle, degrees | Rotation time, s | Number of turns completed | Angle of incomplete turn, degrees | Total angle turned, rad | Average angular velocity $\bar{\omega}$, rad/s | Initial angular velocity $\omega$, rad/s | Deviations |
|---|---|---|---|---|---|---|---|
| 90° | | | | | | | |
| | | | | | | | |
| | | | | | | | |
| 45° | | | | | | | |
| | | | | | | | |
| | | | | | | | |

**Catcher at 90°**

Average value of the initial angular
   velocity $\omega$ _____

Average value of deviation _____

Angular momentum after the ball
   is caught _____

Moment arm $l$ of the ball in flight _____

Angular momentum before the ball
   is caught _____

Percent discrepancy _____

**Catcher at 45°** _____

Average value of the initial angular
   velocity $\omega$ _____

Average value of deviation _____

Angular momentum after the ball
   is caught _____

Moment arm $l$ of the ball in flight _____

Angular momentum before the ball
   is caught _____

Percent discrepancy _____

## CALCULATIONS _____

**1.** Calculate the moment of inertia of the rotating system using Equation 11.1 and the data of Procedures 1 and 2. Remember that $m_{cb}$ is the mass of the catcher complete with its mounting nut and with the ball inside it.

**2.** From the data of Procedure 5, compute the average range of the ball, the deviation of each of your range measurements from the average, and the average deviation (a.d.).

**3.** Knowing that the ball falls a vertical distance of 0.5 m and that $g = 9.8$ m/s$^2$, calculate the velocity of projection $v$ using Equation 11.5. Take the a.d. for the range of the ball found in Calculation 2 as the error in the range and find the resulting error in $v$.

**4.** Calculate the total angle through which the rotating arm turned following each firing of the ball into the catcher. This is done by multiplying the number of complete turns by 360 and adding the number of degrees turned through in the incomplete final turn. The result is converted from degrees to radians by multiplying by $\pi/180$.

**5.** Calculate the average angular velocity $\overline{\omega}$ in each case by dividing the total angle turned through *in radians* by the corresponding time in seconds. Then double each value of $\overline{\omega}$ to obtain the corresponding initial angular velocity $\omega$.

**6.** Find the average of your three values of $\omega$ for the case of the catcher set at 90° to the bar and multiply this result by the moment of inertia found in Calculation 1 to obtain the angular momentum of the system after the ball is caught.

**7.** Find the deviation of each of your values of $\omega$ obtained in Calculation 5 from the average value found in Calculation 6 and then find the average deviation (the a.d.). Treating this as the error in your result for $\omega$, find the error to be associated with the angular momentum determined in Calculation 6, assuming no error in the moment of inertia.

**8.** For the case of the catcher set at 90° to the bar, the moment arm $l$ of the ball in flight is equal to the radius $R$ of the circle in which the catcher moves. Multiply this radius, the mass of the ball, and the ball's velocity from Calculation 3 together to obtain the angular momentum $mvl$ due to the ball before it is caught. Since before the ball is caught the rotating system is at rest, $mvl$ is the total angular momentum of the complete system at that time and should be equal to the system angular momentum afterwards as found in Calculation 6. Attach the proper error to your value of $mvl$ and compare your before and after values of the angular momentum by seeing if they fall within the limits of error established by your a.d.'s. Also calculate their difference and the percent discrepancy.

**9.** Repeat Calculations 6–8 for the case of the catcher set at 45° to the rotating bar. Note that in this case, $l = R \sin 45° = 0.707R$.

## QUESTIONS _____

**1.** Explain in your own words how this experiment demonstrates the principle of conservation of angular momentum.

**2.** Where do you think errors came into your experiment? What could be done to reduce them?

**3.** The counterweight was considered to be a point mass traveling in a circle of radius $R$, but actually it is a cylinder with a diameter of about 5 cm. Since it is mounted on the bar, it turns about its own center of mass with the same angular velocity as that of the rotating system. In other words, its total motion has two components: the motion of its center of mass in a circle of radius $R$ and the rotational motion about its own center of mass. Both contribute to the moment of inertia of the complete rotating system. Why could you neglect the contribution due to the counterweight's own rotation?

**4.** Derive Equation 11.5 from Equations 11.3 and 11.4.

**5.** Calculate the energy $\frac{1}{2}mv^2$ of the system before the ball was caught in the catcher and the energy $\frac{1}{2}I\omega^2$ afterwards. Which is greater? What happened to the difference? Was the difference greater when the catcher was at 45° to the bar or when it was at 90°? Why?

**6.** A wheel having a moment of inertia of 0.01 kg·m² is mounted in a horizontal plane on a frictionless shaft. The wheel has a diameter of 0.5 m and is initially at rest. A 40-g ball is fired in a direction tangent to the rim of the wheel with a velocity of 10 m/s and is caught in a cup mounted on the rim. What is the angular velocity of the wheel after the ball settles down in the cup?

**7.** (a) Calculate the conversion factor between angular velocity in radians per second and in revolutions per minute. (b) Express your answer to Question 6 in revolutions per minute.

# Archimedes' Principle **12**

Buoyancy is the ability of a fluid to sustain a body floating in it or to diminish the apparent weight of a body submerged in it. Archimedes' principle states that this apparent reduction in weight is equal to the weight of the fluid displaced. It is the purpose of this experiment to study Archimedes' principle and its application to the determination of density and specific gravity. In particular, the specific gravities of a solid heavier than water, a solid lighter than water, and a liquid other than water will be measured.

## THEORY

The density of a body is defined as its mass per unit volume. It is usually expressed in grams per cubic centimeter.* The specific gravity of a body is the ratio of its density to the density of water at the same temperature. Since for a given location the weight of a body is taken as a measure of its mass, the specific gravity may be taken as the ratio of the weight of a given volume of a substance to the weight of an equal volume of water. Because the mass of 1 cm$^3$ of water at 4°C is 1 g, the specific gravity of a body at this temperature is also numerically equal to its density in grams per cubic centimeter.

If a body is totally immersed in a fluid, the volume of fluid displaced must be equal to the volume of the body, because if the body weren't there, its volume would be occupied by the fluid. Moreover, the hydrostatic pressure on the bottom of the body is greater than that on the top because hydrostatic pressure rises as depth increases. The situation is illustrated in Fig. 12.1, which shows a beaker of fluid with an object (which we have made rectangular in the interest of simplicity) submerged in it. Let us begin by imagining an area of 1 cm$^2$ on the bottom of the beaker as shown at the right in Fig. 12.1. The dotted lines indicate that we may think of this 1-cm$^2$ patch as supporting a column of fluid 1 cm$^2$ in cross section extending up to the surface a distance $d$ above it, the depth of fluid being $d$ cm. Thus a volume $1 \times d$ cm$^3$ and hence a weight of $\rho g d$ dynes of the fluid sitting on that 1 cm$^2$ of the bottom. Here $\rho$ is the fluid's density and $g$ is the acceleration of gravity in centimeters per second squared. Hence the pressure sustained by the bottom of the beaker is $\rho g d$ dynes weighing on every square centimeter, or $\rho g d$ dynes/cm$^2$. Note that this is the *gauge* pressure. The *absolute* pressure must include the pressure due to the atmosphere on the fluid surface—we could imagine

the dotted lines in Fig. 12.1 going on up to the top of the atmosphere and indicating a column of air 1 cm$^2$ in cross section sitting on top of the column of liquid. Since atmospheric pressure does not affect the present experiment we will not consider it further.

Turning now to the immersed solid, we note that it has height $h$, its bottom end being at depth $h_2$ and its top end at depth $h_1$ where $h_2 - h_1 = h$. The considerations of the last paragraph show that the pressure at any depth $d$ in a fluid of density $\rho$ is $\rho g d$; hence the pressure on the bottom of the body is $\rho g h_2$ and the force exerted on the bottom, which will be an upward force, is $\rho g h_2 A$, where $A$ is the body's cross-sectional area as shown in Fig. 12.1. Similarly, the downward force on the body's top surface is $\rho g h_1 A$, so that there is a net upward or so-called *buoyant force* $\rho g A (h_2 - h_1) = \rho g A h$. But $Ah$ is just the body's volume $V$, so that the buoyant force may be written

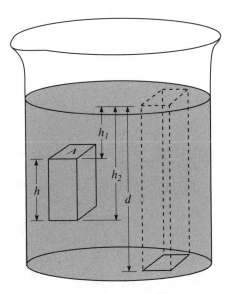

**Figure 12.1** *A beaker of fluid with a submerged rectangular object*

---

*The CGS (centimeter-gram-second) system will be used in this experiment as it is both more usual and more convenient for this work.

as simply $\rho g V$. Then, since $\rho$ is the fluid's density, $\rho g V$ is identical with the weight of the displaced fluid. This result, namely that the buoyant force (which, being an upward force, gives rise to an apparent loss of weight of the submerged body) is equal to the weight of the displaced fluid, is known as *Archimedes' principle*.

The specific gravity of a solid heavier than water may be easily determined by the application of this principle. The body is weighed in air; then it is weighed in water, that is, suspended by a thread from the arm of a balance so as to be completely submerged (see Fig. 12.2). The loss of weight in water is $W - W_1$, where $W$ is the weight in air and $W_1$ is the weight in water. But the loss of weight in water is equal to the weight of the water displaced, or the weight of an equal volume of water. Thus, the specific gravity $S$ will be

$$S = \frac{W}{W - W_1} = \frac{Mg}{Mg - M_1 g} = \frac{M}{M - M_1} \quad (12.1)$$

where $M$ is the body's mass in air and $M_1$ its apparent mass in water.

The specific gravity of a liquid may be found by measuring the loss of weight of a convenient solid body when immersed in that liquid and the loss of weight when immersed in water. The procedure is as follows: A heavy body is weighed in air; this weight is called $W$. Then it is weighed in water; this weight is called $W_1$. Finally, it is weighed in the liquid whose specific gravity is to be determined; this weight is called $W_2$. The specific gravity of the liquid will then be

$$S = \frac{W - W_2}{W - W_1} = \frac{M - M_2}{M - M_1} \quad (12.2)$$

since this expression represents the weight of a certain volume of the liquid divided by the weight of an equal volume of water. Here $W - W_1$ is the loss of weight in water, and $W - W_2$ is the loss of weight in the given liquid.

In order to find the specific gravity of a solid lighter than water, it is necessary to employ an auxiliary body, or sinker, of sufficient weight and density

**Figure 12.2** *Arrangement for weighing a body in water*

to hold the other body completely submerged. The specific gravity of a solid lighter than water, as obtained by the sinker method, is given by

$$S = \frac{W}{W_1 - W_2} = \frac{M}{M_1 - M_2} \quad (12.3)$$

where $W$ is the weight of the solid in air; $W_1$ is the weight of the solid and the sinker, with the sinker alone immersed; and $W_2$ is the weight when both solids are immersed in water.

The hydrometer is an instrument designed to indicate the specific gravity of a liquid by the depth to which it sinks in the liquid. To measure the specific gravity of a liquid by means of a hydrometer, it is only necessary to let the hydrometer float in the liquid and to read the specific gravity directly on the calibrated scale. The reading is taken, if possible, by placing the eye below the liquid surface and seeing where this surface cuts the hydrometer scale.

## APPARATUS

1. Triple-beam balance with hook for suspending weights
2. Metal cylinder
3. Wooden cylinder
4. Lead sinker
5. Distilled water
6. Alcohol
7. Hydrometer
8. Hydrometer jar
9. 1000-cc Pyrex beaker
10. Fine thread

## PROCEDURE

**1.** The triple-beam balance is set up with a fine thread attached to the underside of the pan carrier so

that you can weigh bodies by hanging them on the thread rather than placing them in the pan. Make sure that the thread is of the proper length so that a body attached to its end will hang completely submerged in the fluid in the beaker (see Fig. 12.2). Also check that the beam balance balances with no body attached and adjust it accordingly. Then weigh the metal cylinder in air by suspending it on the end of the thread. Note that the balance is calibrated in grams, so that you will be recording the *mass* of the cylinder.

**2.** Fill the beaker with water and place it on the floor with the metal cylinder submerged in it as shown in Fig. 12.2. Be sure the cylinder is completely submerged and not touching the sides of the beaker. In this way, obtain the weight of the metal cylinder in water. Again you will be reading in grams so that you will be observing an "apparent reduction in mass." When you have completed this measurement, remove and dry off the metal cylinder.

**3.** Empty the beaker, dry it thoroughly, and refill it with alcohol. Then repeat Procedure 2. When you are finished, pour the alcohol into the hydrometer jar and rinse out the beaker with distilled water.

**4.** Weigh the wooden cylinder in air as you did with the metal cylinder in Procedure 1.

**5.** Refill the beaker with distilled water. Attach the sinker to the wooden cylinder and weigh the combination with the sinker alone immersed in water.

**6.** Now weigh the two solids when they are both immersed in water.

**7.** Measure the specific gravity of the alcohol with the hydrometer. See that there is enough alcohol in the hydrometer jar and let the hydrometer float in the alcohol. Read the specific gravity directly.

## DATA

Mass of metal cylinder in air _____
Apparent mass of metal cylinder in water _____
Apparent mass of metal cylinder in alcohol _____
Specific gravity of metal cylinder _____
Specific gravity of metal cylinder from Table IV at the end of the book _____
Percent error _____
Specific gravity of alcohol _____

Specific gravity of alcohol from Table IV at the end of the book _____
Percent error _____
Mass of wooden cylinder in air _____
Apparent mass of cylinder in air and sinker immersed in water _____
Apparent mass of cylinder and sinker, both immersed in water _____
Specific gravity of wooden cylinder _____
Specific gravity of alcohol by using the hydrometer _____

## CALCULATIONS

**1.** From the data of Procedures 1 and 2, calculate the specific gravity of the metal cylinder.

**2.** From the data of Procedures 1–3, calculate the specific gravity of the alcohol.

**3.** From the data of Procedures 4–6, compute the specific gravity of the wooden cylinder.

**4.** Compute the percent error of your measurements by comparing your results for the specific gravity of the metal cylinder and of the alcohol with the accepted values.

**5.** Compare your measurement of the specific gravity of alcohol using the hydrometer with the accepted value and note whether this latter value falls within the limits of precision on the hydrometer reading.

## QUESTIONS _____

**1.** (a) Explain how you can obtain the volume of an irregular solid insoluble in water. (b) How can you obtain the weight of an equal volume of water?

**2.** (a) When a block of wood is completely submerged in water, why does it apparently lose more than its entire weight in air? (b) Given the block of wood's weight and density, find an expression for the minimum weight of a sinker made of a material of density $\rho_s$ (>1) which will completely submerge the combination.

**3.** Suppose there were a bubble of air on the bottom of the metal cylinder immersed in water. How would this affect the calculations of the density of the metal?

**4.** A piece of cork having a mass of 25 g in air and a specific gravity of 0.25 is attached to a lead sinker whose mass is 226 g in air. What will be the apparent mass of the two solids when they are both immersed in water?

**5.** The cork in Question 4 is allowed to float in water. What fraction of its volume is above the surface?

**6.** Derive Equation 12.3.

**7.** Look up Pascal's principle, state it in your own words, and tell how it was applied (without our saying so) in the derivation of Archimedes' principle given in the Theory section.

**8.** (a) Explain the apparent loss of weight suffered by an object when immersed in a liquid. (b) Suggest a modification of the apparatus shown in Fig. 12.2 that will demonstrate your answer to Part (a). (c) If your instructor suggests it and time permits, carry out the experiment you proposed in Part (b) and state your results below.

# Simple Harmonic Motion 13

Simple harmonic motion is one of the most common types of motion found in nature, and its study is therefore very important. Examples of this type of motion are found in all kinds of vibrating systems, such as water waves, sound waves, the rolling of ships, the vibrations produced by musical instruments, and many others. In fact, any time a *linear restoring force* exists, simple harmonic motion results. A linear restoring force is a force that is proportional to the displacement of the body on which the force acts from an equilibrium position (where the force is zero) and is always directed back towards that equilibrium position. The object of this experiment is to study two important examples of such a force, the simple pendulum and the vibrating spring, and to determine the time of vibration (called the period) in each case.

## THEORY

A linear restoring force as defined above may be expressed mathematically by the equation

$$F = -kx \tag{13.1}$$

where $F$ is the force, $x$ represents the displacement from the equilibrium point $x = 0$, $k$ is the proportionality constant, and the minus sign expresses the restoring nature of the force by indicating that the direction of $F$ is always opposite to the displacement $x$ from equilibrium.

If a body to which a linear restoring force is applied is displaced from equilibrium and released, the restoring force will bring the body back to its equilibrium position. But the body's inertia will then carry it beyond this position, and so a restoring force will build up in the opposite direction. This force first brings the body to rest and then accelerates it back towards the equilibrium position again. The action is repeated and the body executes a type of vibration known as simple harmonic motion. In this motion, the maximum displacement from the equilibrium point achieved by the body on either side is called the amplitude, one complete vibration is called a cycle, the time required for one cycle is called the period, and the number of cycles completed per second (the reciprocal of the period) is called the frequency. Clearly, the period is measured in seconds and the frequency in cycles per second ($s^{-1}$). The unit of cycles per second is now commonly called hertz in honor of Heinrich Hertz (1857–1894), who investigated electrical oscillations and first showed (1887) that high-frequency electrical oscillations could produce electromagnetic radiation. Further analysis of the vibrations resulting from a linear restoring force shows that the period is given by

$$T = 2\pi\sqrt{\frac{m}{k}} \tag{13.2}$$

where $T$ is the period, $m$ is the mass of the vibrating body, and $k$ is the proportionality constant from Equation 13.1. If $m$ is expressed in kilograms and $k$ in newtons per meter, $T$ will come out in seconds.

Two physical examples of this type of force are considered in the present experiment. The first is the simple pendulum, which consists of a concentrated mass suspended at the end of a cord of negligible weight. A close approximation to this is a small metal sphere on a long, thin thread. Such an arrangement is diagrammed in Fig. 13.1, which shows a sphere of mass $m$ hung on the end of a thread whose length is such that the distance from the suspension

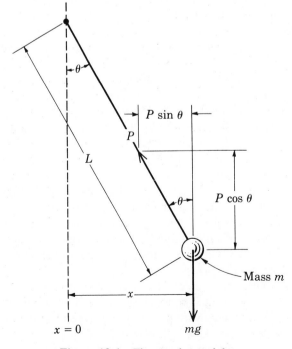

**Figure 13.1** *The simple pendulum*

point to the center of the sphere is $L$. The sphere has been drawn aside to a horizontal distance $x$, the thread consequently making an angle $\theta$ with the vertical. Note that the vertical line through the suspension point is at $x = 0$, for this is the equilibrium position, mass $m$ being in equilibrium when it hangs straight down.

Inspection of Fig. 13.1 shows that the sum of the $x$ components of all the forces acting on mass $m$ is simply $-P \sin \theta$ and the sum of all the $y$ components is $P \cos \theta - mg$, where $P$ is the tension in the thread. If the pendulum is never swung by more than a very small amount so that $\theta$ is always a small angle even at the extremes of the swing, then two simplifying assumptions may be made. The first of these is that for small angles the cosine remains approximately equal to 1, so that $P \cos \theta - mg$ may be approximated by $P - mg$. Secondly, in a small swing there is very little vertical motion of mass $m$, so that zero vertical acceleration may be assumed. Hence,

$$P \cos \theta - mg \approx P - mg \approx 0$$

or $$P \approx mg$$

Further inspection of Fig. 13.1 shows that, by definition, $\sin \theta = x/L$, and therefore by substitution the $x$-directed force $-P \sin \theta$ becomes $-mgx/L$. This qualifies as a linear restoring force with the proportionality constant $k$ equal to $mg/L$, and Equation 13.2 shows that the period of the simple pendulum should then be

$$T = 2\pi \sqrt{\frac{mL}{mg}} = 2\pi \sqrt{\frac{L}{g}} \qquad (13.3)$$

Again the period will be in seconds if the length $L$ is in meters and $g$, the acceleration of gravity, is in meters per second per second ($9.8 \text{ m/s}^2$).

The second example of a linear restoring force presented in this experiment is that of a stretched spring. Ideally, the elongation of a spring is proportional to the stretching force. Therefore, if a mass is hung on a spring, the spring will be stretched by an amount proportional to the weight of the mass. Furthermore, an equilibrium position will be reached at which the spring pulls up on the mass with a force just equal to the mass's weight. The situation is illustrated in Fig. 13.2, where initially the spring hangs relaxed, with no mass attached, and has a length $y_1$. When a weight of mass $m$ is attached, the spring elongates to length $y_0$, and if the elongation is proportional to the stretching force (the weight of the mass), $mg = k(y_0 - y_1)$.

When the weight is at any arbitrary distance $y$ below the spring support, the elongation of the spring is $y - y_1$, the upward pull of the spring on the weight is $k(y - y_1)$ and the downward force of gravity is $mg$. Thus, the net force on the weight is

$$F = mg - k(y - y_1) \qquad (13.4)$$

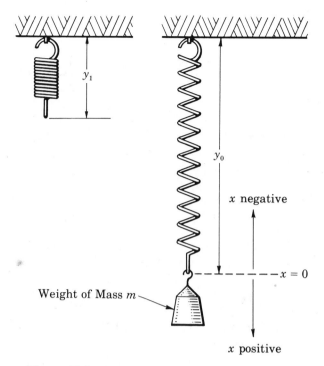

**Figure 13.2**   *Weight suspended on an ideal spring*

If a new coordinate $x$, measured in the same direction as $y$ but having the equilibrium point as its origin, is introduced as shown on the right in Fig. 13.2, then $x + y_0 = y$, for the value of $y$ will always be $y_0$ greater than the value of $x$. Substitution in Equation 13.4 yields

$$F = mg - k(x + y_0 - y_1) \qquad (13.5)$$

But since $mg = k(y_0 - y_1)$, Equation 13.5 becomes

$$F = k(y_0 - y_1) - k(x + y_0 - y_1) = -kx$$

Thus, again we have a linear restoring force, the constant $k$ here being a characteristic of the spring called its force constant. The force constant can be determined by measuring the elongation in meters suffered by the spring when a stretching force of a given number of newtons is applied. A part of this experiment will be devoted to making such a determination.

If the weight in Fig. 13.2 is now pulled down below its equilibrium position and released, it will execute simple harmonic motion with a period given by Equation 13.2. Note, however, that $m$ in this equation will be the mass of the weight only if the spring has negligible mass. In the general case where the spring's mass cannot be neglected, $m$ is the effective mass of the vibrating system, which is made up of the mass of the suspended weight plus a part of the mass of the spring, since the spring itself is also vibrating. It is found by analysis that one third of the mass of the spring must be added to the mass of the suspended weight in calculating the value of $m$ to be used in Equation 13.2.

## APPARATUS _____

1. Metal sphere on a long string (the simple pendulum)
2. Supporting rod with right-angle clamp, short rod, and hook for suspending spring
3. 2-m stick with caliper jaws
4. Vernier caliper
5. Spring (cylindrical type, coils not touching when relaxed)
6. Set of hooked weights (100-, 200-, 200-, 500-g)
7. Stopwatch or stop clock
8. Triple-beam balance

## PROCEDURE _____

**1.** Measure the diameter of the metal sphere with the vernier caliper.

**2.** Attach the string so that the center of the sphere is about 0.6 m below the point of support. Carefully measure the distance from this point to the upper surface of the sphere. Record this distance as the length of string used. The length of the pendulum will be this distance plus the radius of the metal sphere.

**3.** Displace the sphere to one side through an angle of not more than about 5° and let the pendulum oscillate. Record the time it takes the pendulum to make 50 vibrations. In counting vibrations, be sure to start the stopwatch on the count of zero and not on the count of one.

**4.** Repeat Procedures 2 and 3, making the length of the pendulum successively 0.8, 1.0, 1.2, 1.4, and 1.6 m. In the last three cases, record the time for 25 vibrations instead of 50.

**5.** Using a length of about 0.5 m, determine the time for 50 vibrations when the sphere is displaced about 5°, 30°, and 45° from the equilibrium position. Record the length of the pendulum and the time for 50 vibrations at each displacement.

**6.** Weigh the spring and record its mass in kilograms.

**7.** Remove the pendulum from the support and suspend the spring in its place. Using the 2-m stick and a caliper jaw, observe the position of the lower end of the spring and record the reading.

**8.** Suspend 0.1 kg from the spring and again record the position of its lower end.

**9.** Repeat Procedure 8 with loads of 0.2, 0.3, 0.4, and 0.5 kg suspended from the spring. Record the position of the spring's lower end in each case.

**10.** Suspend a mass of 0.2 kg from the spring, displace it about 5 cm downward from its position of equilibrium, and set it into vibration. Using a stopwatch, measure the time for 50 complete vibrations of the mass and record it.

**11.** Repeat Procedure 10, again using a mass of 0.2 kg but displacing it about 10 cm from its position of equilibrium and setting it into vibration. Again record the time for 50 complete vibrations.

**12.** Suspend a mass of 0.5 kg from the spring, displace it about 5 cm from its position of equilibrium, and set it into vibration. Record the time for 50 complete vibrations.

## DATA _____

Diameter of sphere _____     Radius of sphere _____

| Length of string used | Length of pendulum | Number of vibrations | Time | Period | Square of period |
|---|---|---|---|---|---|
| | | | | | |
| | | | | | |
| | | | | | |
| | | | | | |
| | | | | | |
| | | | | | |

Value of $g$ from slope _____     Percent error _____

Length of string used   _____   Length of pendulum   _____

| Initial displacement of sphere | Number of vibrations | Time | Period | Square of period |
|---|---|---|---|---|
| 5° | | | | |
| 30° | | | | |
| 45° | | | | |

| Mass suspended from the spring | Force stretching the spring | Scale reading | Elongation |
|---|---|---|---|
| 0 kg | | | |
| 0.1 kg | | | |
| 0.2 kg | | | |
| 0.3 kg | | | |
| 0.4 kg | | | |
| 0.5 kg | | | |

Mass of the spring   _____   Force constant of the spring   _____

| Mass suspended from the spring | Mass of the vibrating system | Amplitude of vibration | Time for 50 vibrations | Period Experimental value | Period Calculated value | Percent discrepancy |
|---|---|---|---|---|---|---|
| 0.2 kg | | 5 cm | | | | |
| 0.2 kg | | 10 cm | | | | |
| 0.5 kg | | 5 cm | | | | |

## CALCULATIONS

**1.** Calculate the period of the pendulum for each observation.

**2.** Calculate the square of the pendulum's period for each observation. Use only three significant figures in recording the square of the period, rounding off the third figure.

**3.** Plot a curve using the values of the square of the period as ordinates and the lengths of the pendulum as abscissas. Use the entire sheet of graph paper. Read the instructions on plotting graphs in the Introduction.

**4.** Determine your measured value of g from the slope of the curve just plotted and compare it with the known value, $g = 9.8$ m/s$^2$, by finding the percent error. The acceleration of gravity varies slightly with latitude and elevation. The value given is approximately correct for sea level and 45° latitude.

**5.** From the data of Procedures 7–9, determine the elongation of the spring produced by each load by subtracting the zero reading from the reading corresponding to each load. Also multiply each suspended mass load by 9.8 m/s$^2$ to obtain the force stretching the spring in each case.

**6.** Plot a curve using the values of the elongation as ordinates and the forces due to the corresponding loads as abscissas.

**7.** Obtain the force constant of the spring from the slope of the curve plotted in Calculation 6. See the instructions in the Introduction.

**8.** From the data of Procedures 10–12 calculate the experimental value of the period of the oscillating weight for each set of observations.

**9.** Using Equation 13.2, calculate the value of the period for each of the two masses used. The effective mass of the vibrating system will be the mass suspended from the spring plus one third of the mass of the spring.

**10.** Calculate the percent discrepancy between the experimental values of the period and the calculated values.

## QUESTIONS

**1.** Show that if $m$ is in kilograms and $k$ in newtons per meter, $T$ will in fact come out in seconds in Equation 13.2.

**2.** State what you conclude from your work with the simple pendulum.

**3.** Why was it unnecessary to weigh the metal sphere?

**4.** Explain in your own words why the angle through which the pendulum swings must be no more than about 10°.

**5.** (a) At which point on the path of its vibration does the weight suspended on the spring have the greatest acceleration? (b) Where does it have the greatest velocity? (c) Where does it have its least acceleration? (d) Where does it have its least velocity?

**6.** Is the motion of the piston in a one-cylinder steam engine simple harmonic? Explain.

**7.** What does your curve show about how the elongation of the spring is dependent upon the applied force?

**8.** (a) If it takes a force of 0.5 N to stretch a spring 2 cm (0.02 m), what force is needed to stretch the spring 5 cm? (b) How much work is done in stretching the spring 5 cm?

**9.** What percent error is introduced in the calculated value of the period of vibration, for the 0.2 kg load and the 0.5 kg load respectively, if the mass of the spring is neglected?

**10.** If adding a certain weight in Procedure 10 would double the period, what would the mass of the added weight be?

**11.** Rotational simple harmonic motion occurs when there is a linear restoring *torque* given by $\tau_r = -K\theta$, where $\theta$ is the *angular* displacement from equilibrium. The period of the resulting rotational oscillation is given in analogy with Equation 13.2 by $T = 2\pi\sqrt{I/K}$. Derive Equation 13.3 by treating the simple pendulum as a rotating system with the suspension point as the pivot. Note that for small angles, $\sin\theta \approx \theta$.

# Standing Waves on Strings 14

The general appearance of waves is well illustrated by standing waves (so called because they do not seem to move along the medium) on a string. This type of wave is very important because most of the vibrations of bodies, such as piano or violin strings or the air column in an organ pipe, are standing waves. The purpose of this experiment is to observe standing waves on a stretched string, to determine the natural vibrational frequencies of that string, and to study the relation between those frequencies, their associated wavelengths, and the stretching force.

## THEORY

Standing or stationary waves are produced by the interference of two wave trains of the same wavelength, velocity of propagation, and amplitude traveling in opposite directions through the same medium. The necessary condition for the production of standing waves on a stretched string can be met by fixing one end of the string and sending a train of waves down from a vibrating body at the other end. These waves reflect from the fixed end and travel back towards the source, thus interfering with the oncoming waves.

A stretched string has many modes of vibration, all determined by the fact that the two ends must remain fixed and non-vibrating. A non-vibrating point is called a *node,* and we therefore see that we have imposed on our string the constraint that the two ends must be nodes. Because the vibrating string assumes a sinusoidal shape (see, for example, Fig. 14.1 on page 110), a node or zero point occurs every half wavelength, and so our constraint requires that the string's length between its two fixed ends be an integer number of half wavelengths. The string may thus vibrate as a single segment, in which case its total length is equal to one half wavelength, or in two segments with a node in the middle so that the string's total length is two half wavelengths or one whole wavelength. Similarly, vibration may occur with the string's length being equal to three half wavelengths with two intermediate nodes as shown in Fig. 14.1, or in four half wavelengths (two whole wavelengths) with three intermediate nodes, or in five, six, etc. half wavelengths. In short, because in a standing wave the nodes occur every half wavelength, the fact that the ends of the string must correspond to nodes requires that the whole length of the string accommodate an integer number of half wavelengths.

Now since the wavelengths of the various modes are determined by the above considerations, the frequencies at which the string wants to vibrate are also determined, the equation

$$V = f\lambda \qquad (14.1)$$

expressing the relationship between the wavelength $\lambda$, the frequency $f$, and the velocity $V$ with which waves propagate on the string. This velocity is given by the equation

$$V = \sqrt{\frac{F}{d}} \qquad (14.2)$$

where $F$ is the string tension in newtons and $d$ is the string's mass per unit length in kilograms per meter. Thus if we define $L$ as the distance between adjacent nodes as shown in Fig. 14.1, then for any given mode $\lambda = 2L$ and

$$f = \frac{V}{\lambda} = \frac{1}{2L}\sqrt{\frac{F}{d}} \qquad (14.3)$$

We conclude that the frequencies at which the string wants to vibrate are determined by the allowed values of $\lambda$, and hence $L$, and also by the tension and mass per unit length of the string. Consequently, if the frequency of a vibrating body driving one end of the string is set to one of the frequencies given by Equation 14.3, the string will vibrate in the corresponding mode with a relatively large amplitude, whereas it will hardly vibrate at all when driven at other frequencies. Setting the driving frequency equal to one of the mode frequencies establishes a condition called *resonance* between the vibrating body and the string.

In this experiment, standing waves are set up on a stretched string by a string vibrator, which is basi-

**Figure 14.1** *Standing waves on a string*

cally the voice-coil section of a loudspeaker but which has, instead of a speaker cone, a projecting rod that bears on the string so as to move it up and down. The vibrator is driven by a signal generator which provides a sinusoidal output at the frequency selected by the tuning dial. The arrangement of the apparatus is shown in Fig. 14.1. The tension in the string is fixed by the weights suspended over the pulley by means of a weight hanger and can be altered by changing these weights. Thus different sets of modes may be established by varying $F$ in Equation 14.3, and within a particular set, the resonant frequencies corresponding to the various allowed values of $L$ may be determined and measured.

## APPARATUS

1. String vibrator
2. Signal generator
3. Rod mounted pulley and clamp
4. Support rod and clamp for string
5. String
6. Triple-beam balance
7. 2-meter stick
8. Weight hanger
9. Set of slotted weights

## PROCEDURE

**1.** Measure and cut off a piece of string exactly 4 meters long. Weigh this length of string and record its mass. This length is chosen for convenience and also to provide enough string for an accurate weighing. It is enough for two setups.

**2.** Tie one end of the string to the vertical support rod, attach the weight hanger at the other end, and pass the string over the pulley mounted as shown in Fig. 14.1 at the other corner of the laboratory bench about a meter away.

**3.** Place the vibrator under the string as near to the end opposite the pulley as you can get it and adjust the string's height so that it just rests on top of the vibrator's driving rod. Adjust the height of the pulley so that the string is level.

**4.** Connect the vibrator to the signal generator, plug the latter in, and turn it on. Start with a mass of 50 grams on the string (usually the mass of the weight hanger alone) and set the generator frequency to a low value, say 10 Hz. Increase the generator's output amplitude until you can see the vibrator post oscillating up and down, and search for resonance at the lowest frequency mode by slowly increasing the generator frequency until you see the string vibrating in one segment. Adjust the generator frequency to maximize the vibration amplitude and note this frequency.

**5.** Search for resonance in the second mode, i.e., with the string vibrating in two segments with a node in the middle. As this mode has a wavelength half of that associated with the first mode, the frequency, according to Equation 14.1, should be doubled, so that your search will be facilitated by looking for frequencies in the neighborhood of twice the value noted in Procedure 4.

**6.** Tune the signal generator to exact resonance with the second mode by maximizing the vibration amplitude. Record this frequency. Then measure the distance from the point where the string contacts the pulley to the first node beyond the vibrator (see Fig. 14.1). With the string vibrating in only two segments, this node is the one in the middle. The measurement may be made by simply holding the 2-m stick alongside the vibrating string. However, a more accurate result is usually obtainable by grasping the string between your thumb and forefinger with your thumbnail right at the node and measuring the distance from your nail to the pulley. In doing this, be sure not to pull on the string in such a direction as to raise the weight and thus increase the length being

measured. If you pull at all, make sure you pull against the support rod. Record the value of $L$ so obtained.

**7.** Search for resonance in the third mode, which should be at three times the frequency noted in Procedure 4. Determine and record the frequency for this mode as you did for the second mode in Procedure 6. Also measure the distance from the point where the string contacts the pulley to the first node beyond the vibrator as in Procedure 6, but this time observe that your measured length is 2$L$ (see Fig. 14.1). Record your results.

**8.** Repeat Procedure 7 for the fourth and fifth modes, noting that the frequencies for these modes should be close to four and five times that noted in Procedure 4 and that the measured string lengths will be 3$L$ and 4$L$ respectively.

**9.** Repeat Procedures 4–8 with weights of 100, 150, 200, 250, 300, 350, and 400 grams attached to the string. Record all your results.

# DATA _____

Mass of 4 meters of string _____     Mass per unit length _____

### Measurements with 50-Gram Weight

Tension _____     Square root of tension _____

| Number of segments | Frequency | Length to first node beyond vibrator | $L$ |
|---|---|---|---|
| 2 | | | |
| 3 | | | |
| 4 | | | |
| 5 | | | |

### Measurements with 100-Gram Weight

Tension _____     Square root of tension _____

| Number of segments | Frequency | Length to first node beyond vibrator | $L$ |
|---|---|---|---|
| 2 | | | |
| 3 | | | |
| 4 | | | |
| 5 | | | |

**Measurements with 150-Gram Weight**

Tension _____    Square root of tension _____

| Number of segments | Frequency | Length to first node beyond vibrator | L |
|---|---|---|---|
| 2 | | | |
| 3 | | | |
| 4 | | | |
| 5 | | | |

**Measurements with 200-Gram Weight**

Tension _____    Square root of tension _____

| Number of segments | Frequency | Length to first node beyond vibrator | L |
|---|---|---|---|
| 2 | | | |
| 3 | | | |
| 4 | | | |
| 5 | | | |

**Measurements with 250-Gram Weight**

Tension _____    Square root of tension _____

| Number of segments | Frequency | Length to first node beyond vibrator | L |
|---|---|---|---|
| 2 | | | |
| 3 | | | |
| 4 | | | |
| 5 | | | |

**Measurements with 300-Gram Weight**

Tension _____    Square root of tension _____

| Number of segments | Frequency | Length to first node beyond vibrator | L |
|---|---|---|---|
| 2 | | | |
| 3 | | | |
| 4 | | | |
| 5 | | | |

**Measurements with 350-Gram Weight**

Tension _____ Square root of tension _____

| Number of segments | Frequency | Length to first node beyond vibrator | L |
|---|---|---|---|
| 2 | | | |
| 3 | | | |
| 4 | | | |
| 5 | | | |

**Measurements with 400-Gram Weight**

Tension _____ Square root of tension _____

| Number of segments | Frequency | Length to first node beyond vibrator | L |
|---|---|---|---|
| 2 | | | |
| 3 | | | |
| 4 | | | |
| 5 | | | |

| Number of segments | Slope | Average value of L | $1/(2L\sqrt{d})$ | Percent error |
|---|---|---|---|---|
| 2 | | | | |
| 3 | | | | |
| 4 | | | | |
| 5 | | | | |

# CALCULATIONS _____

**1.** Determine the mass per unit length of the string. Express it in kilograms per meter.

**2.** Calculate the tension in newtons and the square root of this tension for each weight used on the string.

**3.** For each observation, calculate the value of $L$, the length of a single vibrating segment.

**4.** Plot the square root of the tension against frequency for the 2-segment mode using your values of $\sqrt{F}$ as abscissas and those of $f$ for that mode as ordinates. Use the entire sheet of graph paper. Draw the best straight line through your plotted points, remembering that the origin should be a point on your line.

**5.** Repeat Calculation 4 using the frequencies for the 3-segment mode, the 4-segment mode, and the 5-segment mode. Use the same sheet of graph paper so that you have four lines of different slope on the one sheet.

**6.** According to Equation 14.3, $f$ and $\sqrt{F}$ are proportional with proportionality constant $1/(2L\sqrt{d})$. This constant (with the appropriate value of $L$) should therefore be the slope of each line drawn in Calculations 4 and 5. Obtain these slopes and record their values in the spaces provided in the last data table.

**7.** Calculate and record an average value of the segment length $L$ from the values of $L$ obtained each time you had the string vibrating in two segments. Similarly find average values of $L$ from the values obtained with the string vibrating in 3, 4, and 5 segments.

**8.** Using the average values of $L$ found in Calculation 7, calculate the value of the proportionality constant $1(2L\sqrt{d})$ in each case and compare with the corresponding values found in Calculation 6 by finding the percent discrepancy.

## QUESTIONS _____

**1.** In this experiment, why did we consider the length of the string between the first node beyond the vibrator and the pulley rather than the entire length between the vibrator and the pulley? What error would have been introduced if the entire length and total number of vibrating segments had been used? With your experimental setup, is this error likely to have been large?

**2.** Why should string that doesn't stretch easily be used in this experiment?

**3.** Why should the curves drawn in Calculations 4 and 5 pass through the origin? What does this mean physically?

**4.** What is meant by resonance?

**5.** A copper wire 1 m long and weighing 0.061 kg/m vibrates in two segments when under a tension produced by a load of 0.75 kg. What is the frequency of this mode of vibration?

**6.** Upon what physical properties do (a) the loudness, (b) the pitch, and (c) the quality of a musical note depend?

**7.** Compute the velocity of the wave on the string for each of the weights used and note how your results compare with the speed of sound in air, which, for air at room temperature, is 343.4 meters/second.

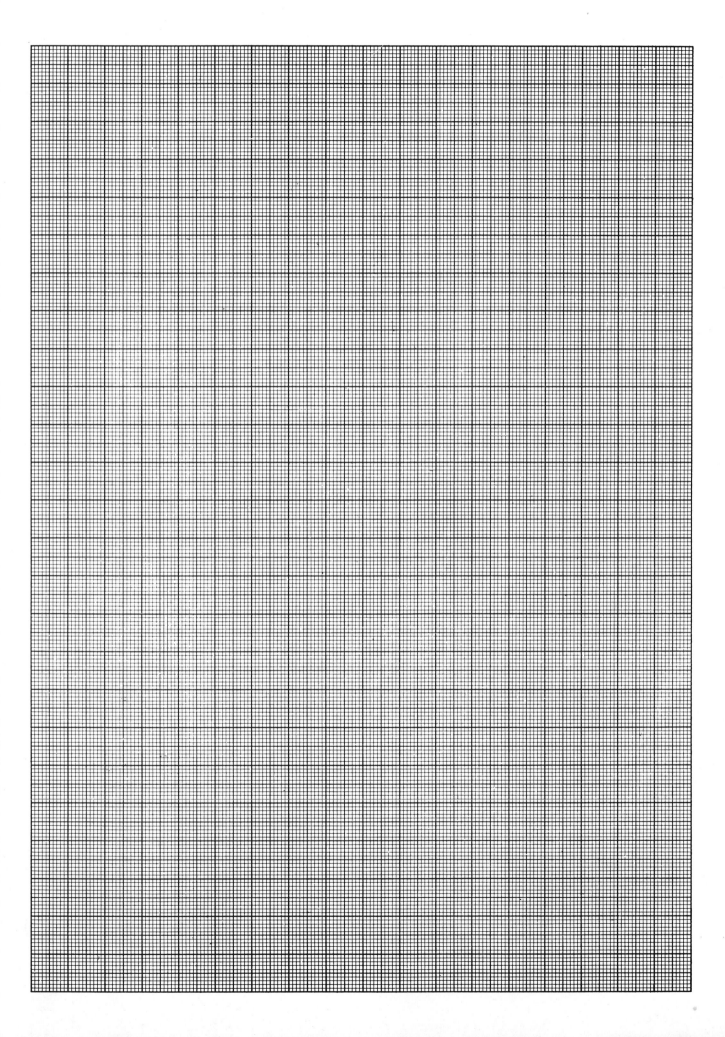

# The Sonometer 15

A stretched string that is set into vibration will emit a musical note, the frequency of which is determined by the length of the string, its mass per unit length, and the tension. By varying any one of these quantities, the frequency may be changed. The purpose of this experiment is to determine (1) how the frequency of a vibrating string varies with its length when the tension is kept constant, (2) how the frequency of a vibrating string varies with the tension in the string when its length is kept constant, and (3) how the frequencies of two strings of the same length and material and under the same tension depend on their cross-sectional areas.

## THEORY

The frequency, or number of vibrations per second, of the fundamental of a stretched string depends upon the length of the string, the tension, and the mass per unit length. The relation between these quantities is

$$f = \frac{1}{2L}\sqrt{\frac{F}{d}} \qquad (15.1)$$

where $f$ is the frequency of the note emitted in vibrations per second or hertz, $L$ is the length of the vibrating segment in meters, $F$ is the tension in newtons, and $d$ is the mass per unit length of the string expressed in kilograms per meter.

In this experiment, the above relation will be tested by the use of a sonometer. This device consists of a resonant case provided with wires that may be stretched. The tension on the wires, or the stretching force, can be easily measured. Adjustable bridges are provided so that the lengths of the vibrating strings may be varied and measured. The general appearance of a sonometer is shown in Fig. 15.1.

The experimental work consists of tuning the sonometer to several known frequencies produced by an audio signal generator and loudspeaker so that the law expressed by Equation 15.1 may be tested. Thus, the variation of the frequency of a vibrating string with stretching force may be studied, as well as the variation of frequency with length. Finally, the effect of the mass per unit length may be shown by using two wires of different diameters.

To detect when the notes produced by the wire and the signal generator are in unison (that is, have the same frequency), the two may be sounded alternately and a rough adjustment made until they appear to have the same pitch. Then the two are sounded simultaneously, the length of the wire and/or the signal generator frequency are adjusted until beats are heard, and the beat frequency is made to get lower and lower until finally the beats disappear. This "zero-beat" condition indicates identical frequencies. Unison may also be obtained by placing the loudspeaker on the resonance board and adjusting the string length for maximum sympathetic vibration of the string. Using a signal generator as the source of the known frequency has the advantage that its tuning dial can serve to make fine adjustments so that the generator and string frequencies agree exactly.

**Figure 15.1** *The sonometer*

## APPARATUS

1. Sonometer with two steel wires of different diameters
2. Audio signal generator covering the range 100–600 Hz
3. Small loudspeaker or telephone receiver
4. Meter stick
5. Micrometer caliper
6. Card giving the mass per unit length of the wires

## PROCEDURE

**1.** Connect the loudspeaker or telephone receiver to the signal generator output, set the generator frequency to 200 Hz, and turn it on. Adjust its output level so that you can comfortably hear the tone. *Note*: Laboratory signal generators do not have a large ouput; thus, you should not expect to get a large volume of sound. If a booster amplifier driving a loudspeaker is used, do not set its volume control and/or the generator output to a high level. You want only enough volume to hear and make the comparison with the sound from the sonometer.

**2.** Set the tension in the sonometer wire of smaller diameter by loading it with 4 kg. Vary the length of the wire by moving the sliding bridge until the frequency produced by this wire is the same as that of the signal generator. Obtaining identical frequencies may be done either by tuning for zero-beat or by looking for maximum sympathetic vibration of the wire. Final fine tuning is most conveniently done with the generator tuning dial. Record the length of the wire, its tension, and the generator's final frequency setting.

**3.** Repeat Procedure 2 starting with generator frequencies of 300, 400, and 500 Hz. Record the length of the wire, its tension, and the final generator frequency in each case.

**4.** Change the load on the small-diameter wire to 5 kg. Set the signal generator frequency to 500 Hz and vary the length of the wire with the sliding bridge until its frequency agrees with that of the signal generator, as you did in Procedures 2 and 3. Record the length of the wire, its tension, and the final generator frequency. Now reduce the generator frequency to 400 Hz and, keeping the length of the wire fixed, gradually reduce the tension until the frequency of the wire matches that of the signal generator. Record the new tension and frequency.

**5.** Repeat Procedure 4 with initial generator frequency settings of 300 and 200 Hz. Record the values of the tension and frequency in each case.

**6.** Load the other (larger) sonometer wire with 5 kg and vary its length until its frequency is the same as that of the first wire. Note that the small wire remains loaded with 5 kg and in unison with a generator frequency near 200 Hz from Procedure 5. You should wind up with your two wires having the same tension and resonating at the same frequency but having different lengths. Record this frequency, tension, and the lengths of both wires.

**7.** Using the micrometer caliper, measure the diameter of each wire and record the values.

## DATA

Diameter of smaller wire _____

Diameter of larger wire _____

Mass per unit length of smaller wire _____

Mass per unit length of larger wire _____

| Signal generator frequency | Length of vibrating string | Tension | Diameter of wire | Square root of tension | Reciprocal of length of the string |
|---|---|---|---|---|---|
| | | | | | |
| | | | | | |
| | | | | | |
| | | | | | |
| | | | | | |
| | | | | | |
| | | | | | |
| | | | | | |
| | | | | | |
| | | | | | |

## CALCULATIONS _____

**1.** Using the values of the tension and length observed in Procedure 2 and the given value of the mass per unit length of the wire, calculate the frequency of the fundamental of the string using Equation 15.1. Compare the calculated value of the frequency with that of the signal generator by finding the percent error.

**2.** From the data of Procedures 2 and 3 calculate the reciprocals of the length of the string. Plot a curve using these reciprocals as abscissas and the corresponding frequencies as ordinates. What does this curve show about the way the frequency of a string under constant tension depends on the length of the string?

**3.** From the data of Procedures 4 and 5 calculate the square roots of the tension. Plot a curve using the square roots of the tension as abscissas and the corresponding frequencies as ordinates. What does this curve show about the dependence of the frequency of a string of constant length on the tension in the string?

**4.** From the data of Procedures 6 and 7 show how the lengths of two strings compare with their cross-sectional areas, if the strings vibrate at the same frequency, are under the same tension, and are made of the same material.

## QUESTIONS

**1.** Upon what physical properties do (a) the loudness, (b) the pitch, and (c) the quality of a musical note depend?

**2.** What is meant by beats?

**3.** A wire of mass 0.0003 kg/m and 0.5m long is vibrating 200 times per second. What must be the tension in newtons? What mass hung on the wire would produce this tension?

**4.** The first overtone of a stretched string 0.75 m long has a frequency of 200 Hz when the string is stretched by a weight of 4 kg. What is the mass per unit length of the vibrating string?

**5.** Two wires equal in length and made of the same material are subjected to tensions in the proportion of 1 to 4; the one under the greater tension is the thicker wire. If they vibrate at the same frequency, how do their diameters compare?

**6.** (a) Find the slope of the curve plotted in Calculation 2 and compare it with the theoretical value obtained from Equation 15.1 using the pertinent values of the tension and mass per unit length of the wire. (b) Repeat (a) for the curve of Calculation 3, comparing the slope with the theoretical value obtained from the values of the length and mass per unit length.

# Resonance of Air Columns 16

The velocity with which a sound wave travels in a substance may be determined if the frequency of the vibration and the length of the wave are known. In this experiment, the velocity of sound in air is found by using a sound wave of known frequency whose wavelength can be measured by means of a resonating air column. The purpose of this experiment is thus to determine the wavelengths in air of sound waves of different frequencies by the method of resonance in closed pipes and to calculate the velocity of sound in air using these measurements.

## THEORY

If a sound wave source is held over a tube open at the top and closed at the bottom, it will send disturbances made up of alternate compressions and rarefactions, that is, a sound wave, down the tube. These disturbances will be reflected at the tube's closed end, thus creating a situation in which identical waves are propagating in opposite directions in the same region. This is the condition for a standing wave, and so we have the possibility of a standing wave being set up in the air column in the tube. However, closing one end of the tube requires that there be a zero-amplitude point or node at that end, while at the other end, which is open, we must have a loop, or maximum-amplitude point. These requirements will be satisfied if the length of the tube is one-quarter of the sound wave's wavelength in air, for this is the distance from a node to an adjacent maximum, or loop. The requirements will also be satisfied if the tube's length is a quarter wavelength plus any integer number of half wavelengths, because half a wavelength is the distance from one node to the next. Whenever the tube has one of these lengths, the tube and the sound source are said to be in resonance, or put another way, the tube resonates at the source's frequency. The condition of resonance is indicated by an increase in the loudness of the sound heard when the air column has the resonant length.

When a sound source such as a loudspeaker or telephone receiver is held over a tube closed at one end, resonance will occur whenever the tube length is an odd number of quarter wavelengths, that is, when $L = \frac{1}{4}\lambda$, $\frac{3}{4}\lambda$, $\frac{5}{4}\lambda$, etc., where $L$ is the length of the tube and $\lambda$ is the wavelength of the sound waves in air. The relation between this wavelength and the frequency of the sound source is

$$V = f\lambda \qquad (16.1)$$

where $V$ is the velocity of sound in air in meters per second, $f$ is the frequency (the number of vibrations or cycles per second), and $\lambda$ is the wavelength in meters. From this relation, the velocity can be calculated from the known source frequency and the wavelength obtained from measurements of $L$ at resonance. It should be noted, however, that the center of the loop at the tube's open end does not fall right at the end but outside it by a small distance that depends on the wavelength and the tube diameter. However, the distance between successive tube lengths at which resonance is obtained gives the exact value of a half wavelength.

In this experiment, a closed pipe of variable length is obtained by changing the level of the water contained in a glass tube. The length of the tube above the water level is the length of the air column in use. The apparatus is shown in Fig. 16.1 and consists of a glass tube 1.2 m long closed at the bottom

**Figure 16.1** *Resonance tube apparatus*

123

with a cup and mounted on a heavy tripod base. A supply tank, which is connected to the cup by a rubber hose, is attached to the support rod by means of a clamp, which permits easy adjustment. The sound source used in this experiment consists of a small loudspeaker driven by a signal generator, and a clamp on the support rod holds this loudspeaker directly above the tube. If the sound level is to be detected by a microphone, it, too, is held just above the tube and next to the loudspeaker by means of a second clamp.

The velocity of sound in air is 331.4 m/s at 0°C. At higher temperatures, the velocity is slightly greater than this and is given by

$$V = 331.4 + 0.6t \qquad (16.2)$$

where $V$ is the velocity in meters per second and $t$ is the temperature in degrees Celsius. Equation 16.2 can be used to get a more accurate value for the speed of sound at the laboratory room temperature.

## APPARATUS

1. Resonance tube apparatus
2. Signal generator
3. Loudspeaker or telephone receiver
4. Meter stick
5. Clamp to mount loudspeaker
6. Celsius thermometer
7. Microphone and mounting clamp (optional)
8. Sound level indicator (optional)

## PROCEDURE

**1.** Using the clamp provided, mount the loudspeaker or telephone receiver so that it faces into the open end of the long glass tube and is about 2 cm above it. If a microphone is to be used to detect the sound level, it, too, should be mounted facing the top of the tube. In this case, the loudspeaker may be mounted a little to one side and slanted to face toward the tube end, thus making room for the microphone beside it.

**2.** Fill the supply tank with water and raise it to adjust the water level in the long tube to a point near the top. Do not overfill the supply tank or it will overflow when you lower it.

**3.** Turn on the signal generator and set it to a frequency of 500 cycles/s (or Hz). Connect the loudspeaker or telephone receiver to the generator output and adjust the level to give a low but audible volume of the 500-Hz tone.

**4.** Determine the shortest tube length for which resonance is heard by slowly lowering the water level while listening for resonance to occur. At resonance, that is, when the air column is adjusted to the proper length, there will be a sudden increase in the intensity of the sound that you can detect either by listening,

or, if you have the equipment, by noting the rise in the reading of the sound level indicator connected to the microphone. When the approximate length for resonance has been found, run the water level up and down near this point to determine the position for which the sound level is maximum. Measure and record the length of the resonating air column to the nearest millimeter. Make two additional determinations of this length by changing the water level and relocating the position of the maximum. Record these two readings also.

**5.** Lower the water level until the second position at which resonance occurs is found and repeat Procedure 4 to determine the air-column length for this case. As in Procedure 4, make three independent determinations of this length and record each to the nearest millimeter. *Do not change either the frequency or the sound level from the settings used in Procedure 4.*

**6.** Repeat Procedures 3–5 with signal-generator frequencies of 400 and 300 Hz. Record your three frequencies in the spaces provided in the data table.

**7.** Record the temperature in your laboratory room.

## DATA

Room temperature _____

Calculated value of the velocity of
    sound in air (average) _____

Known value of the velocity of
    sound at room temperature _____

Percent error _____

| Source frequency | First position of resonance | | | | Second position of resonance | | | | Wavelength | Velocity of sound in air |
|---|---|---|---|---|---|---|---|---|---|---|
| | 1 | 2 | 3 | Average | 1 | 2 | 3 | Average | | |
| | | | | | | | | | | |
| | | | | | | | | | | |
| | | | | | | | | | | |

## CALCULATIONS _____

**1.** For each length of resonating air column, calculate the average of the three readings taken.

**2.** For each frequency, determine the value of the wavelength in air from the lengths of the resonating air column.

**3.** For each frequency, calculate the velocity of sound in air from the wavelength found in Calculation 2. Find the average of these three values.

**4.** Calculate the velocity of sound in air at room temperature from Equation 16.2.

**5.** Compare the average value of the velocity of sound as found in Calculation 3 with the known value as found in Calculation 4 by finding the percent error.

## QUESTIONS _____

**1.** Suppose that in this experiment the temperature of the room had been lower. What effect would this have had on the length of the resonating air column for each reading? Explain.

**2.** How would an atmosphere of hydrogen affect the pitch of an organ pipe?

**3.** A 128-Hz sound source is held over a resonance tube. What are the two shortest distances at which resonance will occur at a temperature of 20°C?

**4.** A sound source is held over a resonance tube, and resonance occurs when the surface of the water in the tube is 10 cm below the source. Resonance occurs again when the water is 26 cm below the source. If the temperature of the air is 20°C, calculate the source frequency.

**5.** An observer measured an interval of 10 s between seeing a lightning flash and hearing the thunder. If the temperature of the air was 20°C, how far away was the source of sound?

**6.** From the spread in your data for this experiment, estimate limits of error to be attached to your measured value of the velocity of sound in air. Does the accepted value fall within these limits? If it does not, can you offer an explanation for the discrepancy?

# Coefficient of Linear Expansion  17

Most substances expand with an increase of temperature. It is found that the change in length of a solid is proportional to the original length and to the change in temperature. The factor of proportionality, which is called the coefficient of linear expansion, depends on the material of which the solid is made. The purpose of this experiment is to determine the coefficient of linear expansion of several metals.

## THEORY

The coefficient of linear expansion of a substance is the change in length of each unit of length when the temperature is changed one degree. The coefficient $\alpha$ may be expressed by the relation

$$\alpha = \frac{L_2 - L_1}{L_1(t_2 - t_1)} \qquad (17.1)$$

where $L_1$ is the original length in meters at the temperature $t_1$ degrees Celsius, and $L_2$ is the length at the temperature $t_2$.

The apparatus consists of a metal rod that is placed in a brass jacket through which steam may be passed. The temperature of the rod is measured by a thermometer inserted in the jacket. One end of the rod is fixed and the other end is allowed to move. A micrometer screw, which may be adjusted to make contact with the movable end, is used to measure the change in length.

In determining the coefficient of linear expansion of a metal rod, the length of the rod is measured at room temperature. The rod is then placed in the brass tube of the apparatus and the initial reading of the attached micrometer screw is taken. Steam is then run through the tube, and the increase in length is measured by means of the micrometer screw. The supports of the apparatus carry binding posts for connecting an electric buzzer or light bulb to indicate the exact point when the screw touches the rod. The coefficient of linear expansion is finally calculated from Equation 17.1.

## APPARATUS

1. Coefficient of linear expansion apparatus
2. Boiler and tripod stand
3. Bunsen burner
4. 0–100°C thermometer
5. Rubber tubing
6. Meter stick
7. 6-V dry cell or power supply
8. Electric buzzer or 6-V light bulb (if not built into apparatus)
9. Three metal rods: aluminum, copper, and steel
10. Paper towels or wipers

## PROCEDURE

**1.** Measure the length of the aluminum rod with the meter stick to 0.1 mm at room temperature. Record the length and the temperature. Avoid handling the rod with your bare hands so as not to raise its temperature above that of the room. Use dry paper towels or wipers.

**2.** Mount the rod in the brass tube and record the setting of the micrometer screw for the rod at room temperature. Make sure that one end of the rod is touching the fixed end of the apparatus and the other end is facing the micrometer screw, but not touching it. Carefully turn the micrometer screw until the buzzer sounds or the light lights, showing that the screw has made contact with the rod. Record the setting of the micrometer screw.

**3.** Turn the micrometer screw back several millimeters to allow room for the expansion of the rod. Insert the thermometer into the brass tube through the port provided for it and pass steam through the tube. Let the steam continue to flow until the temperature has stabilized at a value at or near 100°C. Note that the temperature of the steam at normal atmospheric pressure will be 100°C, but it may take some time before the rod is brought up to this temperature, and in fact heat losses from the apparatus may prevent the rod from ever getting to the full 100°C. In any case, record the final temperature reached. Then, with the steam still flowing, carefully turn the micrometer screw until the buzzer sounds or the light lights, showing that the screw has made contact with the rod. Record the setting of the micrometer screw. Remove the thermometer from the apparatus and let it return to room temperature when you have finished your measurement.

**4.** Repeat Procedures 1–3 for the copper rod.
**5.** Repeat Procedures 1–3 for the steel rod.

## DATA

|  | Aluminum | Copper | Steel |
|---|---|---|---|
| Room temperature |  |  |  |
| Length of rod at room temperature |  |  |  |
| Setting of micrometer screw at room temperature |  |  |  |
| Final temperature of rod |  |  |  |
| Setting of micrometer screw at final temperature |  |  |  |
| Change of length of the rod |  |  |  |
| Coefficient of linear expansion |  |  |  |
| Value of the coefficient of linear expansion from Table VI, at end of book |  |  |  |
| Percent error |  |  |  |

## CALCULATIONS

**1.** Compute the change in length of each rod in centimeters from the two readings of the micrometer screw.

**2.** Calculate the coefficient of linear expansion of aluminum, copper, and steel from your data.

**3.** Compare your results with those given in Table VI at the end of the book, and find the percent error.

## QUESTIONS _____

**1.** Why must the increase in length of the rod be measured so carefully although the length itself can be determined by using an ordinary meter stick?

**2.** A compound bar made of an aluminum strip and an iron strip fastened firmly together is heated. Explain what happens to the shape of the bar.

**3.** When an ordinary mercury thermometer is placed in hot water, its reading drops at first but quickly rises again. Explain why.

**4.** Why is the numerical value of the coefficient of linear expansion different when expressed in Fahrenheit degrees from what it is when expressed in Celsius degrees?

**5.** An aluminum hoop whose inside diameter is 40 cm at room temperature (20°C) is to be "shrunk" on to a disk to form a wheel with an aluminum tire or rim. The process involves heating the hoop to make it expand, placing it on the disk, and letting it shrink down on the disk as it cools. The available furnace can heat the hoop to 300°C. To what diameter should the disk be machined so that the hoop will just fit on when so heated?

**6.** The Golden Gate Bridge over San Francisco Bay is 4200 ft long. Compute the total expansion of the bridge as the temperature changes from $-20°C$ to $+40°C$. The coefficient of linear expansion for steel is given in Table VI at the end of the book.

**7.** If a solid material is in the form of a block rather than a rod, its volume will grow larger when it is heated, and a *coefficient of volume expansion* $\beta$ defined by

$$\beta = \frac{V_2 - V_1}{V_1(t_2 - t_1)}$$

may be quoted. Here $V_1$ and $V_2$ are the initial and final volumes of the block, and $t_1$ and $t_2$ are the initial and final temperatures. Find the relation between the coefficients $\alpha$ and $\beta$. *Hint:* Suppose the block is a rectangular parallelepiped whose volume is its length times its width times its height.

# Specific Heat and Calorimetry 18

To raise the temperature of a body from a given initial temperature $t_1$ to a final temperature $t_2$ requires a total quantity of heat that depends on the mass of the body, the specific heat of the body's material, and the temperature difference. The process of measuring quantities of heat is called calorimetry. The purpose of this experiment is to determine the specific heat of one or more metals by a calorimetric procedure called the method of mixtures.

## THEORY

The measurement of heat quantities by the method of mixtures makes use of the principle that when a heat interchange takes place between two bodies initially at different temperatures, the quantity of heat lost by the warmer body is equal to that gained by the cooler body, and some intermediate equilibrium temperature is finally reached. This is true provided no heat is gained from or lost to the surroundings.

Any given body is characterized by a *heat capacity,* which is the number of calories required to raise the temperature of that particular body 1°C. Clearly, the heat capacity depends both on the mass of the body and the nature of its material. To eliminate the dependence on mass and obtain a characteristic depending only on the material involved, the *specific heat* is defined as the heat capacity of a particular amount of the material in question. Any unit of amount may be used, but it must be explicitly stated in the units of specific heat. Because it is common practice in calorimetry, we shall use the gram of mass as our unit of quantity in this experiment. Our units of specific heat will then be calories per gram per degree Celsius, in contrast to the units of heat capacity, which are just calories per degree. Note that the calorie is defined as the heat required to raise 1 g of water (at 4°C) 1°C. Thus, by definition the specific heat of water is 1 cal/g-°C. The heat capacity of $m$ grams of water is thus numerically equal to $m$, whereas the heat capacity of the same mass of a substance whose specific heat is $c$ cal/g-°C is $mc$. Since $mc/m = c$, the specific heat of any substance may be thought of as the ratio of the heat capacity of a body made of that substance to the heat capacity of an equal mass of water.

In this experiment, the specific heat of two different metals will be measured by studying the heat interchange between a sample of each metal and a mass of water. The vessel in which this interchange takes place is called a calorimeter. Any cup may be used as a calorimeter provided account is taken of the heat given up or absorbed by the cup and care is exercised to prevent heat exchange with the cup's surroundings. A normal way to do this is to use a thin metal cup whose heat capacity is small and easily measured and to mount it in an insulating jacket that prevents external heat exchange. A simpler and quite satisfactory procedure is to use a styrofoam cup, which both provides the needed insulation and has a negligibly small heat capacity.

To insure that an equilibrium temperature is reached in a short time after the metal sample is placed in the water in the calorimeter, the water should be stirred. The thermometer used to measure the equilibrium temperature may also serve as a stirrer. The thermometer itself will absorb some heat in coming to the equilibrium temperature, but its heat capacity is usually so small that negligible error is made by ignoring it.

In the experimental determination of the specific heat of a metal by the method of mixtures, a metal sample of known mass that has been heated to a known high temperature is dropped into a known mass of water at a known low temperature. After equilibrium has been established, the new temperature of the water-metal-calorimeter combination is measured. In arriving at this equilibrium temperature, the water and calorimeter must have gained the heat lost by the metal sample. This situation is expressed by the relation

$$Mc(t_1 - t_2) = (mc_w + m_1c_1)(t_2 - t_3) \quad (18.1)$$

where $M$ is the mass of the metal sample in grams, $c$ is the specific heat of the metal, $t_1$ is the initial temperature of the metal sample, $t_2$ is the final equilibrium temperature, $t_3$ is the initial temperature of the water and calorimeter, $m$ is the mass of water, $m_1$ is the mass of the calorimeter cup, $c_1$ is the specific heat of the material of which this cup is made, and $c_w$ is the specific heat of water, which, in our present set of units, is 1 cal/g-°C. Note that $m_1c_1$ is the heat capacity of the calorimeter cup. This quantity is sometimes called the *water equivalent* of the calorimeter because it is numerically equal to the mass of water having that heat capacity. Equation 18.1 may be used to determine $c$ if all the other quantities in it have been measured.

## APPARATUS

1. Calorimeter
2. Boiler and tripod stand
3. Bunsen burner
4. Triple-beam balance
5. 0–100°C thermometer
6. 0–50°C thermometer
7. Sample cylinder no.1 (metal to be specified by the instructor)
8. Sample cylinder no. 2 (metal to be specified by the instructor)

## PROCEDURE

**1.** Fill the boiler with enough water to cover either sample cylinder with a couple of inches to spare and place it on the tripod. Put the bunsen burner under it and bring the water to a boil.

**2.** Weigh the empty calorimeter cup. Note that you will be determining its *mass in grams*.

**3.** Weigh sample cylinder no. 1 and then lower it into the boiling water by means of a thread. Be sure the cylinder is completely immersed in the water but do not allow it to touch the bottom or sides of the boiler.

**4.** Pour cold water (about 3° below room temperature) into the calorimeter cup until the cup is about half full. Then weigh the cup with the water in it. Note that the temperature of the cold water should be about as much below room temperature as the equilibrium temperature will be above it so as to balance out errors due to transfer of heat by radiation to or from the surroundings.

**5.** Replace the calorimeter in its insulating jacket and measure the temperature of the water with the 0–50°C thermometer. Record this temperature.

**6.** Quickly transfer the sample cylinder from the boiler to the calorimeter without splashing any water. Stir the water and record the equilibrium temperature as indicated on the 0–50°C thermometer.

**7.** Measure and record the temperature of the boiling water using the 0–100°C thermometer.

**8.** Repeat Procedures 3–7 for sample cylinder no. 2.

## DATA

Mass of calorimeter cup _____

Specific heat of calorimeter cup material _____

|  | **Data for cylinder no. 1** | **Data for cylinder no. 2** |
|---|---|---|
| Mass of cylinder |  |  |
| Mass of calorimeter and cold water |  |  |
| Mass of cold water |  |  |
| Initial temperature of cold water |  |  |
| Temperature of boiling water |  |  |
| Equilibrium temperature |  |  |
| Calculated specific heat |  |  |
| Specific heat from Table VII |  |  |
| Percent error |  |  |

# CALCULATIONS _____

**1.** Calculate the specific heat of the sample no. 1 material from the data of Procedures 2–7.

**2.** Calculate the specific heat of the sample no. 2 material from the data of Procedure 8.

**3.** Compare your results for the specific heats with those given in Table VII at the end of the book, and find the percent error.

## QUESTIONS _____

**1.** What is the purpose of starting with the temperature of the water lower than room temperature and ending with it about the same amount above room temperature?

**2.** How would the computed value of the specific heat be affected if some boiling water were carried over with the metal?

**3.** What will be the biggest source of error if too much water is used?

**4.** What is meant by the water equivalent of a body?

**5.** Check that Equation 18.1 is dimensionally correct. The factor $c_w$ (the specific heat of water) is sometimes omitted because its numerical value is 1. What is wrong with doing this?

**6.** A platinum ball weighing 100 g is removed from a furnace and dropped into 400 g of water at 0°C. If the equilibrium temperature is 10°C and the specific heat of platinum is 0.04 cal/g-°C, what must have been the temperature of the furnace? Neglect the effect of the mass of the calorimeter.

# Heat of Fusion and Heat of Vaporization

# 19

The present experiment deals with *changes of phase*, that is, the changing of a substance from the solid to the liquid state and from the liquid to the gaseous state. These three states, or phases, are normally exhibited by all substances, but the temperatures at which changes from one state to another take place and the heat involved vary widely. It is the purpose of this experiment to study the heats involved in the change of water from a solid to a liquid and from a liquid to a gas by the method of mixtures. This method has been described in detail in Experiment 18 on calorimetry, and a review of the Theory section of that experiment will be helpful preparation for the present work.

## THEORY

When a solid body is heated, its temperature rises at a rate determined by the rate at which heat is being supplied and the heat capacity of the body. However, when a certain temperature called the melting point is reached, the body starts to melt, and as melting proceeds the temperature remains constant even though heat is being continually supplied. Increasing the supply rate merely increases the rate at which the solid changes to a liquid. No change in temperature occurs until the body has completely melted, after which further heating will raise the temperature of the resulting liquid at a rate determined by the liquid's heat capacity (its mass multiplied by its specific heat) and the rate at which heat is being added.

Eventually, a temperature called the boiling point will be reached. Here the liquid changes rapidly to a vapor, and again there is no change in temperature while this is going on even though the addition of heat continues. When all the liquid has been evaporated, continued addition of heat to the resulting gas (which must now be contained in a suitably sealed vessel) will once more raise the substance's temperature—this time at a rate determined by the heat supply rate and the heat capacity (mass times specific heat) of the gas.

Upon cooling, the gas begins to condense when the boiling point is reached, and again there is no change in temperature while condensation takes place. The same amount of heat is removed during this condensation process as was added during vaporization; this heat represents the difference in the energy of the molecules in the gaseous state and their energy when bound (although loosely) in the liquid. Similarly, once the temperature has been reduced to the freezing point (which is the same as the melting point), further removal of heat causes the liquid to solidify. No change of temperature occurs during this process, and the heat removed (which is the same as the heat added during melting) represents the difference in the molecular energy when the molecules are loosely bound in the liquid and when they are tightly bound in the solid.

You should note that vaporization takes place at all temperatures, because there are always a few molecules in the liquid with enough energy to break the bonds at the surface and emerge into the space above. One can thus think of a *vapor pressure* due to these escaped molecules above any liquid surface. Naturally, the higher the temperature, the more molecules will have enough energy to break through the surface and the higher the vapor pressure will be. If the liquid surface is exposed to the atmosphere, atmospheric pressure will prevent molecules entering the gaseous state near this surface from forming bubbles until the vapor pressure and atmospheric pressure become equal. At this point, the vapor can form bubbles in opposition to the pressure of the atmosphere and the action we call boiling occurs. Molecules can then break from the liquid in large numbers, energy in the form of heat must be added to the liquid to replace the energy of the departing molecules, and a high rate of heat supply merely permits more molecules to evaporate per second. Note, however, that the boiling point is now seen to be that temperature at which the liquid's vapor pressure becomes equal to atmospheric pressure and is therefore highly dependent on the atmospheric pressure's value. This is why water boils at a lower temperature on top of a mountain, where the atmospheric pressure is less than at sea level. Thus, when the boiling point of a liquid is given without further information, normal sea-level atmospheric pressure is assumed. Because there is usually a small volume change when a solid changes to a liquid or vice versa, the melting point may also be affected by the external pressure, but this effect is usually so small that it can be ignored.

The amount of heat required to change a unit mass of a substance from the solid to the liquid state without a change in temperature is called the *latent heat of fusion* of the substance, or simply its heat of fusion. Similarly, the amount of heat required to change a unit mass of a substance from the liquid to the vapor state without a change in temperature is called the *latent heat of vaporization* of the substance, or simply the heat of vaporization. Since evaporation can go on at any temperature and the heat of vaporization is different at different temperatures, the temperature should be quoted when a heat of vaporization is given. However, heats of vaporization are usually given without this information, it being assumed that the temperature of the boiling point is meant. Since this temperature is dependent on the atmospheric pressure, normal sea-level pressure is also assumed.

In doing heat experiments where the quantity of heat added to or taken from a substance is to be determined, the range of temperature should extend equally above and below room temperature so that the amount of heat absorbed from the surroundings during the course of the experiment will be approximately equal to the amount of heat radiated to the surroundings. Best results will thus be obtained by starting with a calorimeter of warm water when the heat of fusion of ice is to be measured and starting with a calorimeter of cold water for the measurement of the heat of vaporization of steam. This is easily accomplished by measuring first one and then the other of the two latent heats of water. It is the purpose of this experiment to carry out these two measurements.

The experimental determination of the heat of fusion of ice and the heat of vaporization of water is made by the method of mixtures. This makes use of the principle that when a heat interchange takes place between two bodies initially at different temperatures, the quantity of heat lost by the warmer body is equal to that gained by the cooler body, and some intermediate equilibrium temperature is finally reached. This is true provided no heat is gained from or lost to the surroundings.

In determining the heat of fusion of ice, a few small pieces of ice are placed, one by one, into a calorimeter containing water. As the ice melts, heat is absorbed from the water and calorimeter until the mixture comes to a final equilibrium temperature. The heat absorbed by the ice in melting plus the heat absorbed by the ice water thus produced is equal to the heat lost by the warm water and calorimeter. The working equation is

$$ML_f + Mc_w(t_2 - 0) = (mc_w + m_1c_1)(t_1 - t_2)$$

where $M$ is the mass of ice in grams, $L_f$ is the latent heat of fusion of ice, $t_2$ is the equilibrium temperature, $t_1$ is the initial temperature of the water and calorimeter, $m$ is the mass of warm water, $c_w$ is the specific heat of water, $m_1$ is the mass of the calorimeter, and $c_1$ is the specific heat of the material of which the calorimeter is made. Note that if all masses are in grams and specific heats in calories per gram per degree Celsius, $c_w$ has the numerical value 1 and $m_1c_1$ is the water equivalent of the calorimeter.

The heat of vaporization of water is determined in a similar manner by the method of mixtures. A quantity of steam is passed into a known mass of cold water in a calorimeter, where it condenses and raises the temperature of the water. The heat lost by the steam in condensing plus the heat lost by the condensed steam is equal to the heat absorbed by the cold water and calorimeter. The working equation is

$$ML_v + Mc_w(100 - t_2) = (mc_w + m_1c_1)(t_2 - t_1)$$

where $M$ is the mass of steam in grams, $L_v$ is the latent heat of vaporization of water, $t_2$ is the equilibrium temperature, $t_1$ is the initial temperature of the water and calorimeter, $m$ is the mass of cold water, $c_w$ is the specific heat of water, $m_1$ is the mass of the calorimeter, and $c_1$ is the specific heat of the calorimeter material as before.

# APPARATUS

1. Calorimeter
2. Boiler and tripod stand
3. Water trap
4. Rubber tubing
5. Bunsen burner
6. Triple-beam balance
7. 0–100°C thermometer
8. 0–50°C thermometer
9. Ice
10. Paper towels

# PROCEDURE

**1.** Weigh the empty calorimeter. Note that you are finding the calorimeter's *mass* in *grams*.

**2.** Fill the calorimeter half full of water about 10° above room temperature and weigh it. Replace it

in the outer calorimeter jacket and record the temperature of the water.

**3.** Dry some small pieces of ice on a paper towel and add them to the water without touching the ice with your fingers, so as not to melt it. Add the ice until the temperature is about 10° below room temperature, keeping the mixture well stirred with the 0–50°C thermometer. Record the equilibrium temperature when the ice is entirely melted. Weigh the calorimeter with its contents again, but without the thermometer.

**4.** Fill the calorimeter three-quarters full of water about 15° below room temperature and again weigh it. Replace it in the outer calorimeter jacket.

**5.** Set up the steam generator as shown in Fig. 19.1 and adjust it so that the steam passes through the water trap and flows freely from the steam tube. Record the temperature of the cold water and quickly immerse the steam tube. Allow the steam to pass in and condense until the temperature of the water is about 15° above room temperature, stirring continuously with the 0–100°C thermometer. Remove the steam tube and record the equilibrium temperature. Weigh the calorimeter with its contents again, but without the thermometer.

**Figure 19.1**  *Heat of vaporization apparatus*

# DATA _____

Mass of calorimeter _____        Specific heat of calorimeter _____

**Heat of Fusion**

Mass of calorimeter and water _____        Mass of ice _____

Mass of water _____        Calculated heat of fusion of ice _____

Initial temperature of water _____        Accepted value of heat of fusion

Equilibrium temperature _____           of ice _____

Mass of calorimeter, water, and        Percent error _____
   melted ice _____

**Heat of Vaporization**

Mass of calorimeter and water _____        Mass of steam _____

Mass of water _____        Calculated heat of vaporization of

Initial temperature of water _____           water _____

Equilibrium temperature _____        Accepted value of heat of

Mass of calorimeter, water, and           vaporization of water _____
   condensed steam _____        Percent error _____

# CALCULATIONS _____

**1.** From the data of Procedures 1–3, compute the heat of fusion of ice.

**2.** From the data of Procedures 4 and 5, compute the heat of vaporization of water.

**3.** Compare your results for the heat of fusion and the heat of vaporization with the known values by finding the percent error.

## QUESTIONS _____

**1.** Discuss the principal sources of error in this experiment.

**2.** Why is it necessary to stir the mixtures?

**3.** What error would have been introduced if the ice were not dry?

**4.** Suppose the steam generator produced steam so fast that some of it bubbled up through the water and escaped from the calorimeter without condensing. Would this cause an error in your results? Explain.

**5.** (a) What is the purpose of the water trap in determining the heat of vaporization? (b) How would its absence have affected the results?

**6.** A glass tumbler of 200 g mass and 5 cm inside diameter is filled to a depth of 10 cm with tap water whose temperature is 10°C, and 50 g of ice at 0°C are then added. At the end of an hour, it is noted that the ice has just finished melting. What was the average rate, in calories per second, at which heat flowed into the glass from its surroundings?

**7.** The Theory section of this experiment suggests that when a liquid starts to boil, the formation of bubbles should begin near the surface. Why do the bubbles actually form first at the bottom of the containing vessel?

# The Gas Laws 20

When a substance is in the gaseous state, its molecules fly about at random, there being only very small forces of interaction between them (sometimes called Van der Waals forces), and because the molecules are small compared to their average separations, there is only a very small chance of any two of them colliding. A so-called ideal gas is defined as the idealization of the situation just described, that is, a gas in which the molecules are mass points that have *no* chance of colliding with each other and that also have *no* interaction forces, not even weak ones. Actual gases like nitrogen, oxygen, and hydrogen satisfy this definition to a good approximation at ordinary temperatures and pressures; thus, the study of the ideal gas has a practical application. Only at extreme temperatures and pressures do the molecules of real gases get close enough together for their interaction forces to have an appreciable effect and their finite size to become significant.

In order to specify fully the condition of any substance, its pressure, volume, and temperature must be known. These quantities are always interrelated through a relationship characteristic of the substance called its *equation of state,* so that if two of them are known, the equation of state gives the third one. The equation of state for an ideal gas (often called the ideal gas law) is a particularly simple one, derivable theoretically on the basis of the assumptions noted above with regard to no collisions and no interaction forces among the ideal gas molecules. It applies to real gases at ordinary temperatures and pressures and can thus be checked experimentally. The purpose of this experiment is to carry out such a check and incidentally to determine the so-called absolute zero of temperature.

## THEORY

In studying the behavior of a gas under different conditions of pressure, temperature, and volume, it is convenient to keep one of these constant while varying the other two. Thus, if the temperature is kept constant, a relation is obtained between the pressure and the volume; if the volume is kept constant, a relation between the pressure and temperature can be found. Such studies were carried out independently by three different investigators whose names are now associated with the results they obtained. The three possible relations, collectively referred to as the gas laws, are as follows: pressure and volume with temperature constant (Boyle's law); volume and temperature with pressure constant (Charles's law); and pressure and temperature with volume constant (Gay-Lussac's law).

**Boyle's Law** If a gas is kept at constant temperature, we would expect an increase in pressure to "squeeze the gas together," that is, to cause a decrease in volume. Conversely, reducing the pressure allows the gas to expand, that is, to occupy a larger volume. This relation was studied quantitatively by Robert Boyle (1627–1691), who found that for a fixed amount of gas at constant temperature the pressure and volume are inversely proportional over any ordinary pressure range. This result may be stated mathematically as

$$P = \frac{C_1}{V}$$

or

$$PV = C_1 \qquad (20.1)$$

where $P$ is the pressure, $V$ is the volume, and $C_1$ is a proportionality constant whose value depends on the (constant) temperature and the amount of gas involved. The property of gases expressed by Equation 20.1 is called *Boyle's law.*

**Charles's Law** If a gas is heated and the pressure remains constant, it expands so that it occupies an increased volume. The French physicist Jacques Alexandre César Charles (1746–1823) found that under these circumstances a linear relation exists between the volume and the temperature, expressible by the relation

$$V = V_0(1 + \beta t) \qquad (20.2)$$

where $V$ is the volume at temperature $t$, $t$ is the temperature in degrees Celsius, $V_0$ is the volume at 0°C, and $\beta$ is a constant called the coefficient of expansion of the gas at constant pressure. Since the zero of temperature on the Celsius scale is arbitrarily chosen as the freezing point of water at standard atmospheric pressure, negative temperatures on this scale are per-

**147**

fectly meaningful, and it is interesting to note the temperature at which an extrapolation of Equation 20.2 indicates the volume would go to zero. This temperature is clearly $-1/\beta$, and since $\beta$ is very nearly $\frac{1}{273}$ or 0.00366 per degree Celsius for all gases, Equation 20.2 predicts zero volume at a temperature of $-273°C$. If a new temperature scale is now defined having degrees of the same size as Celsius degrees but whose zero is at $-273°C$, we have

$$T = t + 273 \qquad (20.3)$$

where $T$ is the temperature on the new scale, called the absolute or Kelvin scale, and is expressed in degrees Kelvin (K). Substitution of Equation 20.3 in Equation 20.2 with $\beta = \frac{1}{273}$ yields

$$V = \frac{V_0}{273}T = C_2T \qquad (20.4)$$

where $C_2$ is a constant that is seen to depend on $V_0$ and hence on the quantity of and pressure on the gas. Thus, if the temperature is given on the absolute scale, volume and temperature are not just linearly related but are proportional. The linear dependence of volume on temperature at constant pressure is called *Charles's law*.

**Gay-Lussac's Law**  If a gas is heated but not allowed to expand (that is, maintained at constant volume), the pressure will rise. The investigations of the French scientist Joseph Louis Gay-Lussac (1778–1850) showed that, like the volume-temperature relation at constant pressure, the pressure-temperature relation at constant volume is linear and expressible by the relation

$$P = P_0(1 + \beta t) \qquad (20.5)$$

where $P$ is the pressure at $t°C$, $P_0$ the pressure at $0°C$, and $\beta$ the pressure coefficient of the gas at constant volume. Note that the same symbol is used for the pressure coefficient in Equation 20.5 and the coefficient of expansion in Equation 20.2. This is because the same value, $\frac{1}{273}$ per degree Celsius, is found for both and both have the same significance, namely that the volume and therefore also the pressure of an ideal gas should become zero at $-273°C$ or $0°K$. Fig. 20.1 is a plot of pressure versus temperature for a typical gas and shows this extrapolation to zero pressure. If Equation 20.3 is substituted in Equation 20.5 in the same manner as in Equation 20.2, we get

$$P = \frac{P_0}{273}T = C_3T \qquad (20.6)$$

where $C_3$ is a constant that depends on $P_0$ and hence on the quantity of gas involved and the volume in which it is confined. Again, use of the absolute temperature scale makes pressure and temperature proportional. As a result of Gay-Lussac's work, the linear dependence of pressure on temperature at constant volume is usually known by his name.

**The Ideal Gas Law**  Equations 20.1, 20.4, and 20.6 can be combined in the single relation

$$PV = C_4T \qquad (20.7)$$

where $C_4$ is a new constant whose value depends only on how much gas is involved. In view of Avogadro's law (which states that equal volumes of all gases at the same pressure and temperature contain the same number of molecules) stating how much gas we have by giving the actual number of molecules present allows us to write Equation 20.7 as

$$PV = NkT \qquad (20.8)$$

where $C_4 = Nk$, $N$ is the number of molecules in question, and $k$ is a *universal* constant applying to all gases with no dependence on temperature, pressure, volume, or quantity. It is called *Boltzmann's constant* and has the value $1.38 \times 10^{-23}$ joule per molecule-degree absolute. The student should note carefully that the product $PV$ has the dimensions of energy so that the units for Boltzmann's constant are those of energy per molecule-degree. The value of this constant thus depends on the units of energy used. Expressing the amount of gas by the number of molecules present requires inconveniently large numbers, and so the number $n$ of moles is more usually given. Since by definition a mole contains Avogadro's number $N_0$ of molecules, the total number of molecules $N$ is equal to $nN_0$, and Equation 20.8 may be written

$$PV = nRT \qquad (20.9)$$

where $R = N_0k$ and is called the *universal gas constant*. It serves the same purpose as $k$ but has the dimensions of energy per *mole*-degree rather than energy per *molecule*-degree. Equation 20.9 (or 20.8) is a relation between the pressure, volume, and temperature of an ideal gas and is therefore the equation of state for the ideal gas. It is called the *ideal gas law*.

In the present experiment, we will check the laws of Boyle and Gay-Lussac by seeking experimental verification of Equations 20.1 and 20.5. Moreover, if the linear relation predicted by Equation 20.5 is obtained, simple extrapolation to $P = 0$ will give the location of the absolute zero of temperature on the Celsius scale and hence the value of $\beta$. In this way, Equation 20.6 will also be verified. Then, if pressure is found to be proportional to the absolute temperature at constant volume and inversely proportional to volume at constant temperature, the proportionality of volume and temperature at constant pressure (Charles's law), and hence the ideal gas law, is implied. The experimental test of Boyle's law consists in observing a series of different volumes, measuring the corresponding pressures, and plotting $P$ against

**Figure 20.1** *Pressure versus temperature for 1 L of a typical gas*

$1/V$ to see if a straight line is obtained. The apparatus is made up of two glass tubes of uniform bore connected by a piece of flexible rubber tubing and mounted on a supporting frame to which a scale is attached (see Fig. 20.2). The apparatus is partially filled with mercury. One of the glass tubes is provided with a stopcock; for this part of the experiment the air upon which observations are to be made is confined in this tube between the stopcock and the mercury column. The other tube is open to the atmosphere and can be raised or lowered, thus changing the height of the mercury column that confines the air. The difference in height of the mercury in the two tubes causes a pressure to be exerted on the gas trapped in the closed tube. This is called the gauge pressure because it is the pressure that would be measured by a gauge that reads zero when exposed to atmospheric pressure if it were connected to the confined air. The total or absolute pressure on the

confined air is the sum of the gauge pressure and the atmospheric pressure that is applied on top of the mercury column in the open tube. Thus, the total pressure $P$ acting on the confined air is given by

$$P = A + (h_2 - h_1) \qquad (20.10)$$

where $A$ is the atmospheric pressure as read on the barometer; $h_2$ is the level of the mercury in the open tube; and $h_1$ is the mercury level in the tube closed by the stopcock. Notice that Equation (20.10) is consistent if $h_1$ and $h_2$ are in centimeters and all pressures are expressed in centimeters of mercury. Note also that if $h_1$ is greater than $h_2$ the quantity $(h_2 - h_1)$ is negative and subtracts from atmospheric pressure to give the total pressure on the confined air.

For the test of Gay-Lussac's law, a bulb is connected to the stopcock by a short length of rubber tubing as shown in Fig. 20.3. The stopcock is opened so that the volume of confined air includes the vol-

**Figure 20.2** *Boyle's law apparatus*

**Figure 20.3** *Air thermometer*

ume in the glass tube above the mercury column, the volume of the rubber hose, and the volume of the bulb. The total volume is thus very much larger than that used for the test of Boyle's law, and the bulb provides for convenient heating to a known temperature. This is done by immersing the bulb in water contained in a Pyrex beaker, the temperature of the confined air being varied by heating or cooling the water bath. The total volume of confined air is kept constant by adjusting the height of the open glass tube so that the mercury level in the tube connected to the bulb is brought to a particular scale division before each set of readings is taken. Pressure readings are taken by using Equation 20.10 as in the test of Boyle's law and temperature readings by means of a thermometer in the water bath. Note that in the present experiment the relation in Equation 20.5 is being checked and $\beta$ determined by measuring the gas temperature with an external thermometer and obtaining the pressure from the difference in height of the mercury columns. However, if the linear relation postulated by Gay-Lussac is correct and $\beta$ known to high accuracy, the arrangement of Fig. 20.3 could serve as a thermometer, the difference in height of the mercury columns being a direct measure of the temperature of the gas in the bulb. In fact, the linear dependence of gas pressure on temperature at constant volume is so good and the value of $\beta$ has been determined so accurately that the constant volume hydrogen thermometer—an instrument similar to the present setup but using hydrogen instead of air—is the standard thermometer used to calibrate other types of thermometers.

## APPARATUS

1. Boyle's law apparatus
2. Bulb with connecting tube
3. Barometer
4. 0–50°C thermometer
5. 0–100°C thermometer
6. 1000-cc Pyrex beaker
7. Ring stand and clamps
8. Bunsen burner
9. Ice

## PROCEDURE

**1.** Record the reading of the barometer and the room temperature as read on the 0–50°C thermometer. CAUTION: Suspend the thermometer in the room for several minutes to allow it to come to equilibrium and do not touch it during this period.

**2.** Check that the apparatus has the proper amount of mercury in it. To do this, open the stopcock and see if each tube is about half full. The mercury level will, of course, be the same in both tubes when the stopcock is open. Adjust the tube support clamps so that the top of each tube is at about the same height and the clamps are about midway up their support bars, so that the open tube may be both raised above and lowered below the closed one through as great a range as possible.

**3.** Close the stopcock and thereafter do not handle the closed tube or allow the enclosed air to be heated above or cooled below room temperature in any way. Record the level of the top of the closed tube (just below the stopcock) and the level of the mercury columns in the two tubes. Note that atmospheric pressure should continue to exist in the closed tube after the closing of the stopcock, so that the mercury level should remain the same in both tubes.

**4.** Decrease the pressure on the gas trapped in the closed tube by lowering the open tube as far as the instrument will permit. Allow it to remain in this position for a few minutes to test for leaks. If there is any leakage in the apparatus, consult the instructor.

Record the level of the mercury column in the closed tube and in the open tube.

**5.** Vary the pressure in nine approximately equal steps from the smallest to the largest value that can be obtained so as to get ten different observations. Record the level of the mercury column in each tube for each setting. After changing the level, allow a minute or two before taking readings to permit the confined air to reach equilibrium at room temperature. Remember not to handle the closed tube or allow spurious heating or cooling of the enclosed air to take place.

**6.** When the Boyle's law measurements have been taken, return the open tube to the position at which the mercury levels are equal and open the stopcock. Connect the bulb to the stopcock nipple by means of the rubber tube.

**7.** Mount the Pyrex beaker on the ring stand with the bunsen burner under it. Support the bulb inside the beaker by means of a clamp, not allowing the bulb to touch the sides or bottom.

**8.** Fill the beaker with ice and water until the bulb is entirely covered. Watch the motion of the mercury in the closed tube to make sure it does not get up to the stopcock. If the mercury appears to be rising too high, lower the open tube, being careful not to let mercury spill out of it.

**9.** Pick a scale division a few centimeters below the bottom of the stopcock as your constant volume mark, and adjust the open tube so as to bring the mercury column in the closed tube to this level. Wait for temperature equilibrium to be reached, keeping the closed-tube mercury level at your constant volume mark until equilibrium is established. This will be indicated by the pressure stabilizing so that no further adjustment of the open tube is required. Stir the water well and measure its temperature with the 0–50°C thermometer. Note that if you still have a mixture of ice and water in the beaker as called for in Procedure 8, this temperature should be 0°C. Record it and the height of the mercury level in the open tube.

**10.** Remove the ice and add enough water to cover the bulb. Light the bunsen burner and heat the water slowly until its temperature is about 10°C. Remove the bunsen burner and stir the water well with the 0–50°C thermometer. Keep the mercury level in the closed tube at your constant volume mark by

raising the open tube as required. When equilibrium has been reached, record the temperature of the water and the height of the mercury in the open tube.

**11.** Repeat Procedure 10, raising the temperature of the water in steps of about 10° until the boiling point is reached. Use the 0–100°C thermometer in place of the 0–50°C one as soon as the temperature goes above 40°C.

**12.** When the experiment is completed, remove the bulb from the water and lower the open tube progressively as the bulb cools so as to keep the mercury from rising up to the stopcock. Leave the apparatus in the condition it was in at the end of Procedure 6.

**13.** Check the barometer reading to be sure there has been no appreciable change since it was read in Procedure 1. If there is, use the earlier reading for the Boyle's law work and the later one for the test of Gay-Lussac's law.

# DATA

Barometer reading, beginning _____  
                     end _____  
Room temperature _____

Level of the top of the closed tube, $h_3$ _____

| Mercury level | | Gauge pressure $h_2 - h_1$ | Volume of air $h_3 - h_1$ | Total pressure | $PV$ | $1/V$ |
|---|---|---|---|---|---|---|
| Closed tube $h_1$ | Open tube $h_2$ | | | | | |
| | | | | | | |
| | | | | | | |
| | | | | | | |
| | | | | | | |
| | | | | | | |
| | | | | | | |
| | | | | | | |
| | | | | | | |
| | | | | | | |
| | | | | | | |

Height of chosen constant volume
   mark                  _____

Measured value of the absolute
   zero of temperature on the
   Celsius scale            _____

Accepted value of the absolute
   zero                      _____

Percent discrepancy           _____

| Mercury level in the open tube | Gauge pressure | Total pressure | Temperature of the bulb |
|---|---|---|---|
| | | | |
| | | | |
| | | | |
| | | | |
| | | | |
| | | | |
| | | | |
| | | | |
| | | | |
| | | | |
| | | | |

## CALCULATIONS _____

**1.** Compute the value of $h_2 - h_1$ for each setting of the apparatus in Procedures 4 and 5.

**2.** Compute the value of $P$, the total pressure on the confined air, for each observation taken in Procedures 4 and 5. Express the results in centimeters of mercury. *Note:* For these and the following calculations, use three significant figures in the numbers representing the pressure, the volume, and the reciprocal of the volume. In calculating the product, use four significant figures.

**3.** Calculate the volume of the enclosed air for each of the observations of Procedures 4 and 5. *Note:* Since the tube is of uniform bore, the tube's volume is proportional to its length. The length $h_3 - h_1$ of the column of air confined in the closed tube may therefore be taken as a measure of this air's volume.

**4.** Calculate the product *PV* for each corresponding pressure and volume found in Calculations 2 and 3.

**5.** Calculate the value of $1/V$ for each volume found in Calculation 3.

**6.** Plot a curve using the values of the pressure as abscissas and the corresponding values of $1/V$ as ordinates. Use an entire sheet of graph paper for the curve. Read the instructions on plotting graphs in the Introduction.

**7.** For each temperature measured in Procedures 9–11, calculate the gauge pressure exerted on the air in the bulb by subtracting the height of the mercury in the closed tube (height of the constant volume mark) from the height of the mercury in the open tube. Note that this pressure will be negative in many cases. Then add the atmospheric pressure to obtain the total pressure in centimeters of mercury.

**8.** Plot a curve using the values of the temperature as abscissas and the values of the total pressure obtained in Calculation 7 as ordinates. For more accurate results, construct the graph of the data using as large a scale as possible, that is, utilize the whole sheet of graph paper and extend the line only to 0°C.

**9.** From the graph of Calculation 8 determine the position of the absolute zero of temperature on the Celsius scale, that is, calculate the temperature at which this graph predicts that the pressure would be zero. A simple proportion may be used to do this since, according to Equation 20.5, the curve should be a straight line. Compare your result with the accepted value by calculating the percent discrepancy.

## QUESTIONS _____

**1.** Show how your results and the graph from Calculations 4–6 verify Boyle's law. Discuss in particular your results in Calculation 4.

**2.** Explain how an error is introduced into the Boyle's law experiment when the room temperature changes.

**3.** How does the difference in height of the mercury in the two tubes cause a pressure to be exerted on the enclosed air?

**4.** Does Boyle's law hold accurately for a real gas at any temperature or pressure such as a very high or a very low one? Explain.

**5.** Show that Equation 20.7 is in fact a combination of Equations 20.1, 20.4, and 20.6, in the process showing how the constants $C_1$, $C_2$, and $C_3$ are related to $C_4$.

**6.** In the apparatus used to test Gay-Lussac's law, is an error introduced by the fact that the air in the rubber tubing between the bulb and the stopcock and in the glass tube between the stopcock and the mercury column is not in the water bath? Estimate how big and in which direction such an error might be.

**7.** A spherical basketball 10 in. in diameter is pumped up to a gauge pressure of 44.1 lb/in.$^2$ If the normal atmospheric pressure is 14.7 lb/in.$^2$, what volume of air at this pressure has been pumped into the ball?

**8.** In an air thermometer apparatus like that used to test Gay-Lussac's law, the level of the mercury in the open tube is 20 cm higher than that in the closed tube when the bulb is at 0°C. At what temperature will the level in the open tube be 40 cm higher than that in the closed tube? Assume standard atmospheric pressure.

**9.** Use the graph you drew in Calculation 8 to determine $P_0$ and $\beta$ in Equation 20.5 as applied to your setup.

**10.** Calculate the volume occupied by 1 mole of an ideal gas at 0°C and 1 atmosphere pressure. Will your result be true for real gases such as nitrogen or oxygen? What difference will the choice of gas make?

**11.** A constant volume gas thermometer indicates an absolute pressure of 50.0 cm of mercury at 0°C and 68.3 cm of mercury at 100°C. At what temperature will it read an absolute pressure of 56.1 cm of mercury?

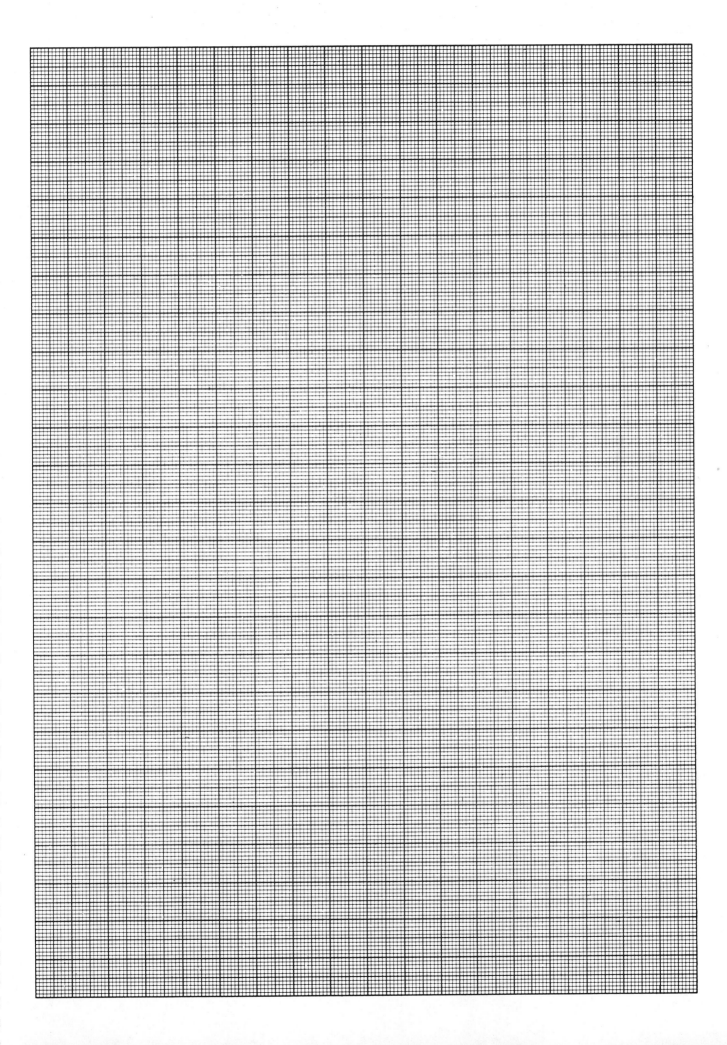

# The Mechanical Equivalent of Heat $21$

Before the work of Benjamin Thompson (Count Rumford, 1753–1814) and James Prescott Joule (1818–1889, after whom the SI unit of energy is named), mechanical energy and heat were considered to be two quite different things. It was Thompson who first noted, while boring cannon for the Elector of Bavaria, that while the large amount of work put into turning the drill seemed to disappear rather than become stored as kinetic or potential energy, the drill and cannon got hot with no apparent source of heat. He suggested that the work was the source of the heat, that mechanical work was transformed into heat by friction and that, therefore, heat was not something separate and apart but just another form of energy. Joule put this idea on a quantitative footing by transforming a known amount of work into a measurable quantity of heat. In this way, he not only showed that heat was a form of energy, so that calories and joules were really different size units of the same thing, but also measured the conversion factor between these units. This conversion factor is called the *mechanical equivalent of heat*. The purpose of the present experiment is to determine the mechanical equivalent of heat by doing a known amount of work against friction and measuring the amount of heat thereby produced. The apparatus used is a highly improved version of Joule's original setup.

## THEORY

When two surfaces slide one over another, a frictional force arises parallel to the surfaces and must be opposed by an equal parallel applied force if motion at constant speed is to be maintained. The applied force will then do a work equal to its magnitude multiplied by the distance through which it moves in pushing one surface over the other; but this work is not stored as potential energy, nor does it appear as an increase in the kinetic energy of the moving body. The sliding surfaces do get hot, however, and Joule's and Thompson's idea was that the friction transforms the mechanical work done into heat, which thus appears as a different form of energy. Evidence in favor of this concept is obtained by showing that the amount of heat that appears is always proportional to the amount of work done against friction. Joule's experiment, like the present one, consisted of a setup by which a predetermined amount of work could be done against friction (in his case, against the viscous friction encountered by a paddle turning in a fluid) and the resulting heat measured.

Figure 21.1 shows the experimental arrangement to be used. A copper band is wrapped around a cylindrical copper calorimeter drum and carries a weight on its lower end. The other end is not attached to the drum but is allowed to go slack. Thus, when the drum is turned (by a hand crank in such a direction as to hoist the weight as if the drum were a windlass), the band slips on the drum surface, and the weight, while supported clear of the floor, is not raised. The value of this weight is thus equal to the frictional force between the band and the drum surface on which it slides, and the distance through which the force acts (the distance the sliding surfaces move with respect to each other) is simply the distance moved by any point on the drum surface. This is just the drum's circumference multiplied by the number of revolutions through which it is turned. The work done is thus

$$W = \pi DNMg \qquad (21.1)$$

where $W$ is the work in joules, $D$ is the diameter of the drum in meters, $N$ is the number of revolutions through which the crank is turned, $M$ is the mass in kilograms of the weight suspended on the copper band, and $g$ is the acceleration of gravity in meters per second squared. Notice that the units of the expression on the right in Equation 21.1 are newton-meters or joules. (By definition a joule is a newton-meter.)

The copper drum serves as a calorimeter. It is filled with water and insulated from the base of the apparatus by the plastic holder and shaft on which it is mounted. The copper band is in thermal contact with the drum and must be included with it. In operation, the drum is filled with water, and a thermometer is inserted through a leakproof seal so that the water temperature may be measured. The relation between the heat delivered to the calorimeter assembly and the observed temperature rise is

$$Q = [m_w c_w + (m_1 + m_2)c + C_1](t - t_0) \qquad (21.2)$$

**Figure 21.1** *Apparatus for measuring the mechanical equivalent of heat*

where $Q$ is the heat delivered, $m_w$ is the mass of the water, $m_1$ the mass of the calorimeter drum, and $m_2$ the mass of the copper band. The specific heats of water and copper are $c_w$ and $c$, respectively, $C_t$ is the effective *heat capacity* of the thermometer, $t$ is the temperature measured after turning the drum, and $t_0$ is the initial temperature. In Equation 21.2, considerable attention must be paid to the units. As the mechanical equivalent of heat is a relation between calories and joules, we will want $Q$ to be in calories, and as masses in this experiment will most conveniently be measured in grams and temperatures in degrees Celsius, the specific heats must be in calories per gram-degree Celsius and the heat capacity $C_t$ in calories per degree Celsius. In these units, the specific heat of water is 1, but the symbol $c_w$ has been included in Equation 21.2 to make it dimensionally consistent.

If, in fact, the work done against friction is converted into heat, then $W$ should be proportional to $Q$ and we may write

$$\pi DNMg = J[m_w c_w + (m_1 + m_2)c + C_t](t - t_0)$$
(21.3)

where $J$ is the proportionality constant and has the dimensions of joules per calorie. $J$ is called the mechanical equivalent of heat and can be computed from Equation 21.3 if all other quantities are known. In the present experiment, the proportionality of $W$ and $Q$ will be checked and the value of $J$ measured.

## APPARATUS

1. Mechanical-equivalent-of-heat apparatus, complete with 5-kg weight and special thermometer

2. Vernier caliper
3. Triple-beam balance

## PROCEDURE

**1.** The apparatus will be set up when you come to the laboratory. Inspect it carefully, noting in particular how the copper calorimeter drum is attached to the plastic mounting disk, so that you will know how to remove it and reattach it. A special thermometer is provided that does not have a very large range but whose expanded scale is arranged to allow small temperature changes near room temperature to be easily read. This thermometer is extremely fragile. Be particularly careful when it is inserted in the calorimeter drum, as a careless knock will break it

off. Also note that it is larger in diameter than most thermometers and rolls easily. Take care not to let it roll off the bench.

**2.** Weigh the empty copper calorimeter drum with the thermometer sealing ferule but without the thermometer. Record the mass $m_1$ thus measured.

**3.** Weigh the copper band; record its mass as $m_2$.

**4.** Measure the diameter of the calorimeter drum with the vernier caliper.

**5.** Fill the drum nearly full of water (about 50 or 60 g) and again weigh it. Then push the thermometer gently through the seal, insert it into the drum, and tighten the sealing ferule. Do not overtighten. Normal tightening with the thumb and forefinger should be sufficient to prevent leaks. Record the thermometer's effective heat capacity $C_t$. This is 0.80 cal/°C for the thermometer supplied by Leybold-Heraeus but may be different for other brands. Your instructor will give you $C_t$ for the particular thermometer you are using.

**6.** Attach the drum to its mounting disk, being careful not to hit the protruding thermometer by accident. Wrap the copper band around the drum until almost its full length is wound up (four or five turns). Attach the 5-kg weight to one end so that it hangs freely and attach the other end to a spring hooked to a peg in the apparatus base. The peg may be placed in any of a number of holes in the base; its position should be chosen so that approximately the same length of copper band extends away from the drum at either end. The cord by which the weight is suspended from the band has a length adjustment in it; this should be set so that the weight clears the floor when the crank is being turned in the direction tending to raise the weight. The band should slide smoothly on the drum surface when the crank is turned at a steady rate, and the end attached to the spring should be slack so as not to detract from the force applied to the band by the suspended weight. Check that there are no leaks around the thermometer.

**7.** The experiment should be started with the temperature of the calorimeter drum and its contents two or three degrees below room temperature. The calorimeter temperature will then be a few degrees above room temperature at the end of the run. In this way, the heat exchange between the calorimeter and its surroundings will just about balance out to zero over the running time. If the initial reading of the thermometer is more than three or four degrees below room temperature due to the calorimeter drum having been filled with cold water, turn the crank until the temperature reaches a better initial value. Notice that a lock is provided on the base that will hold the crank in a fixed starting position and that the thermometer may be turned in its seal so that the scale faces upward for easy reading when the crank is in the starting position just noted. If the thermometer is not adjusted in this way, the apparatus also has a mirror that may be swung out under the thermometer so that its scale may be read when it is facing down. Record your initial temperature value $t_0$ as the temperature corresponding to zero turns of the crank.

**8.** Start your experimental run immediately after recording $t_0$ and work rapidly so as to minimize the opportunity for heat exchange between the calorimeter drum and its surroundings. Turn the crank through 50 turns, then observe and record the temperature. Repeat for another 50 turns, and continue, recording the temperature after each 50-turn interval, until a total of 250 turns have been completed.

**9.** At the end of the experiment, detach the weight and unwind the copper band from the calorimeter drum. Remove the drum from the plastic mounting disk, loosen the sealing ferule, and carefully take out the thermometer. Replace the thermometer in its case and put it away safely. *Do not leave it lying on the laboratory bench.* Empty the calorimeter drum and leave it with the rest of the apparatus.

## DATA _____

Mass of empty calorimeter drum
    and sealing ferule, $m_1$     _____

Mass of copper band, $m_2$     _____

Mass of calorimeter drum, sealing
    ferule, and water     _____

Mass of water, $m_w$     _____

Diameter of calorimeter drum, $D$     _____

Heat capacity of thermometer, $C_t$     _____

| Number of turns | 0 | 50 | 100 | 150 | 200 | 250 |
|---|---|---|---|---|---|---|
| Temperature | | | | | | |
| Temperature rise $t - t_0$ | ✕ | | | | | |

Mechanical equivalent of heat $J$,
    from the graph     _____

Percent error     _____

Value of the mechanical equivalent
of heat from Table I at end of
book     _____

## CALCULATIONS

**1.** Subtract the mass of the empty calorimeter from its mass when full to get the mass of the water.

**2.** Subtract the initial temperature $t_0$ (corresponding to zero turns of the crank) from the temperature observed after 50 turns to get the temperature rise $t - t_0$ due to 50 turns. Repeat using the observed temperatures at 100, 150, 200, and 250 turns.

**3.** Plot a curve using the values of the temperature rise $t - t_0$ as abscissas and the numbers of turns as ordinates. Choose your scales so as to use an entire sheet of graph paper. Note that the origin should be a point on your curve.

**4.** Check that your curve is a straight line and find its slope. According to Equation 21.3, this slope should be $J[m_w c_w + (m_1 + m_2)c + C_t]/\pi DMg$; hence $J$ can be calculated if the other quantities are known. All have been recorded except the specific heat of copper, which is given in Table VII at the end of the book. Enter your calculated value of J on your data sheet.

**5.** Look up the accepted value of *J* and compute your percent error.

## QUESTIONS _____

**1.** Explain how your work in this experiment, and in particular your graph, indicates that heat is a form of energy.

**2.** A clever feature of the apparatus used in this experiment is the fact that the frictional force is developed between the copper band and a *rotating* surface. Why is this a great advantage?

**3.** What problem would be encountered if a standard 0–50°C thermometer were used in place of the special one provided?

**4.** A Prony brake is a device for measuring the power output of an engine. It consists of a brake applied to a drum attached to the engine shaft so that the engine does work against the frictional force provided by the brake. The system is arranged so that the heat produced can be measured. In some Prony brakes, this is done by cooling the brake with water whose rate of flow and temperature rise is measured. If a 50°C rise is noted for the water cooling a Prony brake while the engine under test is developing 200 hp, how many gallons per minute must be flowing through the brake?

**5.** A 2000-lb car is traveling at 60 mi/h when the engine is disengaged and the brakes are applied, bringing the car to a stop. How many calories of heat are produced in the stopping process?

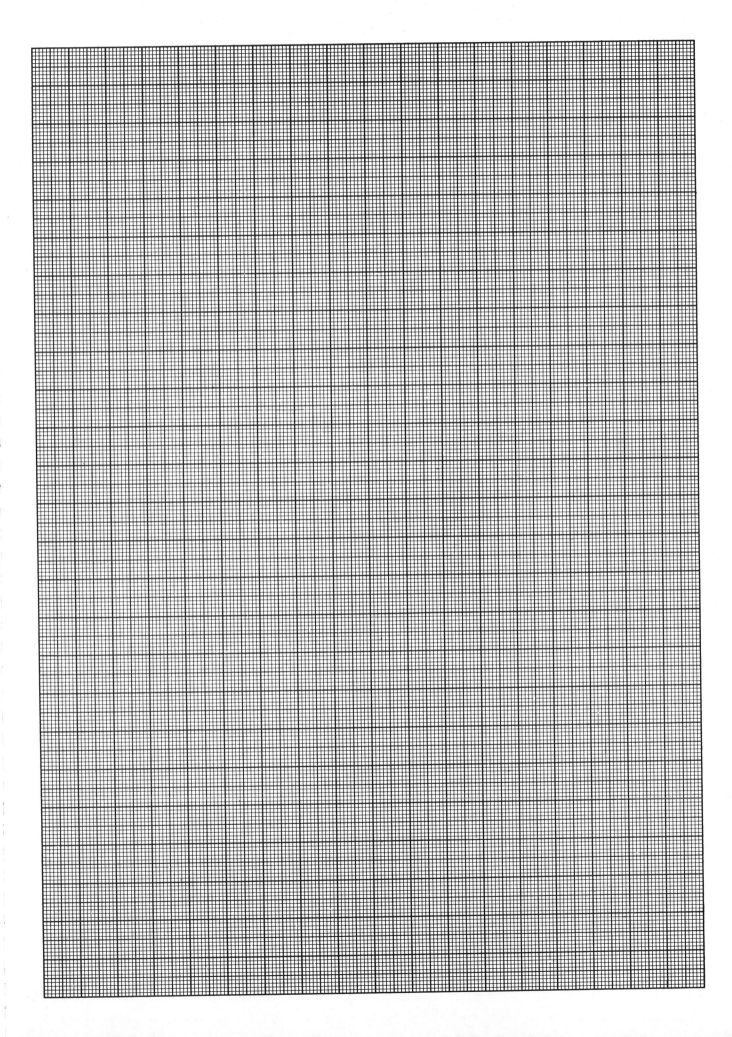

# The Ratio of the Specific Heats of Gases

# 22

Gases, like all substances, exhibit a specific heat, that is, a requirement that a certain amount of heat be added to a unit quantity of the gas in question to raise its temperature one degree. Although it would be possible to quote specific heats in calories per gram-degree Celsius as is usually done with solids and liquids, we shall be interested in a more fundamental unit of quantity, the mole. This is defined as Avogadro's number of molecules, so that we shall be concerned with the amount of heat required to raise the temperature of a collection of this number of molecules of a particular gas by 1°C. The specific heats we will be considering may thus properly be referred to as *molar* specific heats. We will find that a gas exhibits a larger specific heat when it is allowed to expand on heating than it does when it is confined to a fixed volume. In the latter case, the gas undergoes a *constant volume process* and the specific heat is called the *specific heat at constant volume*, designated $c_v$. If, on the other hand, the gas is allowed to expand as it is heated in such a way that the pressure remains constant, a larger specific heat called the *specific heat at constant pressure*, $c_p$, is measured. The ratio $c_p/c_v$ is generally represented by the Greek letter $\gamma$ and is an important quantity in kinetic theory and thermodynamics. The purpose of this experiment is to measure $\gamma$ for a monatomic and a diatomic gas by a method developed by F. E. Rüchhardt and to test the predictions of kinetic theory in this regard.

## THEORY

When a gas is allowed to expand as it is heated, it does work in moving the walls of the containing chamber (for example, pushing out the piston in a containing cylinder), so that some of the energy put in as heat comes out again as this work. Only the remainder stays in the gas to raise its internal energy and thus its temperature. The first law of thermodynamics, which is just a statement of the conservation of energy for this situation, may be written

$$\Delta Q = \Delta U + \Delta W \qquad (22.1)$$

where $\Delta Q$ is the amount of heat added to the gas, $\Delta U$ is the change in the gas's internal energy, and $\Delta W$ is the work done.

Now suppose the increase in internal energy is reflected by a rise $\Delta T$ in temperature. Dividing Equation 22.1 through by $\Delta T$ yields

$$\frac{\Delta Q}{\Delta T} = \frac{\Delta U}{\Delta T} + \frac{\Delta W}{\Delta T}$$

Now $\Delta Q/\Delta T$ is by definition the heat capacity of the particular quantity of gas that you're heating, and if you happen to choose 1 mole for this quantity, then $\Delta Q/\Delta T$ is the molar specific heat. Furthermore, if the heating is done at constant volume (the pressure, of course, rises in such a process), no work is done because the gas doesn't move anything, so that $\Delta W = 0$. Then $\Delta Q/\Delta T = \Delta U/\Delta T$ and we conclude that

$$c_v = \frac{\Delta U}{\Delta T} \qquad (22.2)$$

However, if the gas is allowed to expand at constant pressure, $\Delta W$ is not zero and its value must be calculated. This is easy to do by writing down the ideal gas law (Equation 20.9) with $n = 1$ because we have 1 mole of gas. It is

$$PV = RT$$

where $P$ is the absolute pressure on the gas, $V$ its volume, $T$ its absolute temperature, and $R$ the universal gas constant. Clearly, if the pressure is held constant, a change $\Delta T$ in the temperature will cause a change $\Delta V$ in the volume that is related to $\Delta T$ by the expression

$$P \Delta V = R \Delta T$$

But $P \Delta V$ is the work done in the expansion, for suppose that the expansion is carried out by letting the gas push on a piston whose face has area $A$. Then the force on the piston is $PA$. If the piston moves through a distance $\Delta x$, the work done is by definition $PA \Delta x$. But $\Delta V$ is the volume swept out by the piston as it moves and is thus equal to $A \Delta x$. Hence, we have $\Delta W = PA \Delta x = P \Delta V$ and

$$\frac{\Delta W}{\Delta T} = \frac{P \Delta V}{\Delta T} = R$$

We conclude that the specific heat at constant pressure is greater than the specific heat at constant volume by the universal gas constant, or

$$c_p = c_v + R \qquad (22.3)$$

Note that the dimensions of $R$ are energy per mole-degree Celsius, as they should be.

Classical kinetic theory has shown that a mole of any gas stores an amount of energy equal to $\frac{1}{2}RT$ for each *degree of freedom* of its molecules. A degree of freedom is defined as a way in which a molecule can store either kinetic or potential energy in a way which is independent of other ways. A molecule that acts like a mass point has three degrees of freedom because it can move in each of three mutually perpendicular directions and can thus store kinetic energy in each of these motions. The molecules (actually atoms) of a monatomic gas can be considered mass points, so that a mole of such molecules should have an internal energy $U = \frac{3}{2}RT$. Hence, for monatomic gases such as helium, neon, or argon

$$c_v = \frac{\Delta U}{\Delta T} = \frac{3}{2}R$$

$$c_p = c_v + R = \frac{5}{2}R$$

and

$$\gamma = \frac{c_p}{c_v} = \frac{5}{3} = 1.67$$

A diatomic gas, on the other hand, consists of molecules that may be represented by little dumbbells, the two atoms being spaced apart by the interatomic bond just as the big ends of a dumbbell are by the intervening shaft. Such a structure will have a moment of inertia when rotated about any axis perpendicular to the dumbbell's centerline and can thus store rotational kinetic energy in this motion. Since two mutually perpendicular axes can be constructed perpendicular to the centerline, there are two independent rotational modes (one for rotation about each of these so-called principal axes) and hence two rotational degrees of freedom. There should be a third one corresponding to rotation about the centerline itself (the third principal axis), but if the atoms are really mass points, there is no radius for any of the rotating mass in this mode and thus no moment of inertia and no possibility of storing energy. This means that we have only two rotational degrees of freedom in addition to the three translational ones that any movable object has. It follows that a diatomic molecule has a total of five degrees of freedom, and so a mole of a diatomic gas such as hydrogen, nitrogen, or oxygen should have an internal energy $U = \frac{5}{2}RT$. Consequently, for a diatomic gas

$$c_v = \frac{\Delta U}{\Delta T} = \frac{5}{2}R$$

$$c_p = c_v + R = \frac{7}{2}R$$

and

$$\gamma = \frac{c_p}{c_v} = \frac{7}{5} = 1.40$$

Note that we have the following important results: (a) all monatomic gases have the same *molar* specific heats and thus the same value of $\gamma$, namely 1.67, regardless of their atomic masses or chemical properties, and (b) the same holds for all diatomic gases, except that for them $\gamma = 1.40$. Therefore, the measurement of $\gamma$ for a variety of gases is a very significant test of kinetic theory.

Although it is possible to measure $c_p$ and $c_v$ individually by straightforward methods for any particular gas and then to calculate their ratio, a clever procedure for obtaining $\gamma$ directly was invented by Rüchhardt in 1929. His apparatus is illustrated in Fig. 22.1. It consists of a large flask fitted with a rubber stopper in which is inserted a length of glass tubing having a very smooth bore of precise diameter. A steel ball bearing of mass $m$ is dropped into the tube, in which it fits closely enough to make a gas-tight seal while nevertheless sliding with negligible friction. Fig. 22.1 shows the ball in its equilibrium position. The downward forces are its weight $mg$ and the force due to atmospheric pressure ($P_0$) on its upper surface. This force is $P_0 A$ where $A$ is the ball's cross-sectional area. The balancing upward force is $PA$ where $P$ is the absolute pressure in the flask, so that at equilibrium

$$PA = P_0 A + mg \qquad (22.4)$$

**Figure 22.1**   *Rüchhardt's apparatus for measuring $\gamma$.*

Now let the ball be displaced by a distance $x$ from its equilibrium position as shown in Fig. 22.1. If it is pushed down, the volume $V$ of the gas trapped in the flask and the tubing below the ball will be decreased by a small amount $dV$, the pressure in the flask will be increased by a small amount $dP$, and the ball will be forced back up by the increased pressure. Because of the momentum it gains, it will overshoot the equilibrium position, the pressure will drop below the equilibrium value, and the ball will come to a stop and fall back down again. In short, the ball will bounce up and down, oscillating about its equilibrium position, the pressure and volume of the gas in the flask accordingly varying above and below their equilibrium values.

Because the motion of the ball is rapid, the changes $dP$ and $dV$ in the gas pressure and volume take place in times too short to allow a significant heat transfer into or out of the gas through the walls of the containing vessel. The gas is thus essentially undergoing an adiabatic process, for which the relation between pressure and volume is

$$PV^\gamma = \text{constant}$$

Differentiating to obtain the relation between a small change $dV$ in volume and the resulting small change $dP$ in pressure yields

$$V^\gamma \, dP + \gamma P V^{\gamma-1} \, dV = 0 \qquad (22.5)$$

Now note that the net force $F$ on the ball is $A\,dP$ and that the change in volume $dV$ is $Ax$. Substitution in Equation 22.5 gives

$$V^\gamma \frac{F}{A} + \gamma P V^{\gamma-1} Ax = 0$$

which after a rearrangement of terms becomes

$$F = -\gamma \frac{A^2 P}{V} x \qquad (22.6)$$

We note at once that this is a linear restoring force, so that the ball must oscillate with simple harmonic motion. The proportionality constant in Equation 22.6 is $\gamma A^2 P / V$, and since we know that the period of simple harmonic motion is $2\pi$ times the square root of the ratio of the oscillating mass to this constant, we have

$$\tau = 2\pi \sqrt{\frac{mV}{\gamma A^2 P}}$$

where $\tau$ is the period, $m$ is the mass of the ball, $V$ is the volume of the flask (including the volume in the tube between the ball and the stopper), $A$ is the cross-sectional area of the ball ($\pi/4$ times its diameter squared) and $P$ is the equilibrium pressure in the flask (given by Equation 22.4). The solution for $\gamma$ is

$$\gamma = \frac{4\pi^2 mV}{\tau^2 A^2 P} \qquad (22.7)$$

Thus, by watching the bouncing ball and timing its oscillations to measure $\tau$, $\gamma$ can be obtained from this formula. The method is capable of good results, but note that the requirements that the ball make a gas-tight seal in the glass tube and at the same time slide without friction are highly contradictory, and also note that errors in the period measurement are doubled in $\gamma$ as a result of $\tau$ appearing squared in Equation 22.7. With good technique, however, the difference between the values of $\gamma$ for monatomic and diatomic gases is easily observed.

## APPARATUS

1. Aspirator flask, about 10-L capacity
2. Two single-hole rubber stoppers to fit flask
3. Stopcock tube for flask
4. Rubber pad
5. Precision glass tube
6. Steel ball to fit tube
7. Triple-beam balance
8. Micrometer caliper
9. Wooden ramrod for cleaning tube
10. Lintless cloth patches or tissues
11. Stopwatch or stop clock
12. Barometer
13. Tank of argon with regulator and connecting hose

## PROCEDURE

**1.** Wash your hands and dry them thoroughly. Cleanliness is extremely important in this experiment because the clearance between the steel ball and the bore of the precision glass tube is only about $\frac{1}{10,000}$ in. and no lubricant can be used. The slightest fingerprint, grease spot, or speck of dust can cause the ball to stick in the tube, requiring a thorough cleaning of both tube and ball before a successful run can be achieved. Even when your hands are clean and dry, do *not* touch the ball; handle it using the lintless cloth patches or tissues.

**2.** Record the volume of the aspirator flask. This will be marked on the flask or given to you by the instructor. Be careful of your units.

**3.** Weigh the ball on the triple-beam balance and record its mass in kilograms.

**4.** Measure the diameter of the ball using the micrometer caliper. To do this, first close the caliper jaws to obtain the zero reading. Then make three independent measurements of the diameter, rotating the ball between measurements. Handle the ball only with the cloths or tissues provided and make sure the micrometer's jaws are clean. Record all your readings.

**5.** Inspect the Rüchhardt apparatus, which should already be set up when you come to the laboratory. Make sure that both rubber stoppers are seated firmly in the flask. Note that the flask has a side port (not shown in Fig. 22.1) through which the gas pressure may be released, the flask may be filled with a gas other than air, or the ball may be removed if it falls into the flask at the end of a run. The stopper in this port is fitted with a short tube and stopcock, which you should make sure is closed. Note also that there should be a rubber pad in the flask on the bottom to prevent breakage when the ball drops on it.

**6.** Read the barometer and record the reading.

**7.** Practice your technique and test the apparatus by carefully dropping the ball into the precision glass tube. Remember that the slightest bit of dust or dirt in the tube or on the ball will prevent it from sliding properly. Pick up the ball only with the lintless cloth or tissue; do not touch it with your fingers. If the apparatus is properly clean, you should be able to observe at least seven oscillations of the ball in the tube before the motion damps out. Thereafter, the ball will either stick in the tube or fall into the flask. If it falls into the flask, remove the lower stopper and carefully extract the ball. If it sticks in the tube, remove the tube and upper stopper as a unit and recover the ball either by tipping the tube back and forth until the ball dislodges itself and rolls out (be prepared to catch it in a tissue) or by *gently* pushing it out with the wooden ramrod. CAUTION: Be extremely careful when handling the precision tube. Do not lay it on the bench (it may roll off) or make any rapid motions around it or with it. It is very easily broken, and the sharp ends of a broken glass tube are very dangerous. Also remember that because of the close tolerances to which the bore has been held, this is an expensive piece of apparatus.

**8.** If you cannot get seven oscillations of the ball in the tube, these items will have to be cleaned. The ball may be wiped off with the lintless cloth or tissues. To clean the bore of the precision tube, carefully remove it along with its stopper from the flask and run a cloth patch or tissue through it two or three times with the wooden ramrod. Be extremely careful not to break the tube while doing this. *Note:* No lubricant has ever been found that helps with the contradictory requirements of a gas-tight seal and no friction between the ball and the tube wall. Best results are obtained when the ball and the tube are absolutely dry. This means that the cleaning patches must be dry, and no solvent can be used. If you cannot get the ball and tube clean enough with the dry patches to get at least seven oscillations when you try the experiment, call your instructor.

**9.** When you are sure that the apparatus is working correctly, start your measurement of the period $\tau$ by dropping the ball into the precision tube and timing its oscillations. Do this by starting the stopwatch (or stop clock) at the instant the ball gets to its lowest point and counting a cycle each time the ball returns to this position. Stop the stopwatch at the end of four or five cycles and record the elapsed time and the number of cycles timed. This procedure may be conducted efficiently by two people, one of whom drops the ball in the tube while the other counts the cycles and operates the stopwatch. When the ball has come to rest, remove it as described in Procedure 7, being careful not to get any dust on it or in the tube bore.

**10.** Repeat Procedure 9 twice to get three independent measurements of the period $\tau$.

**11.** Fill the flask with argon by attaching the hose from the argon tank to the glass tube in the lower port and opening the stopcock. The instructor will show you how to operate the regulator on the tank to start the flow of argon. Flush argon through the flask for about 3 minutes or as determined by the instructor. Argon is heavier than air, and so filling the flask from the bottom displaces the air up and out through the precision tube. When the flask has been filled, shut off the regulator first and then the stopcock. Disconnect the line only after you have made sure that both the regulator and the stopcock have been turned off.

**12.** Repeat Procedures 9 and 10 with argon in the flask. Record all your data.

**13.** Read the barometer again at the end of the experiment and record the reading.

# DATA

Mass of ball _____    Volume of flask _____

Diameter of ball:

    Micrometer readings, trials   1. _____   2. _____   3. _____

    Average reading _____    Zero reading _____

    Diameter _____    Cross-sectional area _____

Barometer readings:

Beginning _____   End   _____

| Gas | Run | Time, s | Number of oscillations | Period, s |
|-----|-----|---------|------------------------|-----------|
| Air | 1 | | | |
| | 2 | | | |
| | 3 | | | |
| Argon | 1 | | | |
| | 2 | | | |
| | 3 | | | |

Air:

Average period _____   Value of $\gamma$ _____   Percent error _____

Argon:

Average period _____   Value of $\gamma$ _____   Percent error _____

## CALCULATIONS _____

**1.** Compute an average diameter for the ball by finding the average of your three measurements and correcting the result for your micrometer's zero reading.

**2.** Compute the ball's cross-sectional area from its diameter. Remember that the area of a circle of radius $r$ is $\pi r^2$ and that $r = d/2$ where $d$ is the diameter of the circle.

**3.** Convert your barometer readings to newtons per meter squared (pascals) by using the fact that 1333 pascals is equal to 1 cm of mercury. If there is any difference in the two readings, use the first reading in the calculation of $\gamma$ for air and the second for argon.

**4.** Calculate the equilibrium pressure $P$ in the flask from Equation 22.4. Use the value of $A$ from Calculation 2 and the barometric pressure taken at the beginning of the experiment for $P_0$. Be careful of your units. If the SI system is to be used, you should have recorded the mass of the ball in kilograms despite the fact that the balance reads in grams and should similarly have recorded the volume of the flask in cubic meters and the cross-sectional area $A$ in square meters. The acceleration of gravity in this system is 9.8 m/s$^2$.

**5.** Compute the measured period in each of your three runs for air by dividing the time by the number of oscillations counted in each case. Then compute the average of these three results and record it as the average period for air.

**6.** Calculate your measured value of $\gamma$ for air by using Equation 22.7. Here $A$ is the cross-sectional area obtained in Calculation 2, $P$ is the equilibrium pressure from Calculation 4, and $\tau$ is the average period found in Calculation 5.

**7.** Compare your result from Calculation 6 with the value of $\gamma$ for air predicted by kinetic theory by finding the percent error. Note that because air is composed almost entirely of nitrogen and oxygen, it may be considered a diatomic gas.

**8.** Repeat Calculations 4–7 for argon. Use the barometric pressure taken at the end of the experiment for $P_0$, and compare your measured value of $\gamma$ with that predicted by kinetic theory for a monatomic gas.

## QUESTIONS

**1.** List and discuss briefly the principal sources of error in Rüchhardt's method of measuring $\gamma$. In particular, calculate the a.d. for the period $\tau$ for both air and argon and compute the resulting error to be attached to your measured value of $\gamma$ in each case. Do the theoretical values of $\gamma$ lie within the experimental error of your measured values? If not, can you explain the discrepancies?

**2.** (a) Find the proper relation between $T$ and $V$ for an adiabatic process. (b) Find the proper relation between $T$ and $P$ for this type of process.

**3.** A modification of Rüchhardt's method suggested by Rinkel is to release the ball in the tube from rest at a point where the pressure in the flask is equal to the atmospheric pressure outside, so that the net force on the ball is just its weight $mg$. The distance $L$ through which the ball falls from the release point to the lowest point reached before it stops and starts back up is then measured instead of the period of the subsequent oscillation. (a) Derive a formula for $\gamma$ in terms of $L$ instead of $\tau$. (b) Find the initial amplitude of oscillation, which is the distance between the *equilibrium point* and the lowest position reached.

# Coulomb's Law and the Force Between Charged Plates 23

A fundamental experiment in physical science is the demonstration of Coulomb's law, which states that two electric charges exert a force, one upon the other, that is proportional to the magnitude of each charge and inversely proportional to the square of the distance separating them. Charles Augustin Coulomb (1736–1806) established this law experimentally with the so-called Coulomb balance (Fig. 23.1). Similar to the Cavendish balance for measuring the inverse-square gravitational force, the Coulomb balance consists of two balls on the ends of a stick suspended by a quartz fiber so that this assembly can rotate under the influence of forces due to charges placed on the balls interacting with appropriately placed fixed charges. Quantitatively, this experiment is difficult to carry out, largely because of the complications in measuring the charges involved. In fact, Coulomb worked for years with this apparently simple apparatus before obtaining reliable results. Therefore, we do not attempt to duplicate his work here but instead perform a related experiment that is easier to carry out. Nevertheless, this experiment checks a direct consequence of Coulomb's law and thus constitutes a demonstration of the law itself.

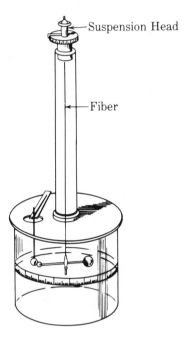

Suspension Head

Fiber

**Figure 23.1** *The original Coulomb torsion balance*

## THEORY

The mathematical expression for the force between two point charges $q_1$ and $q_2$ separated by a distance $r$ is

$$F = k\frac{q_1 q_2}{r^2} = \frac{q_1 q_2}{4\pi\epsilon_0 r^2} \qquad (23.1)$$

where $F$ is the force in newtons, $q_1$ and $q_2$ are in coulombs, $r$ is in meters, and $k$ is a proportionality constant having very nearly the value $9 \times 10^9$ $N \cdot m^2/C^2$. The quantity $\epsilon_0$ is called the *permittivity of free space*. We will shortly see why it is convenient to write the constant $k$ as $1/4\pi\epsilon_0$, but for the moment note only that $\epsilon_0 = 1/4\pi k = 1/(36\pi \times 10^9) = 8.85 \times 10^{-12}$ $C^2/N \cdot m^2$.

Equation 23.1 gives the force between two *point* charges, but in this experiment we deal with two flat, parallel metal plates that are charged, one positively and one negatively, by the simple expedient of connecting a power supply (voltage source) between them so as to "pump" charge from one plate to the other. As a result, the upper surface of the lower

plate and the lower surface of the upper plate become sheets of charge, one positive and one negative, each having a uniform areal charge density $\sigma$ $C/m^2$. We are interested in the force of attraction between these two oppositely charged sheets rather than the force between positive and negative point charges.

It is a direct consequence of Coulomb's law that the electric field near a sheet of charge has the value $\sigma/2\epsilon_0$. Note the simplification that has occurred in this expression as a result of using $\epsilon_0$ instead of $k$. Also, the electric field is by definition the force that a point charge of $+1$ C would feel if placed at the location in question. In this case, the expression $\sigma/2\epsilon_0$, which does not depend on position, shows that the field is constant and uniform. Remember, however, that it is derived on the assumption that the sheet of charge is infinite in extent so that its edges do not enter the picture. Actually, our plates are not infinitely large, so $\sigma/2\epsilon_0$ is an approximation valid only when our test charge is separated from a plate by a distance that is small compared to the plate dimensions. In

173

other words, in this experiment we keep the spacing between the plates small compared to their length and width.

We may now calculate the force on one plate due to the field of the other. Because this field is constant and is defined as the force per unit charge, the force in question is $q\sigma/2\epsilon_0$, where $q$ is the charge on the plate on which the force is being measured. But because this charge has been obtained by "pumping" it from the other plate, the plates have equal and opposite charge, so that $\sigma = q/A$, where $A$ is the plate area. Hence, the force with which one plate attracts the other is

$$F = \frac{q^2}{2A\epsilon_0} \tag{23.2}$$

The question now arises as to how to determine $q$. The advantage of the present experiment is that $q$ is simply related to the voltage or potential difference $V$ imposed between the plates by the power supply. Imagine a point test charge of $+1$ C between the plates. The total field here (due to both plates) is $\sigma/\epsilon_0 = q/A\epsilon_0$, and therefore the work required to carry this test charge across the gap between the plates is $qd/A\epsilon_0$, where $d$ is the plate spacing. But this is just the work per unit charge required to go from one plate to the other, which is by definition the potential difference between the plates. Thus,

$$V = \frac{qd}{A\epsilon_0}$$

or

$$q = \frac{VA\epsilon_0}{d} \tag{23.3}$$

Substituting Equation 23.3 in 23.2 yields

$$F = \frac{A\epsilon_0 V^2}{2d^2} \tag{23.4}$$

This is the result we propose to check. We will do so by measuring corresponding values of $F$ and $V$ and plotting $F$ against $V^2$. The test consists of noting whether this plot is a straight line. Moreover, the slope of this line should be $A\epsilon_0/2d^2$, and, because $A$ and $d$ are easily measured, a value of $\epsilon_0$ can then be found. It should agree with the value already noted in connection with Coulomb's law.

In the present experiment, a sensitive balance is used to measure the force of attraction $F$ between two parallel plates of area $A$. High sensitivity is needed because, although $V$ may be made fairly large, $\epsilon_0$ is quite small; thus the forces to be measured are small. The apparatus, which is shown in Fig. 23.2, consists of a heavy base on which one plate is mounted by means of insulating posts. The second or movable plate forms part of a frame that constitutes one arm of a beam balance. The other arm consists of a threaded rod on which counterweights may be screwed back and forth to balance the beam assembly. This assembly rotates on knife edges, and a small mirror is attached to the central section to allow very small rotations to be observed. This is done by means of a so-called optical lever, as shown in Fig. 23.2. A laser shines a light ray on the mirror, which reflects it to a wall several feet away. This reflected ray is the "lever" because the displacement of the light spot on the wall is twice the angle through which the mirror rotates multiplied by the ray's length (mirror-to-wall distance). As we will see, an actual measurement of the mirror's rotation will be unnecessary, but the mirror-to-wall distance, although it need not be known, should be sufficient to insure detection of even small rotations of the beam balance. In addition, the beam assembly is provided with a damping system consisting of a conducting blade moving in the field of a permanent magnet and a beam lift that raises the assembly off the knife edges to prevent damage during transportation and to

**Figure 23.2** *Electrostatic balance arranged to measure the force between charged plates*

position it correctly on the knife-edge supports. Finally, the electrostatic force between the plates can be measured by placing weights in the center of the movable plate. This process is described in the Procedure section.

## APPARATUS _____

1. Electrostatic balance with calibrated weights
2. Laser with support stand and necessary clamps
3. Level
4. Micrometer caliper
5. $\frac{1}{8}$-in. and $\frac{1}{4}$-in. spacer blocks
6. Vernier caliper
7. Variable power supply capable of delivering 0–1000 V dc at low current (less than 1 mA)
8. Voltmeter, 0–1000 V dc
9. Resistor, 10,000 Ω, 10 W
10. Roll of white calculator paper
11. Masking tape

## PROCEDURE _____

**1.** Set up the electrostatic balance on the laboratory bench and carefully lift off the balance beam. Using the level provided, level the base by means of the two thumbscrews at its front.

**2.** Note the two tapered screws on the lifting mechanism and the matching tapered holes in the bottom of the beam assembly near the knife edges. Rotate the lifting shaft so that the tapered screw points project upward, and carefully place the beam assembly on these screws so that the points enter the tapered holes. Then gently lower the assembly onto the knife-edge bearing surfaces. Bring the movable plate into contact with the fixed plate and insure that these two plates touch over the full extent of their facing surfaces, that is, make sure they are exactly parallel. If they are not, squeeze them together *gently* with the thumb and forefinger of one hand while loosening the set screws retaining the fixed plate to its mounting posts with the other. Note that there are four such set screws, one pair allowing rotation of the plate about a longitudinal axis and the other permitting it to be tipped transversely. Be sure to tighten the set screws after making this adjustment. Lift the beam assembly off the knife edges with the beam lift, noting that an important function of the lifting mechanism is to position the beam assembly and hence the movable plate correctly. Lower the beam onto the knife edges and check the position of this plate. It should cover the lower or fixed plate exactly, the edges matching all the way around. If it does not, loosen the set screws that retain the movable plate support wires to the balance beam and adjust the movable plate position until it is correct. Consult your instructor if the plates are bent or if you have some other type of difficulty in making these adjustments. Be sure all set screws are tight after these adjustments have been completed.

**3.** Use the lifting mechanism to position the balance assembly correctly and adjust the horizontal counterweights until the assembly is balanced and pivots freely on its knife edges. *Note:* The counterweights that hang vertically downward have already been set and should not be touched. Check that the damping blade rides in the middle of the hole cut for it in the base and does not rub on any of the damping magnet poles. If the blade comes too close to one side of the hole, lift the beam assembly off its supports and *gently* press the blade right or left as required. Then replace the assembly, positioning it properly with the lifting mechanism. If either of the damping magnets is too close to the blade, loosen that magnet's retaining screw and adjust its position as necessary.

**4.** Take the beam assembly out of the apparatus and measure the length and width of either plate with the vernier caliper. Record these measurements in meters. The plates used in this apparatus are usually square, so that you may get the same result for both length and width, but measure and record each independently. Then reposition the beam assembly on its supports.

**5.** Wire the apparatus as shown in Fig. 23.3. Notice that there is a connection terminal at each support post holding the fixed plate. The two terminals are needed when the apparatus is used as a current balance (Experiment 25), but in the present case either one may be used for connecting to the fixed plate. A similar consideration holds for the terminals at each knife-edge support, through which connection is made to the movable plate. The 10,000-Ω resistor serves to protect the power supply from being directly short-circuited should the plates come together while high voltage is applied. If this happens during the course of the experiment, however, you should be on the alert to reduce the power supply voltage to zero immediately. Fig. 23.3 shows the voltmeter connected directly across the power supply, although it is supposed to measure the voltage between the plates. However, because the resistance between the plates (when they are not touching) is almost infinite, there should be no current through and therefore no voltage drop across the protecting resistor. We may therefore assume that the voltmeter reading is the same as the actual plate-to-plate voltage.

**Figure 23.3** *Circuit for the electrostatic balance*

**6.** Measure the thickness of the $\frac{1}{8}$-in. spacer block with the micrometer caliper. Note that the $\frac{1}{8}$ in. is a nominal value only. You need much greater precision than this; hence you should make three independent measurements with the micrometer and record the results *in meters* to four significant figures.

**7.** Place the $\frac{1}{8}$-in. spacer between the plates and put some weights on top of the movable plate (at its center) to bring this plate down solidly on the spacer. Lift the balance assembly momentarily with the beam lift to assure that it is correctly positioned. Then check that the plates are truly parallel. If you suspect that they are not, squeeze them together on each side of the spacer with the thumb and forefinger of one hand and loosen the four set screws that hold the fixed plate in position with the other. Then retighten these screws.

**8.** Mount the laser, turn it on, and position it so that its emitted ray strikes the mirror on the apparatus and is reflected back to a wall or screen at least 2m away. Tape a strip of white paper vertically on the wall so that the light spot falls at about its center. Check that the movable plate is weighted down on top of the spacer and mark the position of the light spot on the paper. CAUTION: The low-power laser used in this experiment cannot hurt you even if its light falls directly on your skin, *but do not allow anyone, including yourself, to look directly into the emitted light ray.*

**9.** Now remove the weights on the movable plate, remove the spacer from between the plates, and place weights totalling a mass of 70 mg on the top center of the movable plate. Adjust the horizontal counterweights until the beam assembly balances with the light spot at the marked position on the pa-

per strip. This is a critical adjustment that you should make with great care. Lift the balance assembly and return it to its supports to assure correct positioning several times and check that the spot returns to the marked position. Note that this procedure assures that the plates are spaced by exactly the thickness of the spacer block.

**10.** Remove the 70-mg mass from the top of the movable plate and observe that the balance assembly now balances with the light spot some distance above the position marked on the paper strip. Make sure the power supply's voltage control is turned all the way down and switch the supply on. Gradually increase its voltage while observing the motion of the spot. Note from Equation 23.4 that the force between the plates varies inversely as the square of their separation, so that if you allow the spot to overshoot the mark the plates are likely to be drawn together until they touch. If this happens, reduce the voltage to zero at once, reposition the balance assembly with the beam lift, and try again. With a little practice, you will be able to find a voltage for which the beam assembly balances with the spot right on the marked position. Do this three times, starting from zero voltage each time, and record the voltages obtained.

**11.** Repeat Procedures 9 and 10 using a mass of 50 mg instead of 70 and again with a mass of 20 mg. Record the voltages obtained in three independent measurements with each mass. Satisfy yourself that in each case, when the light spot has been returned to the marked position, the electrostatic force between the plates is exactly equal to the weight of the mass just removed.

**12.** Repeat Procedures 6–11 using the $\frac{1}{4}$-in. spacer, and masses of 100, 70, and 50 mg.

# DATA

Length of plate _____ Width _____ Plate area $A$ _____

Thickness of $\frac{1}{8}$-in. block:

Trial 1 _____

    Zero reading _____ Trial 2 _____

    Thickness $d$ _____ Trial 3 _____

Average _____

| Mass removed from plate | Force on plate | Voltage measurement | | | | |
|---|---|---|---|---|---|---|
| | | **Run 1** | **Run 2** | **Run 3** | **Average** | **V²** |
| | | | | | | |
| | | | | | | |
| | | | | | | |

Thickness of $\frac{1}{4}$-in. block:   Trial 1 _____

    Zero reading   _____   Trial 2 _____

    Thickness $d$   _____   Trial 3 _____

                       Average _____

| Mass removed from plate | Force on plate | Voltage measurement | | | | |
|---|---|---|---|---|---|---|
| | | **Run 1** | **Run 2** | **Run 3** | **Average** | **V²** |
| | | | | | | |
| | | | | | | |
| | | | | | | |

Slope with $\frac{1}{8}$-in. separation   _____   Value of $\epsilon_0$   _____

Slope with $\frac{1}{4}$-in. separation   _____   Value of $\epsilon_0$   _____

Average value of $\epsilon_0$   _____   Percent error   _____

## CALCULATIONS _____

**1.** Multiply the length of your movable plate by its width to obtain the area. Note that this experiment must be done strictly in MKS (SI) units, so that you should express your result in square meters.

**2.** Take the average of your three measurements of the thickness of the $\frac{1}{8}$-in. spacer block. Write your result in meters using as many significant figures as you can justify. Repeat for the $\frac{1}{4}$-in. spacer block.

**3.** Calculate the electrostatic force on the movable plate for each case considered in Procedures 10–12 by finding the weight of the mass removed in each case.

**4.** Take the average of the voltages found in the three runs made for each value of removed mass and plate separation. Then square these averages, entering your results in the proper boxes in the $V^2$ columns on your data sheet.

**5.** Using your data for the $\frac{1}{8}$-in. spacer, plot the forces found in Calculation 3 as ordinates against the values of $V^2$ found in Calculation 4 as abscissas. Choose reasonable scales so as to cover as much of your sheet of graph paper as possible. Draw the best straight line through your plotted points, remembering that the origin (zero force, zero voltage) should be a point on this line.

**6.** Find the slope of the line plotted in Calculation 5, and, by reference to Equation 23.4, calculate your measured value of $\epsilon_0$.

**7.** Repeat Calculations 5 and 6 using your data for $\frac{1}{4}$-in. spacing. This will give you a second straight-line graph from which you can again obtain a value of $\epsilon_0$.

**8.** Calculate the average of your two values of $\epsilon_0$ and compare this result with the accepted value by finding the percent error.

## QUESTIONS _____

**1.** Explain in your own words how your graphs are a test of the theory proposed in this experiment. How well do your plotted points actually support this theory?

**2.** (a) Starting with Equation 23.3, derive an expression for the capacity of the parallel-plate capacitor formed by your two plates. (b) Compute the value of this capacity in picofarads for each of your two plate spacings.

**3.** Check that Equation 23.4 is dimensionally correct. What must be the dimensions of $F$, $A$, and $d$ for this to be true?

**4.** Explain the statement in the Theory section that the displacement of the light spot on the wall is *twice* the angle through which the mirror rotates multiplied by the mirror-to-wall distance. Estimate the smallest angle of rotation of the balance that you could detect by observing the spot.

**5.** While you are bringing the plates to the right spacing as determined by the light spot on the wall, the movable plate may swing up and down a few times so that the plate spacing $d$ varies with time. Suppose that this plate moves so that the spacing changes from $d_1$ to $d_2$ in a short time interval $\Delta t$ while the power supply remains at a fixed voltage $V$. Derive an expression for the current that flows in the connecting wires during $\Delta t$.

**6.** What is the purpose of the damping blade and magnet arrangement? Even if you have not studied magnetic fields yet, find out and explain briefly how this arrangement works.

**7.** Assume that the plates used in your apparatus are made of aluminum sheet about 0.08 cm thick. Suppose one plate, instead of being mounted by supporting wires, is lying on a table and the other is held just above it. Find the plate-to-plate voltage required to just lift the lower plate off the table when the initial spacing between the plates is (a) the spacing you had with the $\frac{1}{8}$-in. spacer and (b) the spacing you had with the $\frac{1}{4}$-in. spacer.

**8.** A parallel-plate capacitor whose plates measure $20 \times 20$ cm and are spaced by 0.5 cm is connected to a power supply delivering 1000 V dc. (a) How much charge does the power supply transfer from one plate to the other? (b) How much work does the power supply do in charging this capacitor?

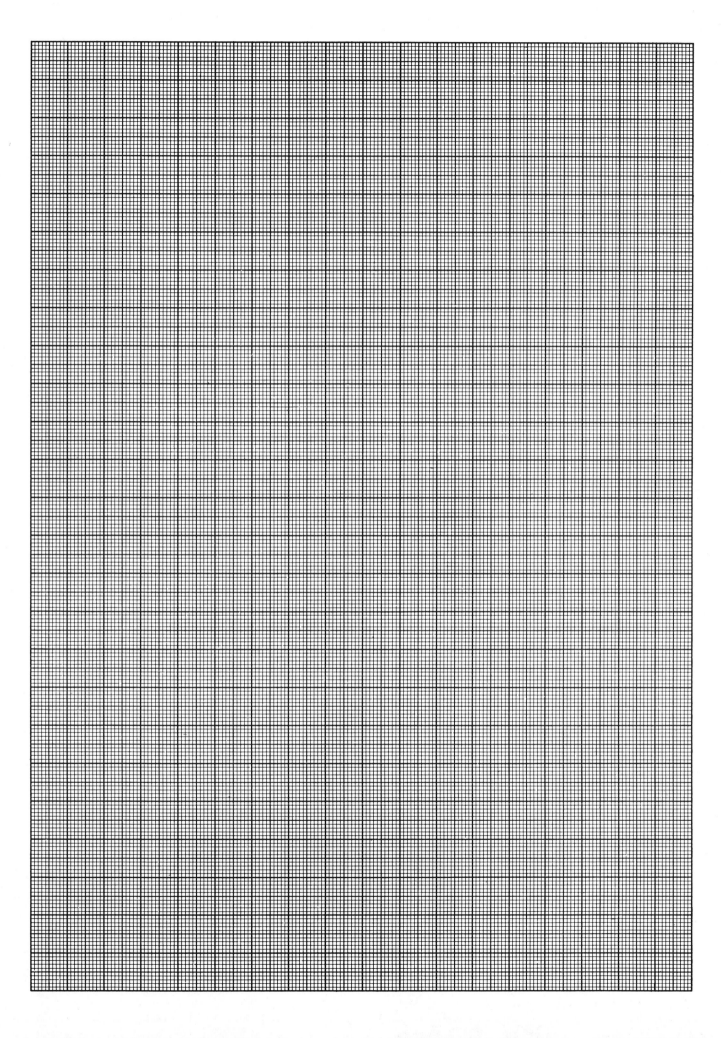

The electrostatic force between electric charges is a so-called action-at-a-distance force, that is, a force between bodies that are not in contact. As a result, the question arises of how one body knows that the other is there so that it should feel a force. For our present purposes, we must assume that the presence of a charged body creates a condition in the surrounding space such that another charged body located in that space feels a force due to it. Qualitatively, this condition is what we mean by an electric field—we introduce the more precise definition of what a physicist means by a "field" in the Theory section. It is the

purpose of this experiment to study the electric field and some aspects of the related problem of electric potential. The latter arises because we are dealing with conservative forces on charges, and motion of those charges will result in work being done and hence the storing of potential energy. We will be interested in drawing two kinds of lines, namely, lines representing the electric field and a second set joining points at the same potential. A result of this experiment will be the determination of these lines for some simple but important configurations of electric charge.

## THEORY

In physics, a field is a region of space in which every point is characterized by the value assumed at that point by some quantity that varies throughout the space. Thus, we could take the space inside the laboratory room, measure the temperature at a large number of points in the room, and draw a three-dimensional map showing our results. We would be able to state the temperature as a function of position in the room and would say that we thus had a *temperature field*. In the electrostatic case, if we have a given arrangement of charged bodies in a certain space and we bring in a test charge (that is, a charged body that we will place at various points in order to measure the force experienced there), we will be able to determine a *force field*, which is a value of the force felt by the test charge at every point in the space. Because the force is proportional to the value of the test charge, we standardize to a unit test charge, by which we mean a charge of $+1$ C. We then define the *electric field* at any given point to be the value of the force exerted on a $+1$-C test charge placed at that point. Notice that this will be a *vector* field because a force has a direction; thus both a direction and a magnitude must be given to specify the field at any point. In contrast, the *scalar* temperature field mentioned earlier needs only one quantity to describe it.

Now if the test charge is moved through a distance $\Delta \mathbf{r}$, an amount of work $\mathbf{E} \cdot \Delta \mathbf{r}$ will be done because the electric field vector $\mathbf{E}$ is the force on $+1$ C, the value of the test charge; hence $\mathbf{E} \cdot \Delta \mathbf{r}$ is simply the usual expression for the work done when a force

moves through a distance.* The potential energy difference between the two ends of $\Delta \mathbf{r}$ is then $-\mathbf{E} \cdot \Delta \mathbf{r}$, and as this is a potential energy *per unit charge*, it is called the *electric potential difference* or *voltage V* between those two points. Its units are clearly joules per coulomb or volts, named in honor of Alessandro Volta (1745–1827), who first produced batteries that would sustain a potential difference between their terminals. Note that as with any example of potential energy, it is the potential *difference* between two points that is of interest. Thus we cannot give the electric potential *at a point* without first stating what point is to be considered at zero potential; as we know, this choice is arbitrary. Because electric forces go to zero when we go infinitely far away from the charge configuration giving rise to them, infinity is often chosen as the zero potential point. In this case, every point of space may be labeled with a potential value giving the work required to bring $+1$ C to that point from infinity, that is, the potential difference or voltage between infinity and the point in question. Such a value is called the *absolute potential*. Obviously, the work required to take $+1$ C from one point to another is simply the difference between the abso-

---

*We assume here that $\Delta \mathbf{r}$ is a small enough distance so that $\mathbf{E}$ is constant over it. In the general case, those who know some calculus will recognize that a $\Delta \mathbf{r}$ that is "small enough" becomes the differential $d\mathbf{r}$ and that the work done in moving the test charge from a point $A$ to a point $B$ may be represented by the integral $\int_A^B \mathbf{E} \cdot d\mathbf{r}$. The electric potential difference (or potential energy per unit charge) is then $V = - \int_A^B \mathbf{E} \cdot d\mathbf{r}$.

lute potentials at the two points, that is, the voltage between them, and the work required to take *any* charge from one to the other is just the value of that charge times the voltage. We can write

$$W_{AB} = qV_{AB} \qquad (24.1)$$

where $W_{AB}$ is the work required to take charge $q$ from $A$ to $B$ and $V_{AB}$ is the voltage between those two points.

We can now construct a surface made up entirely of points that are at the same potential. In the language of geometry, such a surface would be called the *locus* of points characterized by a particular potential. It is called an *equipotential,* and we should note that no work is required to move a charge around on an equipotential surface because by definition there is no potential difference between any two points on it. This implies that there is no component of the electric field lying along such a surface, for if there were, a charge moving along the surface would feel a force parallel to its direction of motion so that the motion would result in work being done and hence a potential difference between points on the surface contrary to our definition. Therefore, *the electric field direction must be perpendicular to an equipotential surface at every point.*

Equipotentials are important in physics, and the simple example of the contour map may help us to understand them. A contour map is a plan view of some area of terrain. Thus, distances measured along the map correspond to horizontal distances over the ground. Elevations are shown by the contour lines, each of which is a line drawn through points all having the same height. A contour line is thus the locus of points having a particular height. Such a line may also be thought of as a gravitational equipotential because gravitational potential energy near the earth's surface is *mgh* and is thus proportional to $h$, the height above some point chosen as the zero of potential energy (which for our map is usually taken to be sea level).

An example of a contour map is shown in Fig. 24.1. Each contour line is marked with a number giving the height above sea level of every point on that line. Clearly, the map shows a hill slightly over 600 ft high, and we note that the right (east) side of the hill has quite a gentle slope whereas the left (west) side is very steep. We know this because close spacing of the contour lines means that the difference in height of the adjacent lines is achieved over a short horizontal distance, whereas wide spacing means a relatively long horizontal distance from one height to the next. We also note that it takes no work to walk around the hill on a contour line because every point of such a line is at the same height, so that as we walk along it we never ascend or descend. Con-

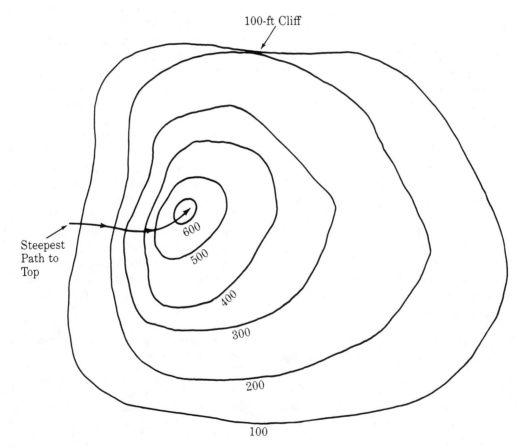

**Figure 24.1** *A contour map of a hill*

versely, the steepest path (one that might be followed by a skier in "schussing" the hill) is at right angles to the contour lines. This is the shortest distance between adjacent lines and therefore the path over which there is the greatest change in height and hence the greatest change in gravitational potential energy per unit distance traversed. This maximum change in potential energy per unit distance is called the *gradient*. It is a vector pointing straight downhill, that is, perpendicular to the contour lines, and its magnitude is a measure of the steepness of the hill and hence of the force urging you downward and against which you must work to climb upward. You need not, of course, climb or descend by the steepest path, that is, along the gradient vector. If you choose a zigzag route, such as a switchback road in the mountains, you will be traveling in a direction having one component along a contour line, with which no force and hence no work is associated, and another along the gradient, which will be responsible for the work done in climbing. Thus, when you reach the top you will have done the same amount of work as you would have by climbing along the shortest path (the gradient), but you would have traveled a lot farther. You would not, however, have had to climb so steeply.

All of these ideas are immediately applicable to the electrical case. The equipotentials are surfaces of constant potential, so that, as already noted, no work is required to move a charge along any one of them. When two equipotentials are close together, there is only the small spacing between them over which the potential must change from its value on one equipotential to its value on the other. Thus, suppose that in the expression $-\mathbf{E} \cdot \Delta \mathbf{r}$ for the work done in traversing $\Delta \mathbf{r}$, this latter were the perpendicular distance between adjacent equipotentials. Then $\Delta \mathbf{r}$ would be in the direction of the gradient and hence of $\mathbf{E}$, and $-\mathbf{E} \cdot \Delta \mathbf{r} = -E\,\Delta r$. Then if the equipotentials are close together so that $\Delta r$ is small, $E$ must be large. Notice that $\mathbf{E}$ is the force urging a unit test charge from one equipotential to the next lower one. It thus represents the "steepness" of the electric potential hill, and in fact, as we just noted, it is the gradient of the potential. Thus, if we could make a map of the equipotential surfaces similar to the contour map of Fig. 24.1, we would have a very good idea of the direction and magnitude of the electric field throughout the mapped region. We could then present this knowledge by drawing lines representing the direction of $\mathbf{E}$ at every point, these lines being constructed perpendicular to the equipotential surfaces. An example of such a construction is given in Fig. 24.2, which shows the field of a point charge $q$. Coulomb's law tells us that the force on $+1$ C in the region surrounding $q$ will be $q/4\pi\epsilon_0 r^2$, where $r$ is the distance from the center of $q$, and as it will be a repulsive force, its direction will be radially outward. The solid

lines thus indicate the direction of $\mathbf{E}$. As the amount of work required to bring a unit test charge in from infinity to a certain distance $r$ from $q$ is quite independent of the direction from which the charge is brought in, all points a distance $r$ from $q$ will be at the same potential, hence a spherical surface of radius $r$, being by definition the locus of points all at a distance $r$ from the center, is an equipotential. The dashed circles in Fig. 24.2 represent several such surfaces.

Fig. 24.2 gives us a good idea of what the field of a point charge looks like and what we would expect a test charge placed anywhere in the region to feel. Note that although the field lines give us a good picture of the field, they are construction lines drawn by us as a convenience in understanding the effects we will observe on test charges placed at various points in the region. Of course there are no actual lines in the space surrounding $q$. Our drawing merely represents the condition in space set up by $q$ and resulting in the action-at-a-distance forces felt by other charges. Moreover, we have only an indirect presentation of the *strength* of the electric field at any point, for the lines of $\mathbf{E}$ simply show its direction. In Fig. 24.2, the equipotentials have been drawn with a constant potential difference between each, and we see that they get closer and closer together as we get nearer $q$. The conclusion is that the field gets stronger as we get nearer $q$, although it is not clear from the drawing that it varies as $1/r^2$. Note, however, that the radial field lines are everywhere perpendicular to the equipotential surfaces, as they should be, and that as a result we can think of a "density of lines," that is, the number of lines crossing unit area of the normal surface. In Fig. 24.2, we have arbitrarily chosen to draw 16 lines from charge $q$, and these lines must all traverse each of the equipotential surfaces at right angles. But any one of these surfaces has area $4\pi r^2$ where $r$ is the radius of the particular surface. Therefore, the line density is $16/4\pi r^2$. Comparing this result with Coulomb's law for the field at distance $r$ from $q$, we see that if $q/\epsilon_0$ is represented by the total number of lines we have selected, the two expressions are identical. Thus, the line density is a direct measure of the field strength, so that in looking at a map of electric field lines we can conclude that the magnitude of $\mathbf{E}$ will be large in regions where these lines are close together.

In the present experiment, we will determine some equipotentials arising from three simple charge configurations and plot some electric field lines using our knowledge that they must be perpendicular to the equipotentials at every point. We will do this in two dimensions using sheets of carbon-impregnated paper that will conduct electricity over its surface, although not very well; that is, it will put up a resistance to the passage of electric current. We will set up fixed

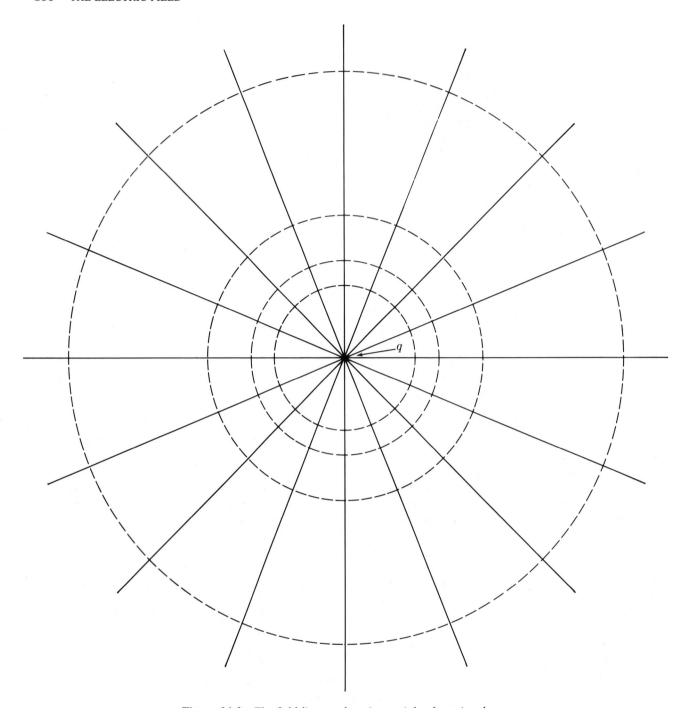

**Figure 24.2** *The field lines and equipotentials of a point charge*

charge configurations by painting the appropriate shapes on the paper with a conducting silver paint that makes the painted area a very good conductor. Two shapes or electrodes will be used in each case with a battery or power supply connected between them. The battery or power supply pumps charge from one electrode to the next, and as the electrodes are conductors, their edges become equipotentials, the potential difference between them being the supply voltage. But because the paper is a resistance, current will flow from one electrode to the other and there will be a fall in potential from the value at the positive electrode to the value at the negative one, which we can call zero to avoid working with negative numbers. Moreover, the current must surely flow in the direction of the electric field at each point so that the flow lines will be in the same direction as the field lines. We cannot see either one, of course, but we can measure the potential at various points by touching the positive probe of a voltmeter to the paper at each of these points. We will then be able to construct equipotential lines on the paper and hence the perpendicular field lines. Notice that since we are working on the two-dimensional surface of the paper, our equipotentials show up as lines just as they do in Fig. 24.2. In that figure, the circles represent cross sections of spherical surfaces. Our experimental work will be applicable to the three-dimensional world if we think of the equipotential lines as being cross sections of cylindrical surfaces that project normally from the paper surface to infinity. In short, our drawings on the paper represent the intersection with the $x$-$y$ plane (the plane of the paper) of surfaces that ex-

tend both up and down in the $z$ direction without change in the $x$ or $y$ directions. Thus, a dot painted on the paper represents the cross section of an infinitely long vertical rod, a circle an infinitely long vertical cylinder, etc. Our results will thus apply to three-dimensional models that can be placed in a three-dimensional *cylindrical* coordinate system with no variation along the $z$-axis.

## APPARATUS

1. Sheets of resistance paper with centimeter grid
2. Conductive ink in dispenser
3. Cork board
4. Metal push pins
5. 15-V dc power supply
6. Electronic voltmeter
7. Connecting leads
8. Masking tape
9. Straightedge and drawing compass

## PROCEDURE

**1.** This experiment uses sheets of a carbon-impregnated paper that has a high but finite resistance across its surface. Electrodes may be drawn on this paper with silver conductive ink that makes the covered areas highly conducting. Such an area then assumes a single potential and may serve as an electrode when connected to the power supply. We will prepare three sheets, each with a pair of electrodes on it, between which we will connect the power supply so that a potential difference of 15 V will be established between them. Fig. 24.3 shows these three configurations. To prepare the sheets, tape each one down flat on the laboratory bench with masking tape. Draw the outline of each electrode using the centimeter grid to establish the dimensions shown in Fig. 24.3, a straightedge to draw the straight lines, and a drawing compass or circle template to draw the circles. Do this with an ordinary pencil and then fill in the outlined areas with the conductive ink. This ink is

supplied in a dispenser resembling a small oil can and must be mixed thoroughly before using. Shake the can vigorously until you hear the stirring ball rattling around inside and then continue shaking for at least 30 s thereafter. Remove the cap and practice drawing with the dispenser on a piece of scrap paper. The ink should flow reasonably freely from the nozzle. If it doesn't, the nozzle can be cleaned out with a pin. If you have trouble getting a proper flow of ink, consult your instructor. Note that the silver pigment tends to settle out whenever you put the dispenser down for a minute, so it is good practice to shake it again before each use.

When you are adept at using the dispenser, fill in your electrode outlines as shown in Fig. 24.3. For the parallel-line configuration of Fig. 24.3(a), fill in the areas between the lines spaced 0.5 cm apart; be careful to bring the ink right up to your penciled lines but do not run over them. You should then have two par-

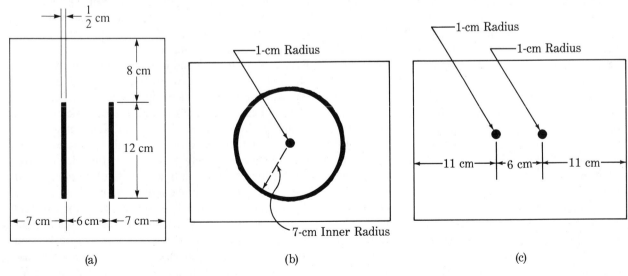

(a)  (b)  (c)

**Figure 24.3** *Electrode configurations*

allel 0.5-cm-wide bands of conductive ink spaced by 6 cm between their facing edges. For the concentric circles, Fig. 24.3(b), fill in the 1-cm-radius central circle so that you have a circular spot of ink 2 cm in diameter there. The outer circle must have a smooth, accurate 7-cm inner radius, but the outer radius doesn't matter and need not be smooth. Paint the ink on the paper approaching your 7-cm penciled circle from the outside and coming carefully up to the line so as to produce the necessary exact 7-cm radius, but do not worry about the smoothness of the outer edge of the circular band you are making. Finally, for the configuration of Fig. 24.3(c), fill in the two circles to produce two circular conductive spots in the same way you did for the central spot in Fig. 24.3(b). When you have finished using the ink, cap the dispenser and leave the three sheets taped down on the bench to dry. This will take 20–30 min, depending on how thick a coat you applied. Go to another bench and start on Procedure 2.

**2.** Copy the three electrode configurations of Fig. 24.3 on the first three sheets of graph paper found at the end of this experiment. This graph paper has centimeter-scale main divisions like the resistive paper grid, so that you can make an exact copy of your electrodes. However, the overall size of the graph paper is not the same as that of the resistive paper sheets, so the dimensions given in Fig. 24.3 for the spacing of the electrodes from the paper edges will not apply. This is of no consequence, however. Simply follow the dimensions for the electrode sizes and spacings, and arrange to center each complete configuration on the graph sheet as closely as you can.

**3.** When you are sure the ink is dry, remove the parallel-electrode configuration (a) from the laboratory bench and tack it down flat on the cork board with a push pin at each corner. Insert push pins into the two conducting bands to make electrical contact with them and wire your setup as indicated in Fig. 24.4. Notice that the negative lead from the electronic voltmeter is connected to the same push pin as the negative lead from the power supply. The electrode contacted by this pin will be considered at zero potential. Potentials at various points on the conductive paper may be determined by touching the point in question with the voltmeter probe.

**4.** When you are ready to make measurements, have your instructor check your setup and then turn on both the power supply and the electronic voltmeter. Allow both to warm up, after which you should set the voltmeter to a scale on which voltages up to +15 V dc are conveniently measured. Check its zero by touching the probe to the zero-potential pin (the one to which you connected the negative voltmeter and power supply leads) and adjusting the zero control until the meter needle is exactly on zero. Then touch the probe to the other pin (the one connected to the other, or positive, electrode) and set the power supply to exactly 15 V.

**5.** You are now ready to investigate the equipotential lines between your two electrodes. Begin by determining a number of points on your resistive paper, all of which are at +12 V. A good way to do this is to look at the horizontal line 4 cm below the bottom ends of your two electrode strips. Run the voltmeter probe along this line, lightly touching the resistive paper until you locate a point where the meter reading is +12 V. Mark this point on the accompanying graph sheet prepared for the parallel-electrode configuration (a) in Procedure 2. Be careful not to try to measure the potential right at a grid mark on the resistive paper, as these marks prevent the voltmeter probe from making good contact. Now look at the horizontal line 3 cm below the bottom ends of your electrode strips and again run the voltmeter probe along this line until you find a point where the potential is +12 V. Plot this point on your graph sheet as before. Repeat this procedure along each horizontal line in the centimeter grid until you reach 4 cm above your electrodes. On your graph sheet, draw a curve through your plotted points and label it "12 V." This is your 12-V equipotential. Note that in the region between the electrodes it should be a straight line, and you may want to use a straightedge to draw the best straight line through these points, but note carefully where your equipotential begins to curve as the ends of the electrodes are approached.

**6.** Repeat Procedure 5 for voltmeter readings of 9, 6, and 3 V. When you have completed this work, your parallel-electrode graph sheet should have four equipotentials plotted on it for 12-, 9-, 6-, and 3-V potentials, respectively.

**7.** Remove the parallel-electrode resistive paper from the cork board and replace it with the circular electrode configuration of Fig. 24.3(b). Use the metallic push pins to make contact with the 7-cm circular band and the central circular spot, and connect the negative leads of the power supply and voltmeter to the spot. The positive power supply lead goes to the 7-cm circle, placing it at a potential of +15 V. Repeat Procedures 5 and 6 for the region between the electrodes, that is, between the circle and the spot. Again, it is convenient to move the voltmeter probe along a horizontal line through the centimeter marks on the grid as in Procedure 5, but symmetry should tell you that the equipotentials will be circles in the present case. You may be guided by this idea in looking for the various points at a particular voltage, and you should be able to use a drawing compass to draw the best circle through your set of points for any particular potential. Be particularly careful in determining the 3-V points, which are quite critical. You should thus obtain four circles as your equipotentials, one for each of the four voltages, 12, 9, 6, and 3 V.

**Figure 24.4** *The experimental setup*

**8.** Replace the resistive paper with the concentric circle configuration of Fig. 24.3(b) with the paper having the two circular spots of Fig. 24.3(c). Con-nect the power supply and voltmeter as before, and find the equipotentials for 12, 9, 6, and 3 V by re-peating Procedures 5 and 6 for this configuration.

## DATA

### Parallel Electrode Configuration (a)

| Equipotential pair | Distance between equipotentials | Deviation |
|---|---|---|
| 15–12 V | | |
| 12–9 V | | |
| 9–6 V | | |
| 6–3 V | | |
| 3–0 V | | |
| Averages | | |

Value of electric field between the
   electrodes        _____

**Concentric Circle Configuration (b)**

| Equipotential | 15-V | 12-V | 9-V | 6-V | 3-V | 0-V |
|---------------|------|------|-----|-----|-----|-----|
| Radius, cm    | 7    |      |     |     |     | 1   |

| Equipotential pair | Distance between equipotentials | Midpoint between equipotentials | Reciprocal of midpoint radius | Field at midpoint |
|--------------------|---------------------------------|---------------------------------|-------------------------------|-------------------|
| 15–12 V            |                                 |                                 |                               |                   |
| 12–9 V             |                                 |                                 |                               |                   |
| 9–6 V              |                                 |                                 |                               |                   |
| 6–3 V              |                                 |                                 |                               |                   |
| 3–0 V              |                                 |                                 |                               |                   |

## CALCULATIONS

**1.** On your graph of equipotentials for the parallel-electrode configuration, sketch several field lines. To draw a field line, start at some point along one electrode edge, draw a line perpendicular to that edge (which is itself an equipotential), and continue the line so that it crosses each successive equipotential at right angles, finally meeting the other electrode edge at right angles. Do your field lines in the central region between the electrodes (that is, not near the ends) go straight across from one electrode to the other?

**2.** Along a field line in the central region noted in Calculation 1, measure the distances separating adjacent equipotentials and record them in the boxes in your data table.

**3.** Your parallel electrodes drawn as two bands on paper represent an edge view of a pair of parallel plates in three dimensions. According to theory, the electric field in the central region between such parallel, oppositely charged plates should be uniform so that equipotentials differing by the same voltage should be equally spaced. See how close you are to realizing this prediction by averaging the five separations recorded in Calculation 2, computing a deviation for each, and averaging the deviations to obtain an a.d. Record your results in the spaces provided in the data sheet.

**4.** Adjacent equipotentials in the present case all differ by 3 V, hence the value of the electric field between the plates is this voltage divided by the average spacing found in Calculation 3. Find this field and record it, attaching the error obtained from the a.d. of the spacing.

**5.** The concentric circle configuration of Fig. 24.3(b) represents a pair of concentric cylinders in three dimensions. According to theory, the equipotentials between such cylinders are cylindrical surfaces (hence circles in our two-dimensional experiment) having radii that increase logarithmically with voltage. Read the radii of the equipotential circles on your concentric-circle graph sheet and record these readings in the pertinent data table. Then plot them against voltage on the semi-logarithmic graph sheet provided. Note that you have a zero-volt equipotential at 1 cm and a 15-V equipotential at 7 cm. Draw the best straight line through your points. Refer to the section on plotting graphs in the Introduction for details on this type of plot.

**6.** On your concentric-circle graph sheet, measure and record the distances between each pair of adjacent equipotentials. Then find and record the radial distance to the midpoint between each equipotential pair and calculate the reciprocal of each of these distances.

**7.** Although the field between concentric cylinders is not uniform, a good approximation over small voltage intervals may still be obtained by dividing the potential difference between adjacent equipotentials by their radial separation and calling that the field at a radius midway between them. Calculate these field values for the radii halfway between each of your five equipotential pairs and enter them in your data table.

**8.** According to theory, the field between oppositely charged coaxial cylinders goes inversely with the radial distance from the axis, hence a plot of $E$ against $1/r$ should be a straight line. Again refer to the instructions on the use of graphs in the Introduction and plot the values obtained for the electric field in Calculation 7 against the reciprocals of the radial locations assigned to them from Calculation 6. Draw the best straight line through your points.

**9.** On your graph sheet with the two circular spots, draw several field lines, constructing each one perpendicular to succesive equipotentials as described in Calculation 1. Do your field lines and equipotentials seem to have any particular shape for this electrode configuration?

## QUESTIONS

**1.** (a) Why was an electronic voltmeter (EVM) used in this experiment as opposed to an ordinary moving coil (d'Arsonval) voltmeter? (b) What would have been the effect of substituting the latter type?

**2.** What is the significance of the slope of the line plotted in Calculation 5? The line in Calculation 8?

**3.** The areal resistivity of the carbon-impregnated paper you used in this experiment is about 10,000 $\Omega$. Explain what this statement means, showing in particular why the units of areal resistivity are the same as the units of resistance between the terminals of a conductor.

**4.** Show why the field between two large but closely spaced parallel plates carrying equal but opposite charges should be uniform in the region not too near the edges. Use any method you can understand and explain.

**5.** Derive a relation between the potential $V$ and the magnitude of the field $\mathbf{E}$ at a radial distance $r$ from the axis of a very long uniformly charged rod of radius $a$ ($r > a$).

**6.** The two circular spots of the configuration of Fig. 24.3(c) represent two long, parallel cylinders in three dimensions. Look up and tell in your own words what the field lines and equipotential surfaces of such a pair of cylinders carrying equal and opposite charges should look like. How does your answer to this question compare with your results in Calculation 9? If your two answers do not correspond, can you explain why?

# The Force Between Electric Currents 25

Although magnets appear to have poles that attract or repel other poles, in reality no magnetic poles are found to exist in the sense that electric charges do. What, then, is the source of the magnetic field if there are no poles or "magnetic charges" to fill this role? Magnetic fields result from the *motion* of electric charges, that is, from electric currents. Permanent magnets produce magnetic fields beause the electrons going around in their orbits in the atoms of the material constitute tiny loops of electric current each of which is a magnetic field source. In materials like iron, nickel, and cobalt, out of which permanent magnets are made, it is possible to line up the atoms so that their fields are collinear and add up to the large fields found around such magnets. This process is called magnetization, and the atoms will remain lined up in permanent magnets provided the temperature is not high enough for the resulting random thermal motion to upset the arrangement (the "Curie temperature").

The most elementary source of a magnetic field is a long, straight wire carrying an electric current. The magnetic lines of force due to such a current are simply concentric circles centered on the wire, as shown in Fig. 25.1. In addition, because there are no magnetic-pole sources of the magnetic field, magnetic forces are not properly described as the action of a magnetic field on a pole. Instead, a force is exerted by a magnetic field on a moving electric charge, that is, on an electric current. Thus, in describing magnetic fields and forces, we consider only moving electric charges and do not think in terms of poles at

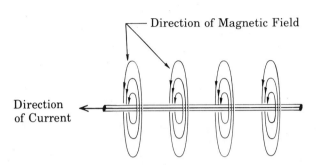

**Figure 25.1** *Magnetic field direction*

all. In particular, when an electric charge moves in a magnetic field, it feels a force in a direction perpendicular both to the direction of the field lines (magnetic lines of force) and to the direction of motion. Clearly, the charge will feel no force if it moves parallel to the field lines, and the force it feels will increase from zero to a maximum as the direction of motion is changed from parallel to perpendicular to these lines. In Fig. 25.1, if a second wire is laid parallel to the one shown, a current in it will be perpendicular to the field and a force pushing it towards or away from the first wire (depending on the current direction) will result. This is the most fundamental possible arrangement for the study of magnetic forces. The force law between parallel current-carrying wires may be regarded as the fundamental law of magnetostatics, just as Coulomb's law is regarded as the fundamental law of electrostatics. The purpose of the present experiment is to study this law.

## THEORY

The magnetic field of a long, straight wire carrying current $I$ is directed as shown in Fig. 25.1 and has the value

$$B = \frac{\mu_0 I}{2\pi r} = 2 \times 10^{-7} \frac{I}{r} \qquad (25.1)$$

where $B$ is the magnetic field strength in teslas, $I$ is the current in amperes, $r$ is the perpendicular distance in meters from the wire to the point where $B$ is being evaluated, and $\mu_0$ is a constant called the *permeability of free space* and has the value $4\pi \times 10^{-7}$ Wb/A·m. Notice that for the electromagnetic work in this experiment, MKS (SI) units must be strictly

adhered to. Therefore, $r$ in Equation 25.1 must be expressed in meters even though it will often be so small that the meter will not be the most convenient unit to use.

If a long, straight wire carrying current $I'$ is in a magnetic field of strength $B$, the force on a length $L$ of the wire is given by

$$F = BI'L \sin \theta \qquad (25.2)$$

where $F$ is the force in newtons, $B$ is the field strength in teslas, $I'$ is the current in amperes, $L$ is the length in meters of the wire on which force $F$ is exerted, and $\theta$ is the angle between the directions of the current and the field. The force is perpendicular to these two directions, but its sense remains to be determined. This is most easily done with the aid of Fig. 25.2, which shows the directions of a current $I'$ and a magnetic field $B$ by means of arrows. If your right hand is then placed so that the fingers curl in the direction needed to rotate the current arrow into the field arrow, the extended thumb will point in the direction of $F$. This procedure, sometimes called "the right-hand rule," is also useful in determining the field direction around the wire of Fig. 25.1. In this case, the wire is grasped with the right hand with the thumb extended in the direction of current flow. The fingers will then curl naturally around the wire in the direction of the magnetic field lines.

According to Equation 25.1, if a second wire carrying current $I'$ is laid parallel to the one shown in Fig. 25.1 at a distance $d$ away from it, the value of $B$ at this wire due to the current $I$ in the first wire will be $2 \times 10^{-7} I/d$. Moreover, Fig. 25.1 shows that the direction of $B$ at the position of the second wire will be perpendicular to the wire and hence to $I'$. Equation 25.2 then says that the force on a length $L$ of the second wire will be $BI'L$ and will be directed either toward or away from the first wire. Consideration of Fig. 25.2 shows that the two wires will attract each other if the currents $I$ and $I'$ are in the same direction and will repel if these currents are oppositely directed.

Combining Equations 25.1 and 25.2 for the case of parallel wires yields

$$F = \frac{\mu_0 I' I L}{2\pi d} = 2 \times 10^{-7} \frac{I' I L}{d} \qquad (25.3)$$

for the force of attraction or repulsion between the wires. Note that it makes no difference which wire is considered the source of the field and which the one on which $F$ acts. Equation 25.3 is obtained in either

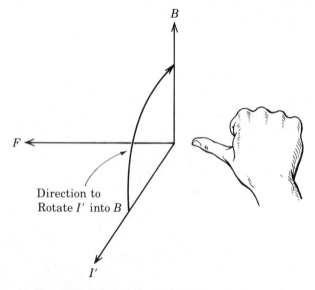

**Figure 25.2** *The direction of the magnetic force*

case. In an actual experiment, one of the two parallel wires will be movable so that the force on it can be measured. The other, which is fixed in position, may then be conveniently taken as the field source.

In the present experiment, a sensitive balance is used to measure the force between two parallel wires. High sensitivity is needed because the factor $2 \times 10^{-7}$ in Equation 25.3 results in very small forces unless inconveniently large currents and/or small values of the separation distance $d$ are used. The apparatus, which is shown in Fig. 25.3, consists of a heavy base on which one straight wire is mounted by means of insulating posts. The second or movable wire forms one side of a rectangular frame which itself forms one arm of a beam balance. The other arm consists of a threaded rod on which counterweights may be screwed back and forth to balance the beam assembly. The beam rotates on knife edges under the side of the rectangle opposite the wire, and a mirror is mounted on this side so that small rotations of the beam assembly may be easily observed. This is done by means of an optical lever as shown in Fig. 25.3. A laser shines a light ray on the mirror, which reflects the ray to a wall several feet away. This reflected ray is the "lever," the displacement of the light spot on the wall being twice the angle through which the mirror rotates multiplied by the ray's length (mirror-to-wall distance). As we shall see, an actual measurement of the mirror's rotation will be unnecessary, but the mirror-to-wall distance, although it need not be known, should be sufficient to insure detection of even small rotations of the beam balance. In addition, the beam assembly is provided with a damping system consisting of a conducting blade moving in the field of a permanent magnet and a beam lift that raises the assembly off the knife edges to prevent damage during transportation and to position it correctly on the knife-edge supports. Finally, the movable wire carries a small pan in which calibrated weights may be placed to measure the force exerted on this wire by the magnetic field of the fixed wire.

The earth's magnetic field is about $5 \times 10^{-5}$ T (0.5 G) and therefore cannot be neglected in comparison with the fields that can be produced at the movable wire by reasonable currents in the fixed one. Although the apparatus can and should be oriented with the wires parallel to the earth's field (roughly north-south) to minimize its effect, we cannot count on eliminating it entirely and steps must be taken to compensate for it. Suppose that the component of the earth's field perpendicular to the plane of the two wires is $B_0$. Then the total field to be used in Equation 25.2 is $B + B_0$, where $B$ is the field due to current $I$ in the fixed wire, and Equation 25.3 becomes

$$F = \left( \frac{\mu_0 I}{2\pi d} + B_0 \right) I' L \qquad (25.4)$$

**Figure 25.3** *Current balance*

The apparatus is now connected with the two wires in series so that $I = I'$. Equation 25.4 then becomes

$$F = \left(\frac{\mu_0 I^2}{2\pi d} + B_0 I\right)L \qquad (25.5)$$

If the current $I$ is then reversed, the force is given by

$$F = \left(\frac{\mu_0 I^2}{2\pi d} - B_0 I\right)L \qquad (25.6)$$

in which the force due to the current is unchanged because $I$ has been reversed in both wires (the sign of $I^2$ doesn't change) but the force due to the earth's field has been reversed by the reversal of the current in the movable wire. Clearly, the average of the two forces measured with the current in each of the two directions will be the force due to the current alone, or that given by Equation 25.3 with $I = I'$.

## APPARATUS

1. Current balance with calibrated weights
2. Laser with support stand and necessary clamps
3. Level
4. Micrometer caliper
5. Meter stick
6. $\frac{1}{8}$-in. and $\frac{1}{4}$-in. spacer blocks
7. Power supply capable of delivering 15 A at 10 V dc
8. Compass
9. Variable line transformer, about 150-W rating ("Variac" or "Powerstat"), if not included in above power supply
10. Ammeter, 0–15 A dc
11. Resistor, 0.5 Ω, 100 W
12. Two double-pole, double-throw knife switches
13. Roll of white calculator paper
14. Masking tape

## PROCEDURE

**1.** Set up the current balance on the laboratory bench with the wires extending in a north-south direction as indicated by the compass. Carefully lift the beam assembly out of the apparatus and lay it aside. Using the level provided, level the base by means of the two thumbscrews at its front.

**2.** While the beam assembly is out of the apparatus, measure the length $L$ of the movable wire with the meter stick. This measurement is made between the centers of the supporting side wires in the beam assembly. Record your result in meters, estimating the measurement to a tenth of a millimeter (0.001 m). Also measure the diameters of both the movable and fixed wires with the micrometer caliper. Make two measurements in different places on each wire and record your results *in meters*.

**3.** Note the two tapered screws on the lifting mechanism and the matching tapered holes in the bottom of the beam assembly near the knife edges. Rotate the lifting shaft so that the tapered screw points project upward and carefully place the beam assembly on these screws so that the points enter the tapered holes. Then gently lower the assembly onto the knife-edge bearing surfaces. Check that the movable wire and the fixed wire below it are in the same vertical plane. If they are not, loosen the set screws holding the wire sides of the rectangle in the knife edges and push the sides in or out as required. Be sure to tighten the set screws after making this adjustment.

Now lift the beam assembly off its supports with the beam lift, noting that an important function of the lifting mechanism is to position the wire rectangle correctly. Again lower the assembly onto its supports and recheck the position of the movable wire. Note that, when viewed from the front, the wires should be straight and parallel. Bring them together by loading some weights onto the pan on the movable wire and check that no light is visible between the wires when a white piece of paper is held behind them. Adjustment for parallelism may be made by loosening the set screws retaining the lower (fixed) wire in its mounting posts and setting it parallel to the upper (movable) wire. However, even after this adjustment has been made, the wires will seldom be so straight as to touch at every point along their full length, but perfect straightness is not essential for good results. Consult your instructor if the wires appear to be seriously bent.

**4.** Use the beam lift to position the balance assembly correctly and adjust the horizontal counterweights until the assembly is balanced and pivots freely on its knife edges. *Note:* The counterweights that hang vertically downward have already been set and should not be touched. Check that the damping blade rides in the middle of the hole cut for it in the base and does not rub on any of the damping magnet poles. If the blade comes too close to one side of the hole, lift the beam assembly off its supports and *gently* press the blade right or left as required. Then replace the assembly, positioning it properly with the lifting mechanism. If either damping magnet is too close to the blade, loosen that magnet's retaining screw and adjust its position as necessary.

**5.** Wire the apparatus as shown in Fig. 25.4. Notice that switch $SW_1$ reverses the current through both wires of the current balance, whereas $SW_2$ reverses the current in one wire with respect to the other in order to change from repulsive to attractive force measurements. The neutral position of either switch opens the circuit, but reducing the current to zero will usually be done with the power supply control. Note that if this control is not built into your supply, you will need to plug the supply into a separate variable transformer. Have your instructor check your setup, and be sure the power supply switch is off and the control or variable transformer is turned

all the way down before plugging it into the power line.

**6.** Measure the thickness of the $\frac{1}{8}$-in. spacer block with the micrometer caliper. Note that the $\frac{1}{8}$ in. is a nominal value only. You need much greater precision than this, hence you should make two independent measurements with the micrometer and record the results *in meters* to four significant figures.

**7.** Place the $\frac{1}{8}$-in. spacer between the wires and put some weights in the pan on the movable wire to bring this wire down on the spacer hard enough to retain it in position. Lift the balance assembly momentarily with the beam lift to assure that it is correctly positioned and check that the spacer block remains properly positioned to separate the wires by just its thickness.

**8.** Mount the laser, turn it on, and position it so that its emitted ray strikes the mirror on the apparatus and is reflected back to a wall or screen at least 2 m away. Tape a strip of white paper vertically on the wall so that the light spot falls at about its center. Mark the position of this spot on the paper. CAUTION: The low-power laser used in this experiment cannot hurt you even if its light falls directly on your skin, *but do not allow anyone, including yourself, to look directly into the emitted light ray.*

**9.** Remove the weights from the pan on the movable wire, remove the spacer from between the wires, and adjust the horizontal counterweights until the beam assembly balances with the light spot at the marked position on the paper strip. This is a critical adjustment and should be made with great care. Lift the balance assembly and return it to its supports to assure correct positioning several times and check that the spot returns to the marked position. Note that this procedure assures that the wires are spaced by exactly the thickness of the spacer block.

**10.** Set switch $SW_2$ for oppositely directed current in the two wires. Turn on the power supply and close $SW_1$ in either direction, the direction you choose to be called "forward" hereafter. Increase the power supply output to a current of about 5 A, noting

**Figure 25.4** *Circuit for the current balance*

that the resulting repulsive force between the wires causes the light spot to move upward along the paper strip. Now carefully add weights to the pan to bring the spot back down approximately to the marked position. Then make a fine adjustment of the current so that the spot comes to the marked position exactly. As the beam assembly is now in the same position it occupied with no weight and no current, the repulsive force between the wires must be just equal to the added weight. Record the mass in the pan and the final current as read on the ammeter.

**11.** Reverse the current through the apparatus by throwing switch $SW_1$ to the opposite or "reverse" position. Readjust the current to return the spot to the exact position of the mark on the paper strip. Record this new current value.

**12.** Repeat Procedures 10 and 11 with additional weight and a current of about 10 A.

**13.** Repeat Procedures 10 and 11 with still more weight and a current of close to 15 A.

**14.** Reduce the current through the apparatus to zero by turning the power supply control all the way down and also opening switch $SW_1$ (neutral position). Adjust the horizontal counterweights to return the light spot to the marked position *with the weights of Procedure 13 still in the pan.* Throw $SW_2$ to give currents in the same direction in the two wires so that the force between them will be attractive.

**15.** Lift the beam assembly momentarily with the beam lift and return it to its supports to insure that it is positioned correctly and balances with the spot on the mark as noted in Procedure 14. Then carefully remove the weights from the weight pan without disturbing the apparatus. Throw switch $SW_1$ to the forward position and slowly increase the current through the wires while watching the spot move back down the paper strip toward your mark. Your aim is to find a current at which the beam assembly just balances with the spot at the marked position. This will be a very delicate procedure, as the spot will tend to overshoot, the movable wire being drawn into contact with the fixed one. If this happens, reduce the current to zero, check the positioning of the balance assembly, and try again. Note that when you have the assembly balanced with the spot on your mark, the attractive force between the wires is equal to the weight you removed from the pan. Record the current read on the ammeter and the mass removed.

**16.** Reduce the current through the wires to zero and throw $SW_1$ to the reverse position. Again carefully increase the current to bring the light spot to the marked position as in Procedure 15 and record the current so obtained.

**17.** Repeat Procedures 14–16 with the weights used in Procedure 12 in the pan for the initial balancing.

**18.** Repeat Procedures 14–16 with the weights used in Procedure 10 in the pan for the initial balancing.

**19.** Repeat Procedures 6–18 using the $\frac{1}{4}$-in. spacer block. The strip of paper you already have taped on the wall may continue to serve with a new mark on it corresponding to the new wire spacing, but label this mark to avoid confusion with the one you were using with the $\frac{1}{8}$-in. spacer.

## DATA _____

| | | | |
|---|---|---|---|
| Length $L$ of movable wire | _____ | Micrometer zero reading | _____ |
| Radius $r_1$ of movable wire | _____ | Thickness of $\frac{1}{8}$-in. spacer | _____ |
| Micrometer readings: | | Micrometer readings: | |
|     Trial 1 | _____ |     Trial 1 | _____ |
|     Trial 2 | _____ |     Trial 2 | _____ |
|     Average | _____ |     Average | _____ |
| Radius $r_2$ of fixed wire | _____ | Thickness of $\frac{1}{4}$-in. spacer | _____ |
| Micrometer readings: | | Micrometer readings: | |
|     Trial 1 | _____ |     Trial 1 | _____ |
|     Trial 2 | _____ |     Trial 2 | _____ |
|     Average | _____ |     Average | _____ |

Wire separation $d$ with $\frac{1}{8}$-in. spacer  _____

| Relative direction | Current | | | Mass on pan | Magnetic force |
|---|---|---|---|---|---|
| | Forward | Reverse | Squared | | |
| Opposite (repulsive force) | | | | | |
| | | | | | |
| | | | | | |
| | | | | | |
| | | | | | |
| | | | | | |
| Same (attractive force) | | | | | |
| | | | | | |
| | | | | | |
| | | | | | |
| | | | | | |
| | | | | | |

Wire separation $d$ with $\frac{1}{4}$-in. spacer  _____

| Relative direction | Current | | | Mass on pan | Magnetic force |
|---|---|---|---|---|---|
| | Forward | Reverse | Squared | | |
| Opposite (repulsive force) | | | | | |
| | | | | | |
| | | | | | |
| | | | | | |
| | | | | | |
| | | | | | |
| Same (attractive force) | | | | | |
| | | | | | |
| | | | | | |
| | | | | | |
| | | | | | |
| | | | | | |

Value of $\mu_0$ with $\frac{1}{8}$-in. spacer  _____   Average value of $\mu_0$   _____

Value of $\mu_0$ with $\frac{1}{4}$-in. spacer  _____   Percent error   _____

## CALCULATIONS _____

**1.** Compute the average of the two diameter readings for each wire. Subtract the zero reading of the micrometer to obtain the diameter in each case and divide your results by 2 to obtain the radii $r_1$ and $r_2$ of the two wires.

**2.** Calculate the actual thickness of the $\frac{1}{8}$-in. spacer by averaging your two measurements and subtracting out the micrometer zero reading. Then find the center-to-center wire separation $d$ for this case by adding $r_1$, $r_2$, and the thickness just determined.

**3.** Compute the square of the current readings obtained in Procedures 10–13 and 15–18.

**4.** From your data on the mass in the weight pan obtained in Procedures 10–13, find the repulsive magnetic force between the wires for each current. This calculation requires multiplication of the mass in the pan *in kilograms* by the acceleration of gravity in meters per second per second.

**5.** From your data on the mass removed from the weight pan in Procedures 15–18, find the attractive magnetic force between the wires for each current. This calculation requires multiplying the mass removed in each case by the acceleration of gravity. Express the masses in kilograms and the acceleration of gravity in meters per second per second to obtain the forces in newtons.

**6.** Plot a curve using the magnetic forces obtained in Calculations 4 and 5 as abscissas and the corresponding values of the square of the current as ordinates. Use an entire sheet of graph paper for this curve. Place the origin near the center of the sheet and consider attractive forces and the corresponding values of $I^2$ to be negative. Distinguish between points representing current through the apparatus in the forward direction and those for current in the reverse direction by using different colored pencils or marking the points in some distinctive way (for example, circles around points for forward current and squares around points for reverse current). Draw the best straight line through your points, remembering that the origin (zero force, zero current) should be a point on this line.

**7.** Calculate a value of $\mu_0$ from the slope of the graph obtained in Calculation 6.

**8.** Repeat Calculations 2–7 for the wire separation obtained with the $\frac{1}{4}$-in. spacer and the associated data called for in Procedure 19.

**9.** Compute an average of your two values of $\mu_0$ and compare this result with the accepted value by finding the percent error.

## QUESTIONS _____

**1.** What conclusions do you draw from your two graphs?

**2.** Explain why reversing switch $SW_2$ changes the sign of $I^2$ but reversing switch $SW_1$ does not.

**3.** The Theory section points out that the effect of the earth's magnetic field can be eliminated by averaging the two force values obtained with opposite flow directions of a particular current. Explain why this was not done in the present experiment and what was done instead. Discuss how the relative positions of the points representing forward and reversed current on your graphs relate to this matter.

**4.** Estimate the smallest angle of rotation of the beam assembly that you could detect by observing the consequent motion of the light spot on the wall. Note that this will require you to estimate the mirror-to-wall distance (you didn't measure it but you must know approximately what it was) and remember that when a mirror turns through a given angle, the reflected ray is turned through twice that angle. Show in your answer why this is so.

**5.** What is the purpose of the damping blade and magnet arrangement? How does it work?

**6.** Suppose the current balance was oriented so that a horizontal component of the earth's magnetic field equal to 0.17 G ($1.7 \times 10^{-5}$ T) remained perpendicular to the movable wire. What mass would have to be added to the weight pan to balance out the effect of this field when a current of 10 A flows in the wire?

**7.** Let the fixed and movable wires be spaced by a distance $d$ between their centers and let $y$ be the coordinate of any point between the wires measured from the center of the fixed wire ($y = 0$). Derive expressions for the total magnetic field between the wires ($r_2 < y < d - r_1$) due to a current $I$ in each wire when (a) the current flows in the same direction in the two wires and (b) the current flows in opposite directions.

**8.** A length of No. 14 bare copper wire (diameter = 0.163 cm) is laid flat on a table and connections are made to its ends with flexible leads. A second wire is mounted 1 cm above and parallel to the first one and is connected in series with it so as to produce an attractive force when a current is passed through the two wires. Find the current necessary to just raise the first wire off the table. Do you think the No. 14 wire could handle this current without overheating?

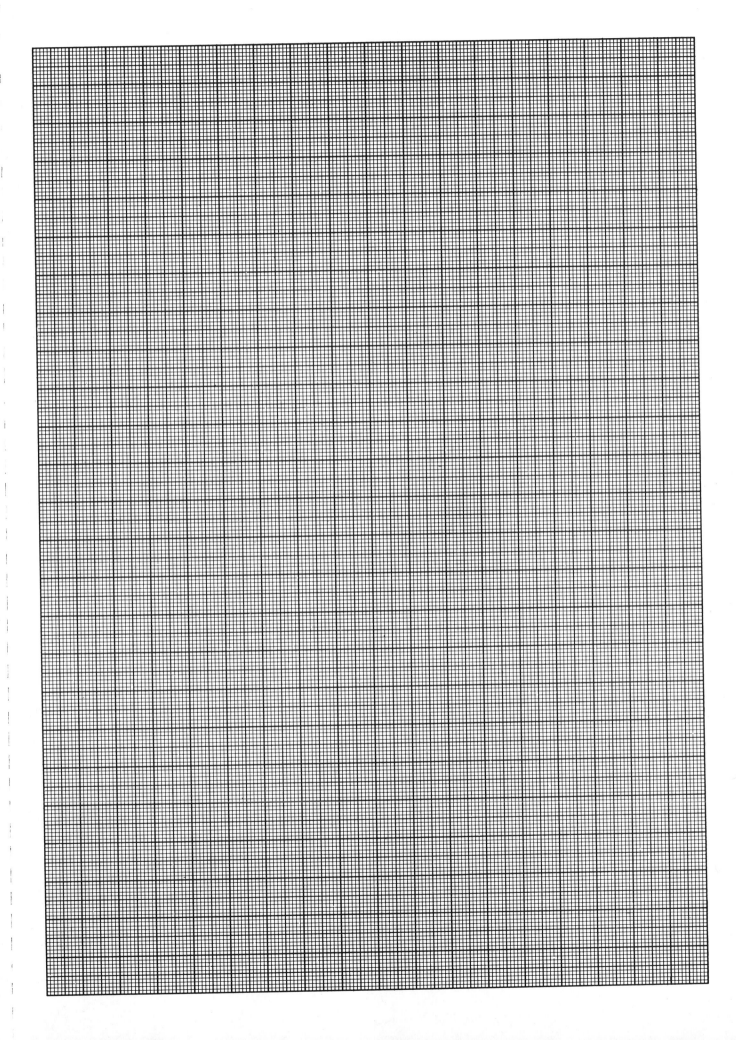

# Ohm's Law 26

The most important principle in direct-current electrical circuit theory is Ohm's law, named after its discoverer, Georg Simon Ohm (1787–1854). This principle states that the current flowing through a conductor (that is, the rate at which electrical charge passes through it) is proportional to the voltage, or potential difference, between that conductor's ends. It is the object of this experiment to study Ohm's law and to derive from it the laws relating to series and parallel direct-current circuits.

## THEORY

Ohm's law states that the voltage across a conductor is proportional to the current through it, the proportionality constant being the *resistance* of the conductor. Although Ohm himself established his law on purely empirical grounds, we can give a theoretical derivation that makes it plausible and helps us understand why most ordinary metals obey it. Consider the bar of conducting material shown in Fig. 26.1. Such a material is a conductor because it contains charged particles, called *charge carriers,* that are free to move under the influence of an electric field. Electrons are the most common charge carriers. It is characteristic of ordinary metals such as copper, silver, and aluminum for the valence or outer atomic electrons to be loosely bound to their respective atoms and thus free to move through the structure of the material with relatively little resistance. This, of course, is a necessary characteristic for a conductor and is what makes metals good conductors. In contrast, insulators, in which all electrons are tightly bound to their respective atoms and cannot move about, do not conduct.

Let us then assume that the conducting bar of Fig. 26.1 contains a charge-carrier density of $n$ such particles per unit volume. In consequence, the slice of thickness $\Delta x$ appearing in this figure will contain $nA\Delta x$ charge carriers, where $A$ is the cross-sectional area of the bar and hence $A\Delta x$ is the volume of the slice. If we now assume that the charge on each carrier is the electronic charge $e$ ($1.6 \times 10^{-19}$ C), the total amount of movable charge in the slice of thickness $\Delta x$ is

$$\Delta q = neA\Delta x \tag{26.1}$$

Now if a battery or power supply maintains a voltage $V$ between the ends of the bar, there will be an electric field $E$ along the bar equal to $V/L$, where $L$ is the bar's length, with the result that each charge carrier will be urged in this direction with a force $eE = eV/L$. The charge carriers will then be accelerated, but although they are free to move in the bar, they do not do so without encountering some resistance. We now make the crucial assumption about the nature of ordinary conductors that leads to Ohm's law: When charge carriers move through the material, they feel an opposing force similar to the force of viscous friction encountered by an object moving through a fluid: for example, a ball bearing sinking in molasses. We further assume that this retarding force is proportional to velocity, so that the net force felt by each charge carrier is $eV/L - Kv$, where $v$ is the so-called drift velocity of the charge carrier and $K$ is the constant of proportionality between this velocity and the retarding force just mentioned. Clearly, a *terminal velocity* will soon be reached as the charge carriers accelerate; at this velocity the net force on each carrier will be zero. The drift velocity $v$ will thereafter be constant at the value

$$v = \frac{eV}{LK} \tag{26.2}$$

required to give a zero net force ($eV/L - Kv = 0$).

Now consider a time interval $\Delta t = \Delta x/v$ in which charge carriers at the back of the slice in Fig. 26.1 just make it to the front. Clearly, in this time *all* the charge carriers that were initially in the slice will have made it out through the front face (to be replaced, of course, by other charge carriers coming in from behind). Thus, the charge $\Delta q$ given by Equa-

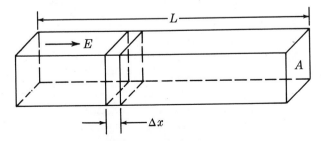

**Figure 26.1** *A conducting bar*

tion 26.1 will have passed through the front face of the slice in time $\Delta t$. The charge passing this point per unit time is then

$$I \equiv \frac{\Delta q}{\Delta t} = neA\frac{\Delta x}{\Delta t} = neAv \qquad (26.3)$$

where $I$ is the current in amperes, which is by definition the amount of charge passing a given point on a conductor per unit time. Substituting Equation 26.2 for $v$ yields

$$I = \left(\frac{ne^2}{K}\right)\left(\frac{A}{L}\right)V \qquad (26.4)$$

This is, in effect, a statement of Ohm's law, for it says that the current through our conducting bar is proportional to the voltage between its ends. The proportionality constant $ne^2A/KL$ is called the *conductance* $G$ of the bar, so that Equation 26.4 may be written

$$I = GV \qquad (26.5)$$

It is more usually written with the voltage as the dependent variable:

$$V = \frac{1}{G}I = RI \qquad (26.6)$$

where $R = 1/G$ is the resistance of the bar and is stated in ohms, the SI unit of resistance. Equation 26.6, which is the mathematical statement of Ohm's law, gives consistent numerical results when $I$ is in amperes, $R$ is in ohms, and $V$ is in volts.

A portion of a direct-current (dc) circuit that contains no voltage sources but only an arrangement of various resistors with wires coming out at two points for connection to the rest of the circuit is called a *two-terminal passive network*. The voltage between the ends (terminals) of such a network is proportional to the current through it, so that the entire network may be replaced by a single resistor whose value is that of the proportionality constant. Two very usual resistor combinations of this sort are the *series* connection shown for the two resistors in Fig. 26.2 and the *parallel* connection shown in Fig. 26.3.

When resistors are connected in series, the current is the same in each because the current in any one must all go on to the next one. The total voltage across the network, however, is the sum of the voltage drops across each resistor. Thus,

$$I = I_1 = I_2$$

and $$V = V_1 + V_2$$

where $I$ and $V$ are the current through and the voltage across the network as a whole, $I_1$ and $V_1$ refer to resistor $R_1$, and $I_2$ and $V_2$ to resistor $R_2$. But by Ohm's law, $V_1 = I_1 R_1$ and $V_2 = I_2 R_2$, so that

$$V = I_1 R_1 + I_2 R_2 = IR_1 + IR_2 = I(R_1 + R_2)$$

**Figure 26.2** *Resistances in series*

Here we see that the voltage across the whole network is proportional to the current through it with proportionality constant

$$R = R_1 + R_2 \qquad (26.7)$$

so that the series combination may be replaced by a single resistor $R$ whose resistance is the sum of those of the individual resistors $R_1$ and $R_2$.

The situation is simply reversed for the parallel connection. In this case, it is the total current that is the sum of the currents through the individual resistors and the voltage across the network as a whole that is equal to that across each resistor:

$$V = V_1 = V_2$$

and $$I = I_1 + I_2$$

This time we apply Ohm's law to the two resistors in the form $I_1 = G_1 V_1$ and $I_2 = G_2 V_2$ so that we get

$$I = G_1 V_1 + G_2 V_2 = G_1 V + G_2 V = (G_1 + G_2)V$$

This result shows that $I$ is proportional to $V$ with proportionality constant

$$G = G_1 + G_2 \qquad (26.8)$$

which looks exactly like Equation 26.7 except that it is a relation among conductances. We conclude that the parallel combination may be replaced by a single resistor whose *conductance* is the sum of the *conductances* of the two individual resistors. However, although this is quite correct, it is unusual to talk about conductances. Consequently, Equation 26.8 is com-

**Figure 26.3** *Resistances in parallel*

monly written in the form

$$\frac{1}{R} = \frac{1}{R_1} + \frac{1}{R_2} \qquad (26.9)$$

which states that the parallel combination of resistors may be replaced by a single resistor such that the re-

ciprocal of its resistance will be equal to the sum of the reciprocals of the individual resistances $R_1$ and $R_2$. In each case, $R$ is the value of the *equivalent* resistor that can replace the complete network without altering the relation between the network current $I$ and its voltage drop $V$.

## APPARATUS

1. Tubular rheostat (about 100 $\Omega$)
2. Tubular rheostat (about 200 $\Omega$)
3. 15-V regulated dc power supply

4. Volt-ohm-milliammeter with 0–500 mA scale
5. dc voltmeter (0–15 V)

## PROCEDURE

CAUTION: This experiment introduces you to a very useful and versatile instrument, the volt-ohm-milliammeter, often called a V-O-M or a multimeter. It consists of a high-quality meter mounted in a case with appropriate circuitry and switching so that by turning a switch and/or plugging the connecting leads into different jacks, voltages, currents, and resistances may be measured on various scales. Of particular interest is the use of this device as an ohmmeter, which requires impressing a known voltage across the resistor under test and measuring the current passing through it. A battery inside the V-O-M provides the voltage, and the circuit is arranged and an appropriate scale on the meter face calibrated to allow a direct resistance readout in ohms. Note, however, that *this meter is extremely sensitive, so that an erroneous connection in your circuit or switching it to the wrong scale can cause it to be instantly and permanently ruined as soon as power is applied.* Do not plug the power supply into the line or turn it on until your circuit has been approved by your instructor. Any changes in the circuit are to be made only after the power supply has been turned off. Have your circuit approved by your instructor every time you make any changes. Remember, you are using expensive equipment; good experimenters always treat their apparatus with great care.

**1.** Connect the two rheostats in series as shown in Fig. 26.2. Use the 100-$\Omega$ rheostat as $R_1$ and set it at about half resistance. Use the 200-$\Omega$ rheostat as $R_2$ and set it at maximum. Switch the V-O-M to its 500-mA scale and connect it at $A$ in Fig. 26.2. The voltmeter is connected across the two rheostats in series as shown. Have your circuit approved by your instructor.

**2.** Plug in the power supply, turn it on, and adjust its voltage control so that the voltmeter reads exactly 15 V. Then, without touching anything else, reconnect the voltmeter across $R_1$ alone. Record the current through the circuit and the voltage across $R_1$ in the first pair of boxes in the first table of your data sheet.

**3.** Increase the current through the circuit in four approximately equal steps by reducing the resistance of $R_2$, the final setting of this rheostat being zero. Record the current through and the voltage across $R_1$ for each setting of $R_2$ in the subsequent boxes in your table. Do not alter the setting of $R_1$ during this procedure.

**4.** Turn off the power supply. Reset $R_2$ to a resistance near its midrange value and reconnect the voltmeter across the two ends of this rheostat without disturbing the rest of the circuit. Turn on the power supply and record the readings of the voltmeter and the V-O-M as $V_2$ and $I_2$, respectively, in the appropriate boxes of the second table in your data sheet. Note that these are *measured* values.

**5.** With the power off, reconnect the voltmeter across the two rheostats in series as shown in Fig. 26.2. Apply power to your circuit and record both meter readings. These are the measured values of $V$ and $I$ for your series circuit.

**6.** Turn off the power supply and disconnect the rheostat used as $R_1$ and the volt-ohm-milliammeter from the circuit. Have your instructor explain the use of the V-O-M as an ohmmeter. Then switch this instrument to the proper scale, zero it, and use it to measure $R_1$. Record your result in the space provided on the data sheet.

**7.** Wire the circuit of Fig. 26.3, which shows the two rheostats in parallel. The three meter symbols marked $A$, $A_1$, and $A_2$ show different possible connections for the V-O-M. Switch this instrument back to its 500-mA scale and connect it at position $A$, the other two positions representing simply direct connections for this step. Do not change the setting of either rheostat when you reconnect them in this circuit.

**8.** Before applying power, have the circuit approved by your instructor. Record the total current through the circuit and the voltage drop across the parallel resistance combination as your measured $I$ and $V$ for the parallel case.

**9.** Switch off the power supply and reconnect the V-O-M at point $A_1$ without disturbing the rest of the circuit. Apply power and record the current through $R_1$ as your measured $I_1$ and the voltmeter reading as your measured $V_2$.

**10.** Repeat Procedure 9 for $R_2$, connecting the V-O-M at $A_2$ instead of $A_1$ and recording your results as $I_2$ and $V_1$. Turn off and unplug the power supply when you have finished your experimental work.

## DATA

| Current, mA | | | | | |
|---|---|---|---|---|---|
| Voltage, V | | | | | |

Resistance value from graph _____   Resistance value from ohmmeter _____

| | | Current, mA | | | Voltage, V | | | Resistance, Ω | | |
|---|---|---|---|---|---|---|---|---|---|---|
| | | $I$ | $I_1$ | $I_2$ | $V$ | $V_1$ | $V_2$ | $R$ | $R_1$ | $R_2$ |
| Series connection | Measured | | ✕ | | | ✕ | | | ✕ | |
| | Calculated | ✕ | | ✕ | ✕ | | ✕ | | | |
| Parallel connection | Measured | | | | | | | | | |
| | Calculated | | ✕ | | ✕ | | | | | |

## CALCULATIONS

**1.** Plot a curve using the values of current obtained in Procedures 2 and 3 as abscissas and the corresponding values of voltage as ordinates. Remember that the origin is a sixth point on this curve.

**2.** From the slope of your graph, find the resistance at which $R_1$ was set in Procedure 1. Remember that your currents are in milliamperes.

**3.** Compare the value of $R_1$ just obtained from your graph with that measured with the ohmmeter in Procedure 6. Decide which value you think is best and enter it as the measured value of $R_1$ for the series connection in the second table of your data sheet.

**4.** Compute the value of $R_2$ and also the value of the total resistance $R$ in the series combination by the application of Ohm's law, that is, from the measured values of the current and the voltage drops in the data of Procedures 4 and 5. Record these results as your calculated values of $R$ and $R_2$ for the series connection. Then calculate $R_1$ from your rule for series resistors (Equation 26.7). Enter this result in the appropriate box in your table and compare it with the measured value recorded in Calculation 3.

**5.** Find the values of $I_1$ and $V_1$ for your series circuit by referring to the discussion of such circuits in the Theory section. Record these values as your calculated $I_1$ and $V_1$ (series connection).

**6.** Calculate the values of the two resistances $R_1$ and $R_2$ in the parallel connection and also of the total resistance $R$ by the application of Ohm's law, that is, from the measured values of the currents and voltage drops found in Procedures 8–10. Although these results are calculated from your voltage and current measurements, record them as your measured values for the parallel connection.

**7.** As you did not change the rheostat settings, the values of $R_1$ and $R_2$ should be the same for the parallel connection as they were for the series. Copy the values found in Calculation 4 as your calculated values for these resistances in parallel, and use Equation 26.9 to obtain a calculated value of $R$. Compare your calculated and measured values of the three resistances for your parallel case.

**8.** Using your knowledge of parallel circuits, determine calculated values of $I$ and $V$ from the values measured for $I_1$, $I_2$, $V_1$, and $V_2$ in your parallel setup. Compare these calculated values with the measured values.

## QUESTIONS _____

**1.** What is an ohm, that is, what combination of fundamental SI units does it represent?

**2.** Explain how your graph demonstrates Ohm's law.

**3.** Show how the comparisons between your measured and calculated values check the formulas for the current, voltage, and resistance in series and parallel circuits.

**4.** In Figs. 26.2 and 26.3, voltmeter $V$ is shown connected so as not to include ammeter $A$ in the network. What error in finding the network resistance from the readings of $V$ and $A$ results from this connection? What error would be made if $V$ were connected to the other side of $A$? How can these errors be minimized?

**5.** The resistance included between two points in a circuit is 10 $\Omega$. How much resistance must be placed in parallel to make the total resistance 4 $\Omega$?

**6.** A coil whose resistance is 60 $\Omega$ and one whose resistance is 40 $\Omega$ are connected in parallel to a 120-V line. (a) What is the resistance of the combination? (b) What is the current in each coil? (c) What would be the current if they were connected in series? (d) What then would be the voltage drop across each coil?

**7.** A lamp is connected to a 120-V line. When the switch is first turned on and the filament is cold, the current through the lamp is 8 A, but this drops quickly to 0.5 A when the filament is hot. By what factor is the lamp's resistance increased due to the change in temperature?

**8.** Prove that the resistance of $n$ identical resistors in series is $n$ times the resistance of one resistor and that the resistance of $n$ identical resistors in parallel is $1/n$ times the resistance of one resistor.

**9.** Express the formula for series resistors in terms of conductances instead of resistances. Does it look familiar? Find out what the unit of conductance is commonly called and write it below.

**10.** You are given three identical resistors, each of 100 $\Omega$ resistance. Show how by using them singly or in series, parallel, or series-parallel combinations you can obtain resistances of $33\frac{1}{3}$, 50, $66\frac{2}{3}$, 100, 150, 200, or 300 $\Omega$.

**11.** Show that in a series combination of resistors the total resistance can never be less than that of the largest resistor whereas in a parallel combination the total resistance can never be more than that of the smallest.

**12.** Referring to the circuit shown below, complete the associated table using Ohm's law and your knowledge of series and parallel circuits. Note that the numbers 1, 2, 3, and 4 in the table refer to the resistance $R$ of, the current $I$ through, and the voltage $V$ across, the four resistors $R_1$, $R_2$, $R_3$, and $R_4$ respectively.

| | R, ohms | I, amperes | V, Volts |
|---|---|---|---|
| 1 | | 0.500 | |
| 2 | | | 4 |
| 3 | 18 | | |
| 4 | | | 10 |

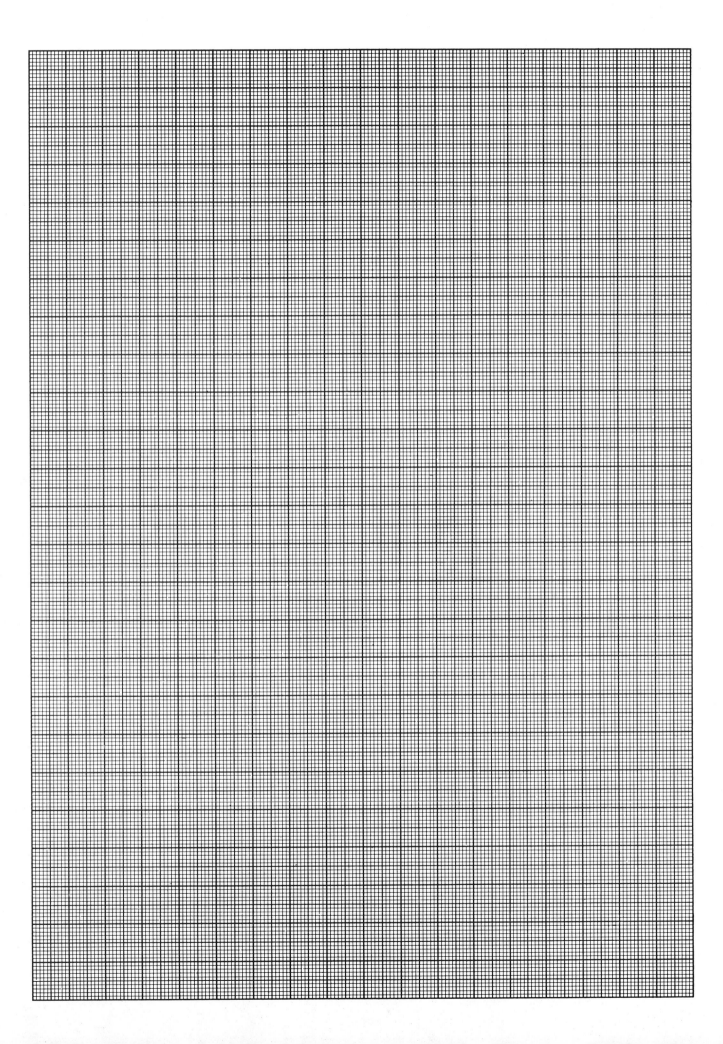

# The Wheatstone Bridge and Specific Resistance

# 27

The most accurate method of measuring resistances of widely different values is by means of the Wheatstone bridge. The purpose of this experiment is to learn to use the slide-wire form of the Wheatstone bridge and to measure the resistance of several metallic conductors with it. Experiment shows that if the temperature of a substance is kept constant, then its resistance depends only on its dimensions. Thus the resistance of a wire varies directly with its length and inversely with its cross-sectional area. For a piece of wire or bar of unit length and unit cross-sectional area, the resistance is called the specific resistance or the resistivity of the material of which the wire is made. Therefore, a further purpose of this experiment is to determine the specific resistance of the metals making up the sample wires used.

## THEORY

A Wheatstone bridge is a circuit consisting of four resistors arranged as shown in Fig 27.1. It is used for finding the value of an unknown resistance by comparing it with a known one. Three known resistances are connected with the unknown resistance, a galvanometer, a dry cell, and a switch, as shown in Fig. 27.1.

For a condition of balance, no current flows through the galvanometer. Hence the current through $R_1$ is the same as the current through $R_2$, and the current through $R_3$ is the same as that through $R_4$. Moreover, because there is no current through $G$, there must be no voltage across it; hence the potential drop across $R_1$ must be equal to that across $R_3$. This requires that

$$i_1 R_1 = i_2 R_3$$

Similarly, the potential drop across $R_2$ must equal that across $R_4$, so that

$$i_1 R_2 = i_2 R_4$$

Dividing the first equation by the second yields

$$\frac{R_1}{R_2} = \frac{R_3}{R_4} \qquad (27.1)$$

Therefore, if three of the resistances are known, the fourth may be calculated from Equation 27.1.

In the slide-wire form of the bridge, as shown in Fig. 27.2, the resistances $R_3$ and $R_4$ are replaced by the uniform wire $AB$ with a sliding contact key at $C$. Since the wire is uniform, the resistances of the two portions are proportional to their lengths; hence the ratio $R_3/R_4$ is equal to the ratio $AC/CB$. If $R_1$ is represented by a known resistance $R$, and $R_2$ by the unknown resistance $X$, Equation 27.1 becomes

$$\frac{R}{X} = \frac{AC}{CB} \qquad (27.2)$$

From this the value of $X$ may be calculated.

It is found that the resistance of a conductor depends on the material of which it is made, its length, its cross-sectional area, and the temperature. For a further discussion of this, refer to the Theory section of Experiment 26 on Ohm's law and in particular to

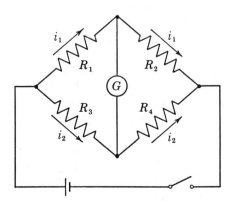

**Figure 27.1** *The Wheatstone bridge circuit balanced*

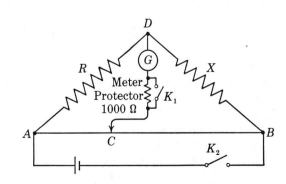

**Figure 27.2** *The slide-wire Wheatstone bridge*

Equations 26.4, 26.5, and 26.6. These show that the resistance $R$ of a conductor is given by

$$R = \left(\frac{K}{ne^2}\right)\left(\frac{L}{A}\right) = \rho\frac{L}{A} \qquad (27.3)$$

where $L$ is the length and $A$ the cross-sectional area of the conducting bar, rod, or wire involved. The proportionality constant $\rho = K/ne^2$ is called the *resistivity* or *specific resistance* of the conducting material and depends only on the temperature and the nature of the material itself. The units of resistivity are clearly ohm-meters, and this quantity may be thought of as the resistance between opposite faces of a 1-m cube of the material in question. Resistivity may be found by determining the resistance of a measured length of a wire made of the given material. The cross-sectional area must also be determined from a measurement of the wire's diameter, after which $\rho$ may be calculated from Equation 27.3. Because $K$ and $n$ in this equation, and hence $\rho$, are somewhat temperature-dependent, the resistivities that we will obtain in this experiment for several common conducting materials will be those corresponding to the laboratory temperature.

## APPARATUS

1. Slide-wire form of Wheatstone bridge
2. Dry-cell (1.5 V, No. 6 size)
3. Galvanometer
4. Standard decade resistance box (1000 Ω)
5. Single-pole, single-throw switch
6. Key
7. 1000-Ω, 1-W carbon resistor
8. Board with eleven lengths of wire of unknown resistance. The board is about 1 m long and the wires are mounted between binding posts. The first nine lengths are a single piece of copper wire. The tenth is a piece of manganin wire and the eleventh a piece of nichrome.
9. Micrometer caliper
10. Meter stick with caliper jaws

## PROCEDURE

**1.** Hook up the slide-wire form of the bridge as shown in Fig. 27.2. Let $X$ be the resistance of the nine lengths of copper wire in series, and let $R$ be the standard resistance box. Use the key at $K_1$ and the single-pole, single-throw switch at $K_2$. Before closing any switches, have the circuit approved by the instructor.

**2.** Measure the resistance of the copper wire with your bridge. To do this, first set the decade box $R$ to 5 Ω and move sliding contact $C$ to the middle of the slide wire. Note that contact $C$ is a knife edge mounted on a slider and equipped with a knob so that it can be pushed down to contact the slide wire at a precise point. Never move the slider while the knife edge is pressed down on the wire.

Now turn your apparatus on by closing switch $K_2$. Notice that key $K_1$ is normally open and should remain so for the moment so that the 1000-Ω protective resistor is in series with the galvanometer. Press contact key $C$ gently down on the slide wire and note whether the galvanometer shows a deflection. If it does, release contact $C$, change the value of $R$, and repeat. Continue this procedure until the least possible galvanometer deflection is observed. Then move the sliding contact to a position at which no deflection occurs when key $C$ is tapped to bring the knife edge into contact with the slide-wire.

**3.** For the most sensitive adjustment of the bridge, the 1000-Ω protective resistor is short-circuited with key $K_1$. When the galvanometer shows no deflection on tapping the contact key $C$, close $K_1$ and hold it down. Adjust the sliding contact very carefully, moving it no more than a fraction of a millimeter at a time, until the galvanometer again shows no deflection when key $C$ is tapped. Record the setting of the standard resistance $R$ and the distances $AC$ and $CB$ from the knife-edge contact to the ends of the slide wire. Open switch $K_2$ when you have finished your measurements.

**4.** Repeat Procedures 2 and 3 with the tenth length of wire (the manganin wire) connected at $X$ in place of the copper wire.

**5.** Repeat Procedures 2 and 3 with the eleventh length of wire (the nichrome wire) substituted at $X$.

**6.** Measure the length of each wire by means of the meter stick and caliper jaws, adjusting the jaws so that they measure the distance between the binding posts where the wire is mounted. Remember that the total length of copper wire is the sum of the nine measured lengths.

**7.** Measure the diameter of each type of wire at four different points with the micrometer caliper. For the copper wire, the four points should be spread out along the total length. Record all four readings of the diameter of each wire, estimating to one tenth of the smallest scale division. The readings should be to 0.0001 cm.

**8.** In a similar manner, take and record four independent readings of the zero setting of the micrometer to 0.0001 cm. Be sure to note the algebraic sign of the zero correction.

Name _____ Section _____ Date _____ EXPERIMENT 27 **219**

## DATA _____

| Wire | R, Ω | AC, cm | CB, cm | X, Ω | Resistivity, Ω·m | Resistivity from Table VIII |
|------|------|--------|--------|------|------------------|------------------------------|
| Copper | | | | | | |
| Manganin | | | | | | |
| Nichrome | | | | | | |

| Wire | Micrometer readings | | | | | Diameter m | Area, m² | Length, m |
|------|---|---|---|---|---------|------------|----------|-----------|
| | 1 | 2 | 3 | 4 | Average | | | |
| Copper | | | | | | | | |
| Manganin | | | | | | | | |
| Nichrome | | | | | | | | |
| Zero reading | | | | | | | | |

## CALCULATIONS _____

**1.** Compute the average of the readings for the diameter of each wire, as well as for the zero reading of the micrometer. Calculate and record the diameter of each wire, properly corrected for the zero reading and converted to meters.

**2.** Compute the cross section of each wire in square meters.

3. Calculate the values of the resistances measured using Equation 27.2.

4. Compute the specific resistance for each metal used.

5. Look up the values of the specific resistances in Table VIII at the end of the book and record them. Why might they differ somewhat from the ones you have just measured?

## QUESTIONS

1. Why is the bridge method not adapted to measuring the hot resistance of an electric lamp?

2. Name three sources of error in using a slide-wire bridge.

**3.** Why is it more accurate to set the sliding contact near the center of the bridge instead of near one end when balance is obtained?

**4.** How is the operation of the Wheatstone bridge affected by changes in the cell voltage? Would there be an advantage in using a higher voltage? What limits the voltage that can be used?

**5.** Assuming that there is an error of 1 mm in locating the position of the contact on the slide wire, calculate the percentage error in determining the value of $X$ when the contact was (a) at 50 cm, (b) at 5 cm. Suppose that $R = 10.0 \ \Omega$ in both cases, and $AB = 1.0$ m.

**6.** Draw a diagram of the bridge as actually used in one of your measurements.

**7.** How does the resistance of a wire vary with the resistivity?

**8.** What would be the effect of a change in temperature on the resistivity?

**9.** Calculate the resistance of 40 mi of No. 22 B&S gauge copper wire. The diameter of the wire is 0.0644 cm. The value of the specific resistance is given in Table VIII.

**10.** What is the resistance of a copper bus bar 10 m long and 0.3 × 0.3 cm in cross section?

**11.** Assume that there was an error of 0.001 cm in measuring the diameter of the copper wire used in the experiment. Calculate the error that this would introduce into your value of the specific resistance.

**12.** The temperature coefficient of resistance for copper is 0.00393 per degree Celsius. Assuming room temperature to be 22°C, compute the specific resistance of copper at 0°C from your measurements.

# Meter Sensitivity: Voltmeter and Ammeter 28

It is extremely important to understand the operation of electrical measuring instruments because they are constantly used in making electrical measurements. A galvanometer is an instrument for measuring very small electric currents. It forms the basis of all non-digital electrical measuring instruments; hence an understanding of its working principle and the type of measurements that can be made with it is essential for quantitative electrical work. In the present experiment the resistance and sensitivity of a galvanometer will be measured, and the instrument will be converted into a voltmeter that gives full-scale deflection for 3 V. In addition, it will be converted into an ammeter giving full-scale deflection for 500 mA (0.5 A).

## THEORY

A galvanometer is an instrument for measuring small currents. It consists essentially of a coil of fine wire mounted so that it can rotate about a diameter in the field of a permanent magnet. When a current flows through the coil, magnetic forces are produced that exert a torque on it. It therefore rotates until equilibrium is established between the torque due to the current and that exerted by a restoring spring. The angle of rotation and hence the deflection of the attached indicating needle is directly proportional to the current flowing through the coil, provided the magnetic field is arranged so that the torque it produces depends only on coil current and does not vary as the coil rotates through its full-scale range. The magnet's poles are shaped so as to bring this about.

The sensitivity of a galvanometer is usually expressed as the current that must be passed through the instrument to produce a deflection of one scale division. The so-called full-scale current sensitivity, which is the current necessary for full-scale deflection, is also often used. If the full-scale sensitivity is known, the sensitivity in amperes per division can be found by dividing the full-scale sensitivity by the number of divisions. This result is called the galvanometer constant $K$. Clearly,

$$i = Kd$$

where $i$ is the current passing through the galvanometer in amperes and $d$ is the galvanometer deflection in scale divisions. Notice that the more sensitive a galvanometer is, the less current is required to cause a given deflection and the smaller the galvanometer constant $K$.

Another important characteristic of a galvanometer is the resistance $r$ of its coil. Both $K$ and $r$ will be measured in the present experiment.

A galvanometer may be converted into an ammeter by placing a suitable shunt in parallel with the coil to permit the flow of a larger current through the complete ammeter than the galvanometer coil will carry alone. This also makes the meter read in amperes with a predetermined full-scale value. Let $S$ be the resistance of such a shunt (see Fig 28.2) and $r$ the resistance of the galvanometer. Now let $I_g$ be the current through the galvanometer and $I$ the total current through the complete ammeter. Then, since $r$ and $S$ are in parallel, the current through the galvanometer is given by

$$I_g = I \frac{S}{r + S} \qquad (28.1)$$

If the ammeter is to display a deflection of $n$ divisions when $I$ amperes pass through it, then $I_g$ must equal $Kn$, and we get

$$I \frac{S}{r + S} = Kn \qquad (28.2)$$

Then if $I$ is the full-scale current reading desired and $n$ the total number of divisions on the meter scale, Equation 28.2 can be used to calculate the necessary shunt resistance $S$. In most cases, where a sensitive galvanometer (small $K$) is being used to measure a large current $I$, $S$ will be much smaller than $r$ and can thus be neglected in the denominator of this equation. $S$ is then given to a very good approximation by

$$S \approx \frac{rKn}{I} \qquad (28.3)$$

A galvanometer may also be converted into a voltmeter by placing a large resistor (called a multiplier) in series with the coil. If $R$ is the multiplier resistance, then

$$I_g = \frac{V}{R + r} \qquad (28.4)$$

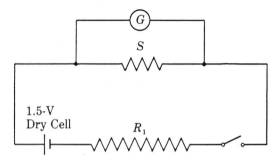

**Figures 28.1 and 28.2** *Circuits for measuring the resistance of a galvanometer*

where $V$ is the voltage impressed across the complete voltmeter (galvanometer and multiplier) terminals. Note that the voltmeter draws current $I_g$ from the source whose voltage is being measured and that because the galvanometer coil and the multiplier resistor are in series, $I_g$ flows through both. Now if $V$ is the desired full-scale voltage reading, the corresponding $I_g$ must equal $Kn$, and Equation 28.4 becomes

$$Kn = \frac{V}{R + r} \quad (28.5)$$

from which the necessary multiplier resistance can be calculated. In most cases $R$ will be much greater than $r$, and so Equation 28.5 may be approximated by

$$R \approx \frac{V}{Kn} \quad (28.6)$$

This simple formula is just Ohm's law, saying that the resistance $R$ is given by the voltage $V$ across it divided by the corresponding current $Kn$ through it, the small additional series resistance $r$ having been neglected.

Much of the above theory requires a knowledge of the galvanometer constant $K$ and coil resistance $r$, and in this experiment a special method will be used to measure these parameters. The problem is to measure them without passing an excessive current through the galvanometer coil. This is done by connecting a known voltage $V$ in series with a large re-

sistance $R_1$ and the coil as shown in Fig. 28.1. The value of $R_1$ is adjusted so that an exact full-scale deflection is obtained. Then by Ohm's law

$$I_g = Kn = \frac{V}{R_1 + r} \approx \frac{V}{R_1} \quad (28.7)$$

so that the galvanometer constant is given by

$$K \approx \frac{V}{nR_1} \quad (28.8)$$

A shunt resistance $S$ is now connected in parallel with $r$ (see Fig. 28.2) and set so that the new deflection is exactly one-half scale. Under these conditions

$$I_g = \frac{1}{2}Kn = \frac{SV}{R_1(r + S)} \quad (28.9)$$

The insertion of the shunt does not alter the fact that $Kn \approx V/R_1$, and substitution of this relation from Equation 28.7 in Equation 28.9 leads to the conclusion that $S = r$. This result is not surprising if we note that the combination of $R_1$ and the voltage source constitutes what is called a "current source," a source of a fixed current $V/R_1$. The current is fixed only to the approximation that resistances such as $r$ or the parallel combination of $r$ and $S$ inserted in series with $R_1$ are so much smaller than $R_1$ that they make a negligible change in the total circuit resistance. This approximation is usually very good, however.

## APPARATUS

1. Portable galvanometer
2. dc voltmeter (0–3 V)
3. Volt-ohm-milliammeter with 0–500-mA scale
4. Dry cell (1.5-V, No. 6 size)
5. Decade resistance box (10,000 Ω)
6. Decade resistance box (1000 Ω)
7. Single-pole, single-throw switch
8. 15-V regulated dc power supply
9. 200-Ω, 50-W rheostat
10. Roll of No. 28 copper wire

## PROCEDURE AND CALCULATIONS

**1.** Measure the voltage of the dry cell with the voltmeter and record it.

**2.** Connect the dry cell in series with the 10,000-Ω resistance box, the galvanometer, and the switch as shown in Fig 28.1. The resistance box, which serves as $R_1$, should be set at 3000 Ω.

**3.** Before closing the switch, have the circuit approved by the instructor. Now close the switch and note the galvanometer deflection. Carefully adjust $R_1$ until the galvanometer reads exactly full scale. Do not allow the galvanometer deflection to exceed the full-scale limit. Record the value obtained for $R_1$.

**4.** Use Equation 28.8 to find the galvanometer constant $K$.

**5.** Open the switch. Connect the 1000-$\Omega$ resistance box in parallel with the galvanometer as shown at $S$ in Fig 28.2, making it equal to 300 $\Omega$ and leaving $R_1$ connected and set as before. Close the switch and adjust $S$ until the galvanometer reads exactly half scale. Record this value of $S$ as your galvanometer resistance $r$.

**6.** Calculate the series resistance $R$ (the multiplier) that will convert the galvanometer into a voltmeter giving full-scale deflection for 3 V.

**7.** Connect the galvanometer in series with the 10,000-$\Omega$ decade resistance box and set the box to your calculated value of $R$. This combination is now a voltmeter. Use it to measure the voltage of the dry cell, making sure that your series multiplier resistance $R$ is included in the circuit. Record this voltage.

**8.** Calculate the shunt resistance $S$ that will convert the galvanometer into an ammeter having a full-scale reading of 0.5 A (500 mA). If you have not already learned how to use the volt-ohm-milliammeter (V-O-M) as an ohmmeter in Experiment 26, have your instructor explain its use to you now. Tell whether you think it could be used to measure the shunt resistance just calculated.

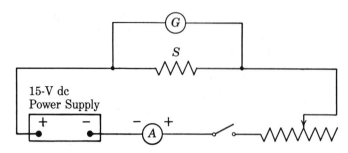

**Figure 28.3**   *Circuit for using a galvanometer as an ammeter*

**9.** Look up the resistance of No. 28 copper wire in Table XIII at the end of the book. Cut slightly over 15 m of this wire from the roll supplied and clean the insulation from the ends. With the ohmmeter, measure the resistance of exactly 15 m by connecting the V-0-M probes to the bare wire ends at points exactly 15 m apart. Compare your result with the value you would expect for 15 m of No. 28 wire from the data of Table XIII.

**10.** When you have established a best value for the resistance of the 15-m length of your No. 28 wire, compute the length needed to give the shunt resistance *S* found in Procedure 8. Cut a length 2 cm greater than this from the supply roll and clean the insulation from 1 cm at each end for connection to the galvanometer terminals. Connect your shunt in parallel with the galvanometer by squeezing the bared 1-cm lengths under the terminal screws. Make sure the full 1 cm is used at each terminal so that the effective length of the shunt is that just computed. Also check that the connections are tight so that no extraneous resistance is introduced. This combination is now an ammeter.

**11.** Connect your ammeter combination in series with the V-0-M set to its 500-mA range, the rheostat set with all of its resistance in the circuit, the switch, and the power supply, as shown in Fig. 28.3. When making connections to your meter, be careful not to disturb the shunt connections made in Procedure 10. Before closing the switch, have your circuit approved by the instructor.

**12.** Turn on the power supply, set it for an output of about 15 V, and close the switch. Adjust the rheostat until the V-0-M reads 0.1 A, and record the reading of your ammeter. Now adjust the rheostat to make the V-0-M read 0.2, 0.3, 0.4, and 0.5 A in turn, recording your ammeter's reading each time. Should your ammeter read off scale (due to your having made *S* a little too large) when the V-0-M indicates 0.5 A, try to estimate the reading, or, if necessary, note that the galvanometer needle is against the stop at the high end.

# DATA _____

### Galvanometer Characteristics

| | | |
|---|---|---|
| Dry cell voltage *V* | _____ | Number of scale divisions *n* _____ |
| Value of $R_1$ for full-scale deflection | _____ | Galvanometer constant *K* _____ |
| | | Galvanometer resistance *r* _____ |

### Voltmeter

Resistance *R* for full-scale
   deflection of 3 V          _____

Reading of this voltmeter for the
   voltage of the dry cell          _____

### Ammeter

Shunt resistance *S* for full-scale
   deflection of 0.5 A          _____

Length of No. 28 copper wire required          _____

| Reading of V-O-M | 0.1 A | 0.2 A | 0.3 A | 0.4 A | 0.5 A |
|---|---|---|---|---|---|
| Reading of galvanometer used as ammeter | | | | | |

## QUESTIONS _____

**1.** Derive Equation 28.1.

**2.** Does your data justify neglecting $r$ in comparison with $R_1$ in Equation 28.7?

**3.** (a) Derive Equation 28.9. (b) Show that when $Kn = V/R_1$ is substituted in this equation, the conclusion $S = r$ follows.

**4.** The sensitivity of a voltmeter is often expressed in "ohms per volt," which means the resistance of the complete meter (galvanometer and multiplier in series) divided by the full-scale voltage reading. Derive a relation between sensitivity in ohms per volt and the galvanometer constant $K$, and find the ohms-per-volt sensitivity of the meter used in this experiment.

**5.** Calculate the series resistance $R$ that will convert your galvanometer into a voltmeter reading 150 V for full-scale deflection.

**6.** Calculate the shunt resistance $S$ that will convert your galvanometer into an ammeter reading 0.3 A for full-scale deflection.

**7.** How should an ammeter be connected in a circuit? A voltmeter?

**8.** Why could you not have measured the resistance $r$ of your galvanometer with the V-0-M set to the appropriate ohms range?

**9.** A certain 3-V voltmeter requires a current of 10 mA to produce a full-scale deflection. How may it be converted into a voltmeter with a range of 150 V?

**10.** Find the sensitivity of the voltmeter of Question 9 in ohms per volt.

**11.** A 1.5-V dry cell furnishes current to two resistances of 2 $\Omega$ and 16,000 $\Omega$ in series. A galvanometer whose resistance is 200 $\Omega$ is connected across the 2-$\Omega$ resistance. The deflection produced is 10 scale divisions. Calculate the current through the galvanometer and the galvanometer constant.

**12.** Design and draw the schematic wiring diagram of a switching circuit that includes the galvanometer, the multiplier resistor $R$ determined in Procedure 6, the shunt resistor $S$ calculated in Procedure 8, and an appropriate switch that allows you to switch your meter from being a 3-V voltmeter to being a 500-mA ammeter. In other words, give the design for a circuit that will make your galvanometer into a simple multimeter having scales of 0–3 V and 0–500 mA.

# Joule's Law 29

Whenever an electric current flows through a conductor, a certain amount of electrical energy is transformed into heat energy by the conductor's resistance, just as mechanical energy may be transformed into heat energy by friction. The object of this experiment is to measure the mechanical equivalent of heat by the electrical method. The electrical energy used may be calculated by measuring the current, the voltage, and the time during which this current flows. The number of calories liberated in the process is measured by means of a calorimeter. The importance of this experiment lies in the fact that heat may be supplied to a body either mechanically (see Experiment 21) or electrically, and the same relation between joules and calories is found in either case.

## THEORY

Energy exists in many forms and is measured in many different units. The purpose of this experiment is to find the ratio between the mechanical unit and the heat unit of energy by using an electric current as a source of heat and equating the energy supplied by the current, expressed in joules, to the same energy expressed in calories.

Whenever a current flows through a conductor, a certain amount of heat is developed. The work done by the electric current is given by

$$W = VIt \qquad (29.1)$$

where $W$ is the work expressed in joules, $V$ is the potential difference between the terminals of the conductor in volts, $I$ is the current in amperes, and $t$ is the time during which the current flows in seconds. But, by Ohm's law

$$V = IR \qquad (29.2)$$

where $R$ is the resistance of the conductor. Equation 29.1 may therefore be written as

$$W = I^2Rt \qquad (29.3)$$

This relation is known as Joule's law of heating. Expressed in words, the law states that when an electric current flows through a pure resistance, the quantity of heat produced in the conductor is proportional to the time of current flow, to the resistance, and to the square of the current. Note that $W$ is the electrical energy dissipated in the conductor expressed in joules. This energy appears as heat, which can be measured in calories by calorimetric methods.

From the principle of conservation of energy, whenever work is done in developing heat, the quantity of heat produced is always proportional to the work done. Thus,

$$W = JH \qquad (29.4)$$

where $W$ is the work expressed in joules; $H$ is the heat energy produced, in calories; and $J$ is the mechanical equivalent of heat, which is the number of joules required to produce one calorie as discussed in Experiment 21.

If the heat energy is used to raise the temperature of the water contained in a calorimeter, then

$$H = (m_w c_w + m_1 c_1 + C_2)(T_2 - T_1) \qquad (29.5)$$

where $H$ is the heat delivered to the water, $m_w$ is the mass and $c_w$ the specific heat of the water, $m_1$ is the mass and $c_1$ the specific heat of the calorimeter, and $C_2$ is the heat capacity of the immersion heater. The temperatures $T_1$ and $T_2$ are the initial and final temperatures, respectively. Note that in this experiment it will be convenient to measure all masses in grams, not only because your balance will be calibrated in these units but also because you can then set $c_w$ equal to 1. The temperatures must of course be in degrees Celsius, the specific heats will be in calories per gram-degree Celsius, and the heat capacity $C_2$ in calories per degree Celsius.

Since the immersion heater is immersed in the water in the calorimeter, all the heat produced in the heater resistance $R$ must pass into the water. Thus Equation 29.4 holds with $W$ given by Equation 29.1 (or Equation 29.3, but we will be measuring $I$ and $V$ in this experiment) and $H$ given by Equation 29.5. Hence,

$$J = \frac{W}{H} = \frac{VIt}{(m_w c_w + m_1 c_1 + C_2)(T_2 - T_1)} \qquad (29.6)$$

As all quantities in the expression on the right in this equation can be measured, $J$ can be found just as it was in Experiment 21. In fact, a comparison of Equations 29.6 and 21.3 shows that the basic difference between this experiment and Experiment 21 is that in

**233**

the latter the heat was provided by the dissipation of mechanical energy by friction, whereas in the present case it is provided by the dissipation of electrical energy in a resistance.

## APPARATUS

1. Immersion heater for 12-V operation
2. Calorimeter
3. 0–50°C thermometer
4. Stopwatch or stop clock
5. dc voltmeter (0–15 volts)
6. dc ammeter (0–5 A)
7. Triple-beam balance
8. Power supply capable of delivering at least 5 A at 10 V dc

## PROCEDURE

**1.** Weigh the empty calorimeter and record its mass $m_1$ in grams. Then fill it about two-thirds full of cold water (as cold as it will run from the tap) and determine the mass of the calorimeter plus the water $(m_1 + m_w)$. Replace the calorimeter in its containing jacket, put on the calorimeter cover, and immerse the heater. Make sure that the heater is fully immersed but take care that neither it nor its connecting leads touch the sides or bottom of the calorimeter cup (see Fig. 29.1).

**2.** Wire the circuit as shown in Fig 29.1. Check that the output control on the power supply is turned all the way down and that its power switch is off. Have your setup approved by your instructor before plugging the power supply in.

**3.** Insert the thermometer into the water through the hole in the calorimeter cover and stir the water with it, being careful not to knock the heater around. Then turn on the power supply, set the heater voltage to exactly 10 V, quickly note and record the ammeter reading, and return the power switch to off.

**4.** Again stir the water gently with the thermometer and read the temperature to the nearest tenth of a degree. Record this reading as your initial temperature $T_1$.

**5.** Turn on the power supply and simultaneously start the stopwatch. Keep the voltage constant at 10 V with the output control and stir the water gently from time to time with the thermometer. When the stopwatch reads 1 min, note and record the ammeter reading.

**6.** Note and record the ammeter reading at 2 min, 3 min, and 4 min, keeping the voltage constant at 10 V with the power supply control and stirring the water occasionally with the thermometer.

**7.** When the stopwatch reads *exactly* 5 min, shut off the power. Continue stirring the water until the maximum temperature is reached. Record this temperature to the nearest tenth of a degree as your final temperature $T_2$.

**Figure 29.1** *Arrangement for electrical measurement of the mechanical equivalent of heat.*

## DATA

Mass of the empty calorimeter _____

Mass of the calorimeter with water _____

Mass of water _____

Heat capacity of the immersion heater _____

Initial water temperature $T_1$ _____

| Time | Initial | 1 min | 2 min | 3 min | 4 min | | Average |
|------|---------|-------|-------|-------|-------|---|---------|
| Current | | | | | | | |

Final water temperature $T_2$ _____    Temperature rise $T_2 - T_1$ _____

Mechanical equivalent of heat from
    Equation 29.6 _____

Mechanical equivalent of heat from
    Table I at end of book _____

Percent error _____

## CALCULATIONS _____

**1.** Subtract the mass of the empty calorimeter from the mass of the calorimeter with water to obtain the mass $m_w$ of the water alone.

**2.** Compute your value of the mechanical equivalent of heat from Equation 29.6. The value of $C_2$, the heat capacity of the immersion heater, will be given to you by the instructor. Remember that $V = 10$ V and the time $t$ is 5 min but must be converted to seconds. Your instructor will tell you the material of which the calorimeter is made so that you can look up its specific heat in Table VII at the end of the book.

**3.** Compare your value of the mechanical equivalent of heat with the accepted one by calculating the percent error.

## QUESTIONS _____

**1.** Discuss the principal sources of error in this experiment and justify the difference between your value of the mechanical equivalent of heat and the correct value.

**2.** Using your data, calculate how long it would take to heat the water from 20°C to 100°C by setting the power supply to send a current of 2 A through the same immersion heater.

**3.** In Question 2, how long would it take if a current of 4 A were used? Remember that the resistance $R$ of the immersion heater is fixed at a value you should be able to calculate from your data.

**4.** State the factors on which the heating effect of an electric current depends and how it varies with each.

**5.** Give three examples of desirable heating effects of an electric current and three examples of undesirable ones.

**6.** How long after it is turned on will a 100-W electric heater take to bring a quart of water to a boil from room temperature (20°C)? Assume normal sea-level atmospheric pressure. Watch your units.

**7.** What must be the resistance of a coil of wire in order that a current of 2 A flowing through the coil may produce 1000 cal of heat per minute?

**8.** What is the purpose of starting with the temperature of the water lower than room temperature and ending, as you probably did, at about the same amount above room temperature?

One of the most important instruments used in the electrical laboratory is the potentiometer, an instrument for measuring voltages, or potential differences. The potentiometer possesses a great advantage over the voltmeter because under operating conditions it draws no current from the source whose voltage is being measured and thus does not disturb the value of that voltage. The potentiometer is capable of very high precision and is easy to handle. One of its most

important uses is the calibration of electrical measuring instruments such as voltmeters and ammeters. The purpose of this experiment is to study the slide-wire form of the potentiometer and to measure several voltages with it. Note that modern electronic voltmeters also draw little or no current from the source being measured, but many of them have the equivalent of a potentiometer circuit built in.

## THEORY

The so-called open-circuit voltage of a dry cell, generator, or other voltage source is the potential difference that the source exhibits between its terminals under "no-load" conditions, that is, when no current is being drawn from it. Practical voltage sources are never "ideal" in the sense that they cannot maintain their output voltage constant at the open-circuit value regardless of the current drawn from them. Their terminal voltage always falls somewhat below this value whenever a load (some circuit drawing current) is connected. Fig. 30.1 illustrates what happens. A less-than-ideal voltage source is represented by the series combination of an ideal voltage source giving the open-circuit voltage $V_{oc}$ and a resistor $R_t$ called the *internal resistance* of the source. Ohm's law tells us that the current $i_0$ around the circuit will be $V_{oc}/(R_t + R_L)$ and that therefore the actual voltage

delivered by the source to the load $R_L$ will be

$$V_o = i_o R_L = V_{oc} \frac{R_L}{R_t + R_L} = V_{oc} \frac{1}{1 + R_t/R_L} \quad (30.1)$$

Clearly, $V_o$ is less than $V_{oc}$ by a fraction that approaches unity as the internal resistance $R_t$ of the source is made small compared to the load resistance $R_L$.

Although in many cases the condition $R_t \ll R_L$ is realized, the open-circuit voltage of a voltage source cannot be measured accurately with an ordinary voltmeter because the voltmeter itself draws some current. This problem may be minimized by using a voltmeter of high sensitivity or even an electronic voltmeter (EVM). In the latter, the actual meter movement is driven by a separate power source, and only the very high input resistance (of the order of 10 million $\Omega$ in ordinary EVMs) of the associated electronic amplifier is connected across the voltage to be measured.* All such devices require calibration, however. The potentiometer measures the voltage of a given source by a null method, so that no current is drawn from the source when a balance is reached and the reading is being taken. Standard cells are available that produce an accurately known open-circuit voltage. The potentiometer allows precise measurements of other voltages to be made by comparing them with the known standard voltage.

Basically, a potentiometer is a voltage divider with a circuit such as the ones shown schematically in Fig. 30.2. Battery voltage $V_{in}$ causes current $i = V_{in}/(R_a + R_b)$ to flow through the two series resistors in Fig. 30.2(a) so that the output voltage $V_o$ is

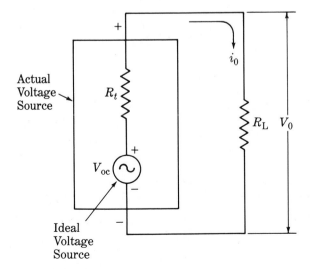

**Figure 30.1** *Schematic diagram of a practical voltage source and load*

---

*See also the discussion of EVMs in the Theory section of Experiment 33.

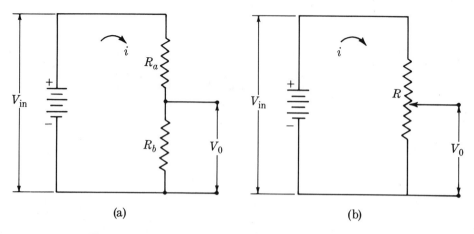

**Figure 30.2** *Voltage dividers (a) fixed and (b) variable*

given by

$$V_o = iR_b = V_{in}\frac{R_b}{R_a + R_b} \qquad (30.2)$$

and is thus the fraction $R_b/(R_a + R_b)$ of $V_{in}$. We can say that the circuit has divided the voltage $V_{in}$ by the factor $(R_a + R_b)/R_b$. In Fig. 30.2(b) this factor has been made continuously variable by making $R_a$ and $R_b$ into a single resistor $R$ with a sliding contact. Thus $R$ is always equal to $R_a + R_b$, and the fraction by which $V_{in}$ is multiplied to give $V_o$ is just $R_b/R$ and is continuously variable from 0 to 1. Thus, any voltage from 0 to $V_{in}$ may be selected.

In the slide-wire form of the potentiometer, resistor $R$ is a uniform bare wire provided with a sliding contact and through which a steady current is sent from an appropriate power supply. The voltage source to be measured is connected through a galvanometer and a protective resistance between one end of the wire and this contact in such a way that its polarity opposes that of the wire's potential difference. For balance, the contact is moved along the wire until no current flows through the galvanometer. The voltage $V_o$ across this portion of the wire is now equal to that of the source being measured. This pro-

cess is repeated with a standard cell. Because the wire is uniform, the length of the wire spanned is proportional to $R_b$ and hence to the potential drop $V_o$. Thus, the open-circuit voltage (sometimes called the *electromotive force* or emf) of any voltage source can be compared with that of a standard cell.

The circuit used in this experiment is shown in Fig. 30.3. $E_s$ is the standard cell, $E_x$ is a cell whose emf is to be measured, $K_2$ is a double-throw switch that can connect either cell in the circuit, $G$ is the galvanometer, $R_1$ is the protective resistance, and $R_2$ is a rheostat used to adjust the current in the slide wire. Because the electromotive force of the standard cell is equal to the potential drop in the length of the wire spanned for a condition of balance and the same is true for the unknown cell, the emf's are proportional to the lengths of wire spanned. Thus,

$$\frac{E_x}{E_s} = \frac{L_x}{L_s}$$

The unknown emf is then given by

$$E_x = E_s\frac{L_x}{L_s} \qquad (30.3)$$

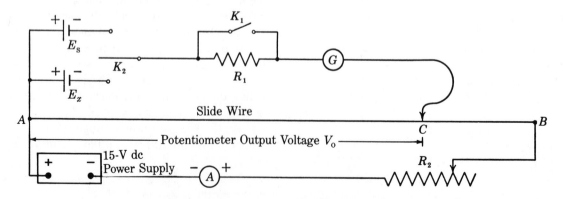

**Figure 30.3** *The potentiometer circuit*

**Figure 30.4** *Typical student potentiometer (Cenco)*

where $E_x$ is the unknown emf, $E_s$ is the emf of the standard cell, $L_x$ is the length of the wire $(AC)$ used for the unknown cell, and $L_s$ is the length of the wire used for the standard cell for a condition of balance.

The scale measuring length along slide wire $AB$ can be made to read voltage directly by proper adjustment of the rheostat $R_2$. Thus, if the total length $AB$ of the slide wire is 1 m and $R_2$ is set so that the emf across this length is 1 V, output voltage $V_o$ will be given by

$$V_o = V_{in} \frac{R_b}{R} = V_{in} \frac{kAC}{kAB} = 1 \frac{kAC}{k(1)} = AC \quad (30.4)$$

where $k$ is the proportionality constant between the length and resistance of the uniform slide wire, and we have inserted $V_{in} = 1$ V and $AB = 1$ m. An unknown voltage can then be read directly from the measured position of contact $C$ at balance. To estab-

lish exactly 1 V across $AB$, the standard cell is selected with switch $K_2$, contact $C$ is set at a scale reading equal to the known standard cell voltage, and balance is obtained by adjustment of $R_2$.

If the potential across $AB$ is set at 1 V, the potentiometer is limited to measuring voltages equal to or less than 1 V. Since in this experiment voltages up to 1.5 or 1.6 V are to be measured, the voltage across $AB$ will be set at 2 V and unknown voltages read from the scale by taking the scale reading at contact $C$ and doubling it.

Commercial potentiometers are available with slide wires calibrated directly in volts and with various auxiliary resistors designed to give the unit several voltage ranges included in a single instrument assembly. A simple example of such an instrument is shown in Fig. 30.4.

## APPARATUS

1. Slide-wire potentiometer
2. Galvanometer
3. 4700-$\Omega$, 1-W carbon resistor
4. 100-$\Omega$, 50-W rheostat
5. 15-V regulated dc power supply
6. Standard cell

7. Good dry cell, or dry cell that tests 1.5 V
8. Old dry cell, or a dry cell that tests about 1.0 V
9. dc ammeter (0–1 A)
10. dc voltmeter (0–3 V)
11. Single-pole, single-throw switch
12. Single-pole, double-throw switch

## PROCEDURE

**1.** Connect the apparatus as shown in Fig. 30.3 using the good dry cell for $E_x$. $R_1$ is the 4700-$\Omega$ resistor and $R_2$ the rheostat. Keep all switches open while

connections are being made. Have your circuit approved by the instructor.

**2.** *Leave switch $K_1$ open*. Set $R_2$ for maximum resistance and then turn on the power supply. Set its output to about 15 V and adjust the current as read on the ammeter to about 0.6 A, keeping it constant at this value for the next four steps by means of the control rheostat. Read the exact value chosen.

**3.** Close switch $K_2$ in the direction to put the standard cell $E_s$ in the circuit and find the setting of the contact point $C$ for a condition of balance. For this initial adjustment, the galvanometer is connected in series with a high protective resistance. Contact $C$ is pressed and the galvanometer deflection observed. This contact is then moved along the wire until there is no galvanometer deflection when the key is pressed. CAUTION: Do not slide contact $C$ along the wire while pressing it into contact. This will cause the knife edge to scrape the wire and thereby damage it.

**4.** Final adjustments are made with the protective resistance $R_1$ short-circuited by switch $K_1$. When the galvanometer shows no deflection on pressing $C$, close $K_1$. Again locate $C$ so that the galvanometer shows no deflection. Only a *very small* displacement of the contact point will be needed to do this. Closing $K_1$ increases the sensitivity of the galvanometer many times and allows point $C$ to be located more accurately. Record the final setting of the contact point and the known value of the standard cell voltage.

**5.** *Open switch $K_1$.* Move $K_2$ to put the good dry cell $E_x$ in the circuit instead of the standard cell. Obtain a balance as described in Procedures 3 and 4 to measure the emf of the good dry cell. Record the final setting of contact point $C$.

**6.** Again *open switch $K_1$.* Replace the good dry cell with the old dry cell and repeat Procedure 5 to measure the old dry cell's emf. Record the final contact point setting.

**7.** *Open switch $K_1$* and return switch $K_2$ to its original position putting the standard cell in the circuit. Set contact point $C$ to a scale reading exactly equal to one half the known emf of the standard cell and adjust the rheostat for zero galvanometer deflection. Close $K_1$ for final adjustments of the rheostat but *be very careful* not to drive the galvanometer off scale when the protective resistance is short-circuited. Record the ammeter reading and make sure it does not change for the remainder of the experiment.

**8.** *Open switch $K_1$.* Move $K_2$ to put the old dry cell back in the circuit. Obtain a balance as described in Procedures 3 and 4. Read contact point $C$'s position on the scale and record it.

**9.** Again *open switch $K_1$.* Replace the old dry cell with the good dry cell and repeat Procedure 8. Record the contact point setting so obtained.

**10.** Measure the terminal voltage of the good dry cell and of the old dry cell with the voltmeter and record the values. Do not measure the voltage of the standard cell with the voltmeter because the current taken by the meter will cause the cell to change so that it will no longer be standard.

## DATA

Slide-wire current, Procedure 2 _____    Slide-wire current, Procedure 7 _____

| Cell used | emf from Equation 30.3 | | emf direct from scale | | Voltmeter reading |
|---|---|---|---|---|---|
| | AC, cm | emf, V | AC, cm | emf, V | |
| Standard cell | | | | | |
| Old dry cell | | | | | |
| Good dry cell | | | | | |

## CALCULATIONS

**1.** Compute the emf of the good dry cell from the data of Procedures 4 and 5.

**2.** Compute the emf of the old dry cell from the data of Procedure 6.

**3.** Obtain the emf's of the good and the old dry cells again by multiplying the distances *AC* measured in Procedures 8 and 9 by 2.

## QUESTIONS _____

**1.** Compare the emf of the good dry cell with that of the old dry cell. Compare their voltages as measured with the voltmeter. Explain these results.

**2.** If you use a 1-m wire of uniform cross section having a resistance of 5 $\Omega$ with a 6-V storage battery whose internal resistance is 0.2 $\Omega$, what resistance must be placed in series with the wire in order that the potential drop per millimeter will be exactly 1 mV?

**3.** The power supply voltage was set only approximately and was not measured or used in the calculations. Why is the value of this voltage not needed?

**4.** A cell whose emf is 1.52 V is connected to a voltmeter that records 1.48 V. If the resistance of the voltmeter is 200 $\Omega$, what is the internal resistance of the cell?

**5.** Find the resistance of your slide wire from data taken in this experiment.

**6.** Eight small flashlight cells in series will give 12 V, just as the storage battery in your car does. Could you start your car with the flashlight cells? Explain.

**7.** Suggest modifications to the circuit of Fig. 30.3 that would increase the voltage range of the potentiometer.

**8.** Find, by any method you can think of, what value $R_L$ should have in Fig. 30.1 so that the power delivered to it by the actual (non-ideal) voltage source will be maximized.

# Measurement of Capacitance by the Bridge Method 31

Electrical charge can be stored in a device called a capacitor in a manner completely analogous to the storing of water in a tank. Many uses for capacitors in providing short but intense pulses of electric current for various purposes are encountered in practice. Practical capacitors exist in many forms, but each basically consists of two conducting sheets or plates separated by an insulator. If a potential difference is applied between the plates, these will acquire equal charges of opposite sign, the value of each charge being proportional to the applied voltage. The constant of proportionality, which is characteristic of the particular capacitor, is called its capacitance. The object of this experiment is to study the bridge method of measuring capacitance and to check the formulas for the capacitance of capacitors connected in series and in parallel.

## THEORY

A capacitor basically consists of two metal plates separated by an insulating material called a dielectric. Such an arrangement stores electric charge by reason of the fact that, if a voltage source is connected between the plates, positive charge flows onto one plate from the other, leaving the latter negative. This process continues until the field in the dielectric is strong enough to stop the further flow of charging current. At this point, a certain amount of charge (positive in one case and negative in the other) is stored on each plate, and a potential difference equal to the source voltage exists between them. The ratio between this charge and this potential difference is a constant for a given capacitor and is called its capacitance. Thus,

$$C = \frac{Q}{V} \tag{31.1}$$

where $C$ is the capacitance in farads, $Q$ is the charge in coulombs, and $V$ is the potential difference in volts. Note that the unit of capacitance is actually a coulomb squared per newton-meter, but it is called a farad both to honor Michael Faraday and for the sake of convenience. Because of the size of the coulomb, the farad is too large a unit for practical use; hence the microfarad (a millionth of a farad) is the unit commonly encountered. The picofarad ($10^{-12}$ F) is also often used.

When a capacitor is connected in an ac circuit, the alternate charging and discharging make it appear to pass an ac current. A linear relationship between the ac voltage applied and the apparent current passed that is very similar to Ohm's law for resistance is found to hold. The proportionality factor, $X_C$, between this voltage and current is called the *capacitative reactance* of the capacitor and is measured in ohms, as is resistance. However, it is related to the capacitance by

$$X_C = \frac{1}{2\pi f C} \tag{31.2}$$

where $X_C$ is the reactance of the capacitor in ohms, $C$ is the capacitance in farads as before, and $f$ is the frequency in cycles per second (or hertz) of the applied ac. Thus the reactance, unlike resistance, depends on frequency and becomes infinite as the frequency goes to zero, an expression of the fact that for dc ($f = 0$), a capacitor is in fact an open circuit. However, for ac of a particular frequency, a capacitor behaves in many ways like a resistor. It can be measured by a bridge circuit similar to the Wheatstone bridge of Experiment 27 except that capacitors replace the resistors on one side of the bridge, an ac source is used instead of a battery, and an appropriate detector of ac (in this case a telephone receiver) replaces the galvanometer (see Fig. 31.1). Comparing this figure with Fig. 27.1, we note that if $C_1$ and $C_2$ are substituted for $R_1$ and $R_2$, then reactances $X_{C1} = 1/(2\pi f C_1)$ and $X_{C2} = 1/(2\pi f C_2)$ appear in place of

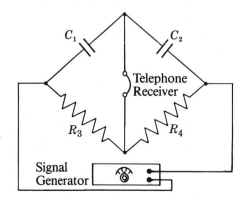

**Figure 31.1** *Bridge method of comparing capacitances*

**Figure 31.2**   *Capacitors in series*

the respective resistances in Equation 27.1, which thus becomes

$$\frac{2\pi f C_2}{2\pi f C_1} = \frac{C_2}{C_1} = \frac{R_3}{R_4} \qquad (31.3)$$

This relation, as described in Experiment 27, holds when the balance condition, that is, a null indication of the ac detector that replaces the galvanometer, has been set up. With $R_3$ and $R_4$ known, $C_2$ can be found in terms of $C_1$, a standard capacitor whose value is known to reasonably high precision. Notice the inverse relationship between the capacitances and the resistances, due to the inverse proportionality of capacitance and its associated reactance.

This inverse proportionality can be used to obtain the formulas for the capacitance of a series or parallel arrangement of capacitors. Suppose three capacitors are connected in series as shown in Fig. 31.2. In this ac circuit, they exhibit reactances $X_{C1}$, $X_{C2}$, and $X_{C3}$, and the total reactance of the combination is the sum of these individual reactances. Thus,

$$X_C = X_{C1} + X_{C2} + X_{C3}$$

Substitution of the expression for reactance, Equation 31.2, and multiplying through by $2\pi f$ leads to the conclusion that the capacitance $C$ exhibited by the three capacitors in series is given by

$$\frac{1}{C} = \frac{1}{C_1} + \frac{1}{C_2} + \frac{1}{C_3} \qquad (31.4)$$

Similarly, with capacitors in parallel as in Fig 31.3, we should find that

$$\frac{1}{X_C} = \frac{1}{X_{C1}} + \frac{1}{X_{C2}} + \frac{1}{X_{C3}}$$

and that there is an equivalent capacitance

$$C = C_1 + C_2 + C_3 \qquad (31.5)$$

Note that these equations are just like those for series

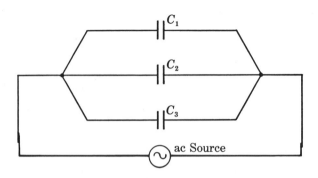

**Figure 31.3**   *Capacitors in parallel*

and parallel resistors but are reversed with respect to which equation goes with which combination, again because of the inverse relationship between capacitance and reactance.

The capacitance bridge of Fig. 31.1 can also be used to measure the so-called dielectric constant of an insulating material. The capacitance of a parallel-plate capacitor is given by the expression

$$C = \frac{\epsilon_0 \kappa A}{d} \qquad (31.6)$$

where $C$ is the capacity in farads (multiply by 1 million or $10^6$ for microfarads), $A$ is the area of either plate in square meters, $d$ is the plate separation in meters, $\epsilon_0$ is the permittivity of free space (see Table I at the end of the book), and $\kappa$ is the dielectric constant of the material between the plates. Note that $C$ varies directly with $\kappa$. If in our bridge we use two parallel-plate capacitors of identical dimensions, one of which ($C_1$) has air ($\kappa$ very nearly equal to 1) or, even better, a vacuum ($\kappa \equiv 1$) between the plates while the other ($C_2$) has the space between the plates filled with the insulator under test, $\kappa$ for that material will be equal to $C_2/C_1$ and will be given directly by the ratio $R_3/R_4$ at balance.

In this experiment, a standard capacitor will be used as $C_1$ and an unknown capacitance inserted at $C_2$. Resistors $R_3$ and $R_4$ will be decade resistance boxes, which will be adjusted to obtain the balance condition. The unknown capacitance can then be calculated from the known value of $C_1$ and the values of $R_3$ and $R_4$ read from the boxes. Two unknown capacitors will be measured in this manner, after which series and parallel combinations of these capacitors will be measured to check the validity of Equations 31.4 and 31.5.

## APPARATUS

1. Standard capacitor ($C_1$) in the range 0.1–1.0 $\mu$F
2. Two unknown capacitors ($C_2$) in the above range
3. Decade resistance box (1000 $\Omega$)

4. Decade resistance box (10,000 $\Omega$)
5. Audio signal generator
6. Telephone receiver or high-impedance headphones

## PROCEDURE

**1.** Begin by connecting the telephone receiver directly to the signal generator output terminals. Turn the signal generator on, set the output level for comfortable listening, and tune the generator frequency to the optimum value. This will be the setting for which the sound heard in the receiver is loudest. Telephone receivers have a limited frequency response range and are made to be most sensitive in the range occupied by the human voice. What you have just done is to find the frequency for which your receiver is most sensitive and will thus perform best as a null detector. Remember that you are going to adjust $R_3$ and $R_4$ to give no audible sound in the receiver, and the more sensitive the receiver is the more precise that adjustment can be. For ordinary telephone receivers and communication (not high-fidelity) headsets, the optimum frequency is around 800–1000 Hz.

**2.** Wire your capacitance bridge as shown in Fig. 31.1. Use the 10,000-$\Omega$ decade box for $R_3$ and set it to 200 $\Omega$. Use the 1000-$\Omega$ decade box for $R_4$, the standard capacitor for $C_1$, and one of the unknown capacitors for $C_2$. Set the signal generator output at maximum, check that it is still tuned to the optimum frequency found in Procedure 1, and listen for the tone in the telephone receiver. Leave $R_3$ set at 200 $\Omega$ and adjust $R_4$ for minimum audible signal. Record the value so obtained along with the value of $R_3$ and that of the standard capacitor ($C_1$).

**3.** Repeat Procedure 2 with the other unknown capacitor substituted at $C_2$. Record $R_4$ for this case.

**4.** Repeat Procedure 2 with the two unknown capacitors in series at $C_2$. Record $R_4$ as before.

**5.** Repeat Procedure 2 with the two unknown capacitors in parallel at $C_2$. Again record $R_4$.

## DATA

Value of $R_3$ _____    Value of standard capacitor ($C_1$) _____

| Capacitors measured | $R_4$ | Measured capacitance ($C_2$) | Calculated or given capacitance | Percent difference |
|---|---|---|---|---|
| Unknown no.1 | | | | |
| Unknown no. 2 | | | | |
| Series combination | | | | |
| Parallel combination | | | | |

## CALCULATIONS

**1.** Compute the capacitance of each of the unknown capacitors from the data of Procedures 2 and 3, respectively. *Note:* If ordinary commercial capacitors are used for the unknowns, their values will be marked on them and can be compared with your measured values. Do not expect close agreement, however, because these capacitors have a tolerance of $\pm 10\%$ or more from the marked values.

**2.** From the data of Procedure 4, find the capacitance of the series combination of the two unknown capacitors.

**3.** Use Equation 31.4 and your measured values of the unknown capacitors from Calculation 1 to obtain a calculated value for the capacitance of the series combination. Compare this value with that obtained in Calculation 2 by finding the percent difference.

**4.** From the data of Procedure 5, find the capacitance of the parallel combination of the unknown capacitors.

**5.** Use Equation 31.5 and your measured capacitance values to obtain a calculated value for the capacitance of the parallel combination. Compare this value with that obtained in Calculation 4 by finding the percent difference.

## QUESTIONS _____

**1.** Do the results you obtained experimentally for the series and parallel connections of capacitors check the laws for such connections within experimental error?

**2.** What would be the effect on the capacitance of a parallel-plate capacitor if a sheet of glass were introduced between the plates? Assume the dielectric to be air (dielectric constant $\kappa \approx 1$) before the glass was inserted.

**3.** Assuming that the charge remains the same, what would be the effect on the potential difference between the plates of a parallel-plate capacitor if a sheet of glass were introduced between the plates? Take $\kappa = 1$ before insertion of the glass as in Question 2. See Table XI at the end of the book for the dielectric constant of glass.

**4.** Which factor in Equation 31.6 is varied in the tuning capacitors used in radio sets? Explain how this is done.

**5.** Three capacitors with capacitances of 2, 3, and 5 $\mu$F, respectively, are joined in parallel. The combination is connected in series with a 5-$\mu$F capacitor. What is the capacitance of the final arrangement?

**6.** Why did you use the maximum output of the signal generator to power your bridge? Discuss the factors affecting the choice of the voltage used to supply a bridge.

**7.** A 4-$\mu$F and a 6-$\mu$F capacitor are connected in parallel and charged from a source of 120 V dc. They are then disconnected from the source and from each other and are reconnected to each other in parallel but with the leads from the 4-$\mu$F capacitor interchanged. Find the final voltage across the parallel combination.

**8.** A parallel-plate capacitor is made up of two circular plates separated by a sheet of mica. The plates have a diameter of 20 cm, and the mica is 0.1 mm thick and has a dielectric constant of 3. Calculate the capacitance so obtained in microfarads.

**9.** Derive Equation 31.5, the formula for the capacitance of a combination of capacitors in parallel, using Equation 31.1 and the fact that in any parallel combination of circuit elements the voltage across the combination is the same as that across each element. Compare your derivation with that given for parallel resistors in the Theory section of Experiment 26.

**10.** If a number of uncharged capacitors are connected in series and a charging current is sent into the combination, each capacitor must receive the same charge. Use this fact and the basic definition of capacitance given by Equation 31.1 to derive Equation 31.4, the formula for the capacitance of capacitors in series. Compare your derivation with that given for series resistors in the Theory section of Experiment 26.

# Measurement of Capacitance by the Ballistic Galvanometer

# 32

There are two general methods of measuring capacitance. One involves the comparison of the capacitance of the unknown capacitor with that of a standard capacitor in a manner similar to the comparison of resistances by a Wheatstone bridge. This method is the subject of Experiment 31. The other involves the use of the relation defining capacitance, that is, the ratio between the quantity of charge on a capacitor and the potential difference between the plates.

The quantity of charge is determined by the use of a ballistic galvanometer. The purposes of this experiment are to measure the ballistic constant of such a galvanometer, to determine the capacitance of a capacitor from the measurement of the charge with a ballistic galvanometer, and to check the laws of series and parallel connections of capacitors by measuring the charge on such combinations.

## THEORY

A capacitor consists of two metal plates separated by an insulating material, which is called a dielectric. If the plates are connected to two points between which there is a difference of potential, the two plates will acquire equal and opposite charges, the magnitude of each charge being directly proportional to the applied voltage. This relation may be expressed in the form

$$Q = CV \qquad (32.1)$$

where $Q$ is the quantity of electricity in coulombs, $V$ is the potential difference in volts, and $C$ is a constant of proportionality known as the capacitance of the capacitor. Thus the capacitance of a capacitor is the ratio between the quantity of charge on one plate and the potential difference between the plates. In the above expression, $C$ is the capacitance in farads. The farad is a unit named after the great experimental scientist Michael Faraday.

A capacitor has a capacitance of 1 F if it acquires a charge of 1 C when 1 V is impressed across it. The farad is too large a unit for practical purposes; hence the microfarad, or millionth of a farad, is used for greater convenience. The picofarad ($10^{-12}$ F) is also often encountered.

When capacitors are connected in series (see Fig. 32.1 and the discussion in the Theory section of Ex-

periment 31), the reciprocal of the capacitance exhibited by the combination is equal to the sum of the reciprocals of the individual capacitances. Thus,

$$\frac{1}{C} = \frac{1}{C_1} + \frac{1}{C_2} + \frac{1}{C_3} \qquad (32.2)$$

where $C$ is the capacitance of the series combination.

When capacitors are connected in parallel (see Fig. 32.2 and the discussion in the Theory section of Experiment 31), the total capacitance is equal to the sum of the individual ones. Thus,

$$C = C_1 + C_2 + C_3 \qquad (32.3)$$

where $C$ is now the capacitance of the parallel combination.

The capacitance of a parallel-plate capacitor is given by the expression

$$C = \frac{\epsilon_0 \kappa A}{d} \qquad (32.4)$$

where $C$ is in farads, $A$ is the area of one plate in square meters, $d$ is the distance between the plates in

**Figure 32.1**  *Capacitors in series*

**Figure 32.2**  *Capacitors in parallel*

meters, $\epsilon_0$ is the permittivity of free space (see Table I at the end of the book), and $\kappa$ is the dielectric constant of the material between the plates. To get the capacitance in microfarads, the above result has to be multiplied by 1 million ($10^6$).

When a capacitor is connected to a source of voltage, a charge is transferred from one plate to the other and the potential difference between the plates rises from the initial value zero to the final value of the impressed voltage. In transferring this charge, work has to be done. Thus, suppose that at a certain instant during the charging process the voltage between the plates is $V_i$. Then to take a very small piece of charge $\Delta Q$ from one plate to the other requires work $\Delta Q V_i$, for the potential difference $V_i$ is by definition the work required to carry unit charge across the gap. But every time such a transfer is made, we raise the voltage, as shown by Equation 32.1 in which the voltage is proportional to the total charge transferred. Hence, the total work required to charge the capacitor from zero to the final voltage $V$ is

$$W = \Delta Q V_1 + \Delta Q V_2 + \cdots + \Delta Q V_n$$
$$= \Delta Q (V_1 + V_2 + \cdots + V_n) \qquad (32.5)$$

where $V_1, V_2, \ldots V_n$ are the voltages obtained after each charge transfer. Multiplying and dividing Equation 32.5 by $n$, the total number of charge transfers made, yields

$$W = n \Delta Q \frac{V_1 + V_2 + \cdots + V_n}{n} = Q\bar{V} \quad (32.6)$$

where the $n$ pieces of charge $\Delta Q$ make up the total charge $Q$ on the capacitor and $\bar{V}$ is the average of the voltage that varied from 0 to $V$ as the capacitor was charged from 0 to $Q$. But since $V$ is proportional to $Q$, $\bar{V} = \frac{1}{2}V$ and Equation 32.6 becomes

$$W = \frac{1}{2}QV \qquad (32.7)$$

where $W$ is the work done in joules, $Q$ is the charge on the capacitor in coulombs, and $V$ is the potential difference between the plates in volts. This work is stored as electrical energy in the capacitor. In view of Equation 32.1, this energy can also be expressed as

$$W = \frac{1}{2}CV^2 = \frac{1}{2}\frac{Q^2}{C} \qquad (32.8)$$

where $W$, $V$, and $Q$ are the same as before and $C$ is the capacitance of the capacitor in farads.

In this experiment, the capacitance of a capacitor is determined by measuring the charge on it and the potential difference between its plates. The capacitor is charged to a known voltage and then discharged through a ballistic galvanometer to measure the quantity of charge on the plates.

A ballistic galvanometer is used for the measurement of quantity of electricity. This type of galvanometer is so constructed that its suspended coil has a relatively large moment of inertia and a long period of oscillation. As a result, when a capacitor is discharged through it, the pulse of current produced rises through a maximum and decreases to zero in a time that is short compared to the period of oscillation of the suspended coil. Thus, all the charge passes through the coil before it has been able to move appreciably, and the coil receives an impulsive torque. Instead of producing a steady deflection, this type of torque produces a throw; that is, the coil swings through an angle and then comes back to rest. The reading of the galvanometer, therefore, is the maximum deflection obtained and not a steady deflection as in the ordinary type of galvanometer.

We can easily see that this maximum deflection is proportional to the quantity of electric charge that has been passed through the coil. Any galvanometer is made so that the torque on the coil about its axis of rotation is proportional to the current through it. Thus,

$$\tau = Ki \qquad (32.9)$$

where $\tau$ is the torque, $i$ is the coil current, and $K$ is a proportionality constant that is characteristic of the galvanometer. In our present case the current $i$, which discharges the capacitor, flows for a very short time $\Delta t$, so that its average value is just $Q/\Delta t$ where $Q$ is the capacitor charge and hence the total quantity of electricity passed through the coil. Substitution in Equation 32.9 and multiplication by $\Delta t$ then gives

$$\tau \, \Delta t = KQ \qquad (32.10)$$

But since $\Delta t$ is the time during which the current flows and hence the time during which the torque $\tau$ acts, the quantity $\tau \, \Delta t$ is the angular impulse applied to the galvanometer's rotating assembly. Because $\Delta t$ is small, we can set this impulse equal to the initial angular momentum of the rotating assembly and thus write

$$\tau \, \Delta t = KQ = I\omega_0 \qquad (32.11)$$

where $I$ is the rotating assembly's moment of inertia and $\omega_0$ its initial angular velocity.

The rotating assembly in this type of galvanometer is suspended on a torsion fibre that exerts a linear restoring torque on the assembly. Hence, if this assembly is allowed to swing freely, it will execute rotational simple harmonic motion at some low frequency $f$. Letting $\theta$ represent the angle through which the rotating assembly has turned from the equilibrium position it occupied at $t = 0$, we can write

$$\theta = \theta_m \sin 2\pi ft \qquad (32.12)$$

where $\theta_m$ is the amplitude of the rotational oscillation and thus represents the maximum angle of rotation from equilibrium. The angular velocity of such motion as a function of time is

$$\omega = 2\pi f \theta_m \cos 2\pi f t \qquad (32.13)$$

and hence the initial velocity $\omega_0$ (that is, $\omega$ at $t = 0$) is just $2\pi f \theta_m$. Replacing $\omega_0$ by this expression in Equation 32.11 gives

$$KQ = 2\pi f I \theta_m \qquad (32.14)$$

and solving for $Q$ then yields

$$Q = \frac{2\pi f I}{K} \theta_m \qquad (32.15)$$

Thus, $Q$ is proportional to $\theta_m$, the maximum angular swing of the galvanometer's rotating assembly. In the usual ballistic galvanometer, this assembly includes a mirror, the observed deflection of the galvanometer being that of a light beam reflected from the mirror to a scale a distance $D$ away. If $\theta_m$ is in radians, the maximum deflection $d_m$ of the spot on the scale is $2D\theta_m$; hence Equation 32.15 becomes, in terms of the observed deflection $d_m$,

$$Q = \frac{\pi f I}{DK} d_m \qquad (32.16)$$

which establishes the proportionality between $Q$ and $d_m$.

It would, of course, be possible to calculate the proportionality constant $\pi f I / DK$ by measuring $D$, the distance from the mirror to the scale, and determining the pertinent galvanometer properties, that is, the natural frequency of the rotating assembly, its moment of inertia, and the electrical constant $K$. However, it is more convenient to write the complete constant as $k$ so that Equation 32.16 becomes simply

$$Q = k d_m \qquad (32.17)$$

Here $k$ is called the *ballistic constant* of the galvanometer and can be measured directly by passing a known charge through the coil. Note that the ballistic constant is numerically equal to the quantity of electricity necessary to produce a unit scale deflection. If $Q$ is expressed in microcoulombs and $d_m$ in millimeters, then $k$ will be in microcoulombs per millimeter.

## APPARATUS

1. Ballistic wall galvanometer
2. Laser light source and scale as required by the galvanometer
3. Two unknown capacitors in the range of 0.2 to 2.0 $\mu$F

4. Capacitor decade capable of giving 0.5, 1.0, 1.5, 2.0, 2.5, 3.0, 3.5, and 4.0 $\mu$F
5. Single-pole, double-throw switch
6. Two 1.5-V dry cells
7. dc voltmeter (0–3 V)

## PROCEDURE

**1.** Connect the apparatus in the circuit shown in Fig. 32.3, using the decade capacitor at $C$ and one dry cell at *BAT*. Mount the laser or other light source so that it shines on the galvanometer mirror with the beam oriented to strike the scale at a convenient point. CAUTION: *If you are using a laser, be careful not to look into the beam or allow anyone else to do so.* The galvanometer itself has already been adjusted and should be ready for use. If it is not, consult the instructor. Do not try to adjust it yourself, and have your instructor check your circuit before connecting the dry cell.

**2.** Observe that when switch *SW* is closed in the direction $A$, the capacitor is being charged by the dry cell, and when it is closed in the direction $B$, the capacitor discharges through the galvanometer. Close the switch in the direction $B$ to insure that $C$ is discharged and set the capacitor decade to 0.5 $\mu$F. Then throw the switch to position $A$, thus charging the capacitor. Wait a few seconds for it to become fully charged and record the voltage read on the voltmeter. Now close the switch in the direction $B$, thus discharging the capacitor through the galvanometer. Observe the maximum deflection of the galvanometer, estimating the reading to 0.5 mm. You may want to practice doing this a few times before recording any data in order to gain experience in catching the light spot at its true maximum position on the scale. When you feel you can do this with high precision, repeat the procedure three times, recording the maximum deflection observed each time in the appropriate box in the data sheet.

**3.** Repeat Procedure 2 with decade capacitor settings of 1.0, 1.5, 2.0, 2.5, 3.0, 3.5, and 4.0 $\mu$F.

**Figure 32.3**  *Circuit for the measurement of capacitance*

**4.** Replace the decade capacitor with one of the unknown capacitors. Then repeat Procedure 2, recording your results in the appropriate boxes in the second table in the data sheet.

**5.** Repeat Procedure 2 using the same capacitor as in Procedure 4 but with two dry cells in series instead of just one at *BAT*. If the dry cells are new and thus give slightly more than 1.5 V each, the voltmeter will read slightly off-scale. If this happens, either estimate the reading by noting the approximate number of small scale divisions by which the pointer has passed the high end of the scale, or disconnect the voltmeter from the circuit and measure the voltage of each dry cell separately. Add these two measurements to get the total charging voltage. In either case, record the charging voltage in the space provided.

**6.** Repeat Procedures 4 and 5 using the other unknown capacitor.

**7.** Connect the two unknown capacitors in parallel at *C* and charge them with only one dry cell at *BAT*. Repeat Procedure 2 with this arrangement, recording your data in the appropriate boxes.

**8.** Connect the two unknown capacitors in series and charge them, using two dry cells in series at *BAT*. Repeat Procedure 2 with this arrangement, measuring the charging voltage as discussed in Procedure 5. Record your results as before.

## DATA

| Capacitance used | Charging voltage | Galvanometer deflection | | | | Total charge |
|---|---|---|---|---|---|---|
| | | Trial 1 | Trial 2 | Trial 3 | Average | |
| 0.5 μF | | | | | | |
| 1.0 μF | | | | | | |
| 1.5 μF | | | | | | |
| 2.0 μF | | | | | | |
| 2.5 μF | | | | | | |
| 3.0 μF | | | | | | |
| 3.5 μF | | | | | | |
| 4.0 μF | | | | | | |

Ballistic constant of galvanometer _____

| Capacitance used | Charging voltage | Galvanometer deflection | | | | Total charge | Value of capacitance |
|---|---|---|---|---|---|---|---|
| | | Trial 1 | Trial 2 | Trial 3 | Average | | |
| Capacitor no. 1 | | | | | | | |
| Capacitor no. 1 | | | | | | | |
| Capacitor no. 2 | | | | | | | |
| Capacitor no. 2 | | | | | | | |
| Capacitors 1 and 2 in parallel | | | | | | | |
| Capacitors 1 and 2 in parallel | | | | | | | |

Average value of capacitor no. 1 _____
Average value of capacitor no. 2 _____

Calculated value of capacitors in parallel _____

Calculated value of capacitors in series _____

## CALCULATIONS _____

**1.** Calculate the average galvanometer deflection obtained with each of the capacitances used in Procedures 2 and 3. Then use Equation 32.1 to find the total charge observed in each case. Note that if $V$ is the charging voltage in volts and $C$ is the capacitance in microfarads, the charge $Q$ will come out in microcoulombs.

**2.** Using the data of Procedures 2 and 3 and the results of Calculation 1, plot the average galvanometer deflections along the $x$ axis and the corresponding values of the charge along the $y$ axis. Draw a straight line through the plotted points in such a way as to fit them as closely as possible. Note that the origin should be a point on your line. This graph will be the calibration curve for the galvanometer. Calculate its slope. This will be the value of the galvanometer's ballistic constant in microcoulombs per scale division.

**3.** Find the average of the three galvanometer deflections obtained with each capacitor setup in Procedures 4–8. Then, using these results, read the corresponding values of the total charge in microcoulombs from your graph and record them in the appropriate boxes in the second data table.

**4.** From the data of Procedure 4, calculate the value of unknown capacitor no. 1. Then obtain and record the value of this same capacitor from the data of Procedure 5. How closely do your two values agree? Compute their average and enter this result in the space provided.

**5.** Repeat Calculation 4 using the data of Procedure 6 to obtain the value of the second unknown capacitor.

**6.** From the data of Procedure 7, determine the capacitance of the two unknown capacitors in parallel by dividing the total charge on them by the charging voltage.

**7.** Calculate the capacitance of the two unknown capacitors in parallel by using Equation 32.3 and the values of the individual capacitances obtained in Calculations 4 and 5. Compare this value with the one obtained in Calculation 6.

**8.** From the data of Procedure 8, determine the capacitance of the two unknown capacitors in series by dividing the measured charge by the charging voltage.

**9.** Calculate the capacitance of the two unknown capacitors in series by using Equation 32.2 and the values of the individual capacitances obtained in Calculations 4 and 5. Compare this value with the one obtained in Calculation 8.

## QUESTIONS _____

**1.** Express the farad in terms of the fundamental SI units: meters, kilograms, seconds, and coulombs.

**2.** Do the results you obtained experimentally for the series and parallel connections of capacitors check the laws for such connections within experimental error?

**3.** A capacitor is charged by a 45-V battery and is then discharged through a galvanometer whose ballistic constant is 1.5 $\mu$C per scale division. The galvanometer deflection produced is 20 divisions. What is the capacitance of the capacitor?

**4.** Which factor in Equation 32.4 is varied in the tuning capacitors used in radio sets? Explain how this is done.

**5.** Two capacitors with capacitances of 4 and 5 $\mu$F respectively are connected in parallel. The combination is connected in series with a 3-$\mu$F capacitor. What is the total capacitance of the final arrangement?

**6.** Compute the energy in joules in your 4.0-$\mu$F standard capacitor when it is charged with one dry cell. What happens to this energy when the capacitor is discharged through the galvanometer?

**7.** A parallel-plate capacitor is made up of two circular plates separated by a thin sheet of mica. The plates have a diameter of 18 cm, and the mica is 0.2 mm thick and has a dielectric constant of 5.7. Calculate the capacitance of this capacitor in microfarads.

**8.** A 10-mF capacitor is charged to a potential difference of 1000 V. It is then disconnected from the voltage source and connected to an uncharged 4-mF capacitor. (a) What is the final voltage across the terminals of the parallel combination? (b) What energy was initially stored in the 10-mF capacitor and what is the final energy stored in the parallel combination? (c) Are these energies equal? If not, what happened to the difference?

**9.** A parallel-plate capacitor has square plates 24 cm on a side separated by 1 mm. The dielectric can be air, or a 1-mm-thick sheet of glass can be slipped between the plates without friction. The capacitor is connected to a 1000-V dc source with the glass sheet between the plates. How much work is needed to withdraw the sheet (a) if the capacitor is first disconnected from the voltage source, and (b) if the capacitor is left connected to the voltage source during the withdrawal process?

**10.** What is the voltage across and the charge on the capacitor of Question 8 after the glass sheet has been withdrawn in each case?

# The *RC* Circuit 33

When a resistor is connected in parallel with a charged capacitor, a current flows through the resistor, draining charge out of the capacitor. The capacitor is thus discharged over a period of time. The relation between the voltage remaining across the capacitor and the time is a very useful one in electronics and electrical engineering, and the purpose of this experiment is to study it. In doing so, we will find that this relation depends on the resistance $R$ of the resistor and the capacitance $C$ of the capacitor only through their product $RC$ and that another method of measuring capacitance is presented.

## THEORY

Fig. 33.1 shows a capacitor $C$ and a resistor $R$ connected in parallel. Let the capacitor be initially charged to a voltage $V_0$ by some external voltage source which does not appear in this figure and has been removed at time $t = 0$. The capacitor then discharges through resistor $R$, the voltage $V$ across the capacitor at any particular time causing a current $i = V/R$ to flow at that instant according to Ohm's law. Current $i$ is the rate in amperes at which charge is leaving the capacitor (recall that an ampere is a coulomb per second), so that voltage $V$ decreases steadily. Note, however, that the smaller $V$ gets, the smaller $i$ is and hence the slower the rate of discharge. Situations like this in which the rate at which something changes is proportional to its value at every instant are common in physics and are always described by the exponential function, which is characterized by the property of having its rate of change proportional to its value. The exponential function is simply the base of natural logarithms $e$ raised to a power proportional to the time. For example, in the absence of inhibiting effects, the rate of growth of the population of any given species is proportional to the number already present. The population as a function of time is then described by an exponential in which the exponent is $t$ times a constant expressing the probability that any particular specimen will reproduce. In the present case, the voltage across the capacitor *decreases* as time goes on, which means that its rate of change is negative. Consequently, the exponent in the exponential must be negative. It can be shown that the voltage across the capacitor is given as a function of time by

$$V = V_0 e^{-t/RC} \qquad (33.1)$$

where $V$ is the voltage in volts at time $t$, $V_0$ is the initial voltage (at $t = 0$), $R$ is the resistance of the resistor in ohms, $C$ is the capacitance of the capacitor in farads, and $t$ is the time in seconds measured from the instant at which $V$ was equal to $V_0$. It is interesting to note that $R$ and $C$ occur only as the product $RC$. Moreover, the exponent as a whole cannot have a dimension, and so $RC$ must have the units of time. Indeed, multiplication of ohms by farads gives seconds. The product $RC$ is therefore called the circuit's *time constant*. Note that when $t = RC$, $V = V_0/e = 0.368V_0$ or somewhat less than half of the original voltage. Equation 33.1 shows that theoretically $V$ never gets to zero (the discharge rate getting slower and slower the smaller $V$ becomes), but in a practical case $V$ becomes negligibly small after a period equal to a few time constants has elapsed.

The validity of Equation 33.1 will be investigated in this experiment by plotting the discharge of a capacitor by a parallel resistor as a function of time. Since practical capacitors have capacitances measured in microfarads, the discharging resistor must have a value of several megohms (millions of ohms) if a time constant of a few seconds is to be realized. Although electronic equipment often uses $RC$ circuits with time constants in the microsecond range, we are not now prepared to measure such short intervals and must therefore have time constants sufficiently large to time the voltage decrease with a stopwatch. To measure $V$, however, a voltmeter must be connected

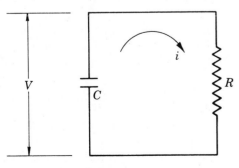

**Figure 33.1**  *The RC* circuit

across the capacitor, and an ordinary voltmeter will itself have a resistance less than the several megohms desired. To overcome this difficulty, an electronic voltmeter or EVM (sometimes called a VTVM for vacuum-tube voltmeter) will be used. An electronic voltmeter contains an amplifier that drives the actual meter with power obtained either from batteries or from the power line. Ideally this means that the meter actuating current does not have to come from the circuit under test so that the EVM input terminals can have an infinite resistance between them. Practical EVMs usually have an input resistance of about 10 MΩ, but this is already so large that most circuits are not affected by it. Thus, the EVM, like the potentiometer, has the advantage of drawing negligible current from the circuit under test, but it is much more convenient for general voltage-measuring purposes.

In this experiment, an EVM will be used both to measure the voltage across a capacitor and to serve as the parallel resistor. A plot of *V* against *t* will be made on semilogarithmic paper and should turn out to be a straight line if Equation 33.1 is correct. The time constant *RC* can be obtained from the slope of this line, as discussed in the section of the Introduction on plotting exponentials. The values of *R* and *C* cannot, however, be found individually. Another plot will therefore be made using a different value of *R* obtained by connecting a large known resistance in parallel with the EVM, and a new value of *RC* will be calculated from its slope. Since the two values of *R* are related by the formula for resistances in parallel, both the value of *C* and the EVM input resistance can be found from the data. The nominal value of *C* is marked on the capacitor and serves as a check on your results.

## APPARATUS

1. Power supply capable of delivering 300 V dc
2. Electronic voltmeter with 0–300 V or higher scale
3. 4-μF capacitor (may be the capacitor decade of Experiment 32 set to this capacity)

4. 10-MΩ precision resistor (1% tolerance)
5. Single-pole, single-throw switch
6. Stopwatch or stop clock

## PROCEDURE

**1.** Wire the power supply, capacitor, electronic voltmeter, and switch in the circuit shown in Fig. 33.2. Omit the 10-MΩ resistor $R_s$ in this step. Be sure the EVM is set to an appropriate range (0–300 V or higher). Leave switch $K_1$ open and do not plug in or turn on the power supply until your circuit has been approved by the instructor. CAUTION: The output of the power supply is high enough to give a painful shock. Be careful around the apparatus and do not make changes in your circuit with the power on. Notice that opening switch $K_1$ is not enough. The power supply itself must be turned off with its own

switch or unplugged from the power line before any wiring changes are attempted.

**2.** After the instructor has checked your circuit, turn on the power supply, close switch $K_1$, and adjust the power supply output to 300 V as read on the EVM.

**3.** Open $K_1$ and at the same instant start the stopwatch. Stop the stopwatch at the instant the EVM passes through 250 V. Record the time thus measured.

**4.** Close switch $K_1$ and readjust the power supply if necessary to bring the voltage across the capacitor to 300 V. Then again open $K_1$ and start the stopwatch simultaneously. Again stop the stopwatch when the EVM passes through 250 V and record the indicated time. Repeat this procedure a third time to give you three independent measurements of the time required for the capacitor voltage to fall from 300 to 250 V.

**5.** Repeat Procedures 3 and 4, stopping the stopwatch at 200, 150, 100, and 50 V.

**6.** Connect the 10-MΩ resistor at $R_s$ in Fig. 33.2. Repeat Procedures 2–5 with this setup.

**7.** Be sure to turn the power supply off and unplug it when you have finished the experiment.

**Figure 33.2** *Circuit for measuring time constants*

## DATA _____

| Voltage across the capacitor | Ratio $V/V_0$ | Time, EVM alone | | | | Time, EVM in parallel with $R_s$ | | | |
|---|---|---|---|---|---|---|---|---|---|
| | | Trial 1 | Trial 2 | Trial 3 | Average | Trial 1 | Trial 2 | Trial 3 | Average |
| 300 | 1.00 | 0.00 | 0.00 | 0.00 | 0.00 | 0.00 | 0.00 | 0.00 | 0.00 |
| 250 | | | | | | | | | |
| 200 | | | | | | | | | |
| 150 | | | | | | | | | |
| 100 | | | | | | | | | |
| 50 | | | | | | | | | |

Value of $R_m C$ from graph  _____          Calculated value of capacitor $C$  _____
Value of $R_1 C$ from graph  _____          Nominal value of $C$  _____
Value of meter resistance $R_m$  _____

## CALCULATIONS _____

**1.** Divide each of your values of the capacitor voltage $V$ by the initial voltage $V_0$ to obtain the ratio $V/V_0$ in each case. Note that this will allow you to plot Equation 33.1 in the form

$$\frac{V}{V_0} = e^{-t/RC}$$

**2.** Calculate the average of each set of three time measurements made in Procedures 3–5 and enter your results in the appropriate boxes in the data sheet.

**3.** On semilogarithmic paper, plot the average times obtained in Calculation 2 against the corresponding values of the voltage ratio $V/V_0$. Plot this ratio on the logarithmic scale and the time on the linear one, choosing your scale so as to spread your graph over as much of the sheet as possible. Read the instructions for plotting exponential functions in the Introduction. Draw the best straight line through your plotted points.

**4.** From the slope of your graph, calculate $R_m C$ where $R_m$ is the resistance of the EVM.

**5.** Calculate the average of each set of three time measurements made in Procedure 6 and enter your results in the appropriate boxes in the data sheet.

**6.** On the second sheet of semilogarithmic graph paper, plot the average times obtained in Calculation 5 against the corresponding values of $V/V_0$. Use the logarithmic scale for the voltage ratio and the linear scale for time, spreading your graph out as much as you can, as before. Draw the best straight line through this set of plotted points.

**7.** From the slope of the graph drawn in Calculation 6, compute $R_1 C$, where $R_1$ is the resistance of $R_m$ and $R_s$ in parallel.

**8.** You now have values for $R_m C$ and $R_1 C$, and you know that $(1/R_1) = (1/R_m) + (1/R_s)$ where the value of $R_s$ is given. On this basis, calculate the capacitance $C$ and the meter resistance $R_m$.

**9.** Compare your calculated value of $C$ with the nominal value marked on the capacitor. *Note:* Commercial capacitors have a tolerance of $\pm 10\%$ or more from the nominal value. You should therefore not expect remarkably good agreement between this and your measured capacitance. The instructor may, however, give you a more precise value of $C$ with which to compare your results.

## QUESTIONS _____

**1.** Explain how your graphs check the validity of Equation 33.1.

**2.** Why was it better to plot the ratio $V/V_0$ rather than simply the voltage $V$ against $t$ in Calculations 3 and 6?

**3.** Show that the product $RC$ has the dimensions of seconds when $R$ is in ohms and $C$ is in farads.

**4.** Let $R_m C = A$ and $R_1 C = B$, where $A$ and $B$ are the measured values of these time constants obtained from the slopes of your graphs in Calculations 4 and 7. Recalling that $R_1$ is the resistance of $R_m$ and $R_s$ in parallel, derive expressions for $R_m$ and $C$ in terms of the known quantities $A$, $B$, and $R_s$.

**5.** When the capacitor in this experiment is initially charged to voltage $V_0$, the energy stored is $\frac{1}{2} C V_0^2$. What happens to this energy as the capacitor discharges? Can you show quantitatively that your answer is correct?

**6.** A charged 2-$\mu$F capacitor is connected in parallel with a 500,000-$\Omega$ resistor. How long after the connection is made will the capacitor voltage fall to (a) 50% of its initial value; (b) 30% of its initial value; (c) 10% of its initial value; (d) 5% of its initial value?

**7.** (a) Show by any method you know that Equation 33.1 may be approximated by

$$\frac{V}{V_0} \approx 1 - \frac{t}{RC}$$

if $t \ll RC$. (b) The above equation is linear, and a plot of $V/V_0$ against $t$ on ordinary graph paper would therefore be a straight line as long as this equation was valid. To what fraction of $V_0$ can $V$ fall before the error made in using this approximation reaches 10%?

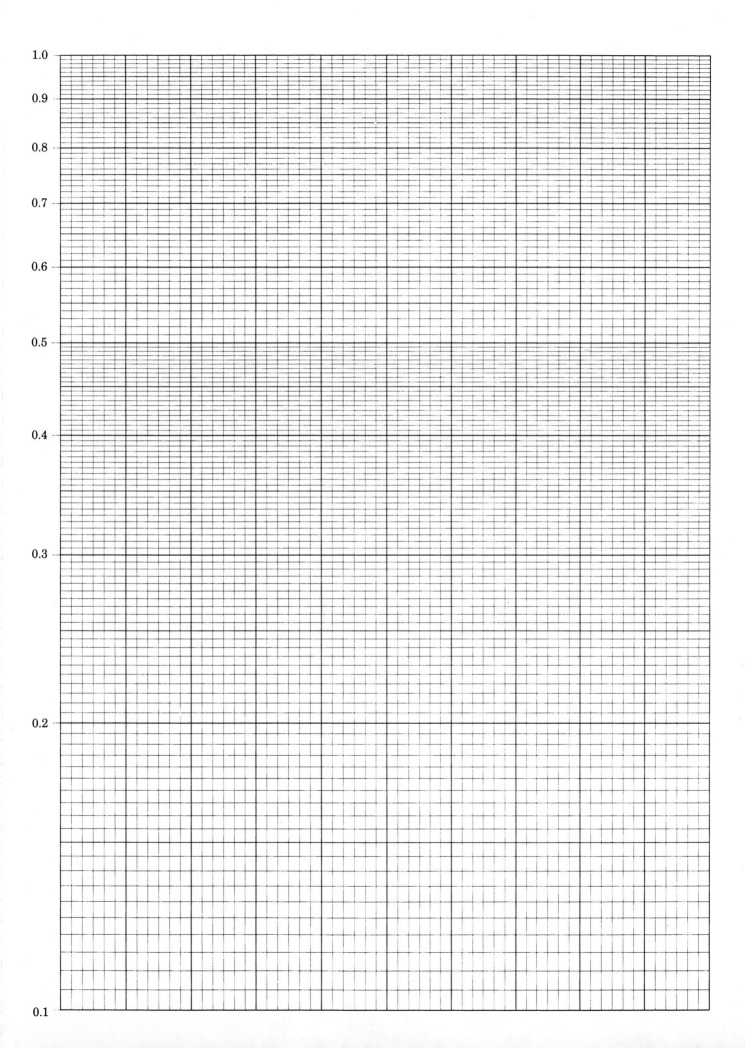

# Electromagnetic Induction: The Transformer and the Generator

# 34

The production of induced currents is one of the most important branches of the study of electricity, for it is the basis of operation of both the transformer and the generator and thus forms one of the foundations of the electrical industry. The electric generator has revolutionized both modern industry and home life by furnishing us with a source of cheap electricity. Generators convert mechanical energy directly into electrical energy by electromagnetic induction. Thus, the enormous power of steam engines, gas engines, and water wheels can be transformed into electricity that can be conveniently used in motors to turn all kinds of machinery. The principle of electromagnetic induction is also fundamental to the operation of the transformer, which, by providing a simple and efficient means of stepping voltage up and down, has made long-distance transmission of electric power possible.

The present experiment deals with some aspects of electromagnetic induction. Much of the work is qualitative, so that a complete record of the observations must be kept. The object of the experiment is to study Faraday's law of electromagnetic induction, Lenz's law, and the general features of a magnetic circuit. In particular, simple examples of a transformer and an alternating current generator will be studied. In each case, the voltage induced in a coil of wire by a varying magnetic flux will be investigated.

## THEORY

A voltage is induced in a loop of wire whenever there is a change in the magnetic flux passing through the loop. Magnetic flux is the product of the magnetic field intensity $B$ and the area bounded by the loop through which this field passes. If the direction of $B$ is not perpendicular to the plane of the loop, then the perpendicular component must be taken. Mathematically the flux is given by

$$\Phi = B_\perp A \qquad (34.1)$$

where $\Phi$ is the flux in webers, $B_\perp$ is the perpendicular component of the magnetic field in teslas, and $A$ is the area bounded by the loop. Faraday's law of electromagnetic induction then states that the voltage induced in a loop by a changing magnetic flux is given by

$$V_\ell = -\frac{\Delta\Phi}{\Delta t}$$

where $V_\ell$ is the average value of the induced voltage in volts; $\Delta\Phi$ is the change in flux in webers; and $\Delta t$ is the time interval in seconds during which the change in flux takes place. If instead of a one-turn loop we have a coil of $N$ turns, voltage $V_\ell$ is induced in each turn, and since the coil is in effect $N$ turns in series, the total voltage at the coil terminals is

$$V = NV_\ell = -N\frac{\Delta\Phi}{\Delta t} \qquad (34.2)$$

From the definition of flux, we note that a change in $\Phi$ may be produced in any of three different ways: the area of the loop may be changed, thus changing the factor $A$ in Equation 34.1; the strength of the magnetic field $B$ may be altered; or the coil may be rotated so that the field's perpendicular component changes. All of these methods of changing $\Phi$ are used in practical applications of Faraday's law.

The minus sign in Equation 34.2 gives the polarity of the coil voltage relative to the sign of the change in flux. It is best understood in terms of Lenz's law, which states that the polarity of $V$ will be such as to oppose the change in $\Phi$ that produced it. Thus, suppose the terminals of a certain coil were connected together so that an induced voltage would cause a current to flow around the coil. Now suppose that a magnetic field directed through the coil from an external source were increased. Lenz's law requires that the current induced in the coil flow in such a direction as to result in a magnetic field opposing the applied one, that is, bucking the increase in $B$ that induced it. Similarly, should the externally applied field decrease, the induced current will flow around the coil in the other direction so as to produce a field in the same direction as the applied field and thus oppose the decrease. Note that the relation between the magnetic field produced by a current flowing around a coil and the direction of this current is given by the right-hand rule. When the fingers of the right hand curl around the coil in the direction of the current flow, the extended thumb points in the direction of the resulting magnetic field.

An important application of electromagnetic induction is the transformer. A transformer consists of two interwound coils called the primary and secondary. These coils are often wound on a core of magnetic material, and this core is extended to form one or more complete loops of the material passing through and around the coils. Such a loop forms a magnetic circuit that concentrates the magnetic field due to current in either coil in the loop of material passing through both coils.

If a current is now sent through the primary coil, a magnetic field passing through the plane of both coils and hence a magnetic flux through these coils will be produced. Any change in the current will result in a change in this flux. The primary will therefore have induced between its terminals a voltage voltage

$$V_1 = -N_1 \frac{\Delta \Phi}{\Delta t} \qquad (34.3)$$

and a voltage

$$V_2 = -N_2 \frac{\Delta \Phi}{\Delta t} \qquad (34.4)$$

will be induced between the terminals of the secondary, where $V_1$ and $V_2$ are in volts, $N_1$ is the number of turns in the primary, and $N_2$ the number of turns in the secondary. Note that even though the primary is wound of heavy copper wire so that it has negligible resistance, a voltage nevertheless develops across it when a current is sent through it, *but only when the current is changing*. Hence with direct current, once the current is on, $\Delta \Phi$ and therefore both $V_1$ and $V_2$ will be zero. This explains why transformers work only on alternating current, which is by definition a current that is continually changing. The voltage $V_1$ appearing across the primary as a result of the changing primary current is sometimes called a *back emf* because of the opposing polarity given it by Lenz's law as expressed by the minus sign in Equation 34.3.

The most important feature of the transformer is revealed when we divide Equation 34.3 by Equation 34.4 to obtain

$$\frac{V_1}{V_2} = \frac{N_1}{N_2} \qquad (34.5)$$

This result says that the ratio of the primary to the secondary voltage is equal to the turns ratio. Thus, in an ac circuit a transformer may be used to change the voltage to any desired value by simply winding the coils with the appropriate numbers of turns. This simple arrangement is used throughout the electrical industry from household appliances to huge transmission-line substations. Much of the electrical equipment in common use today would not work without it, and the fact that the transformer works on ac only is a principal reason for the widespread use of ac rather than dc power.

Another example of an important application of Faraday's law of induction is the electric generator. A generator is a machine that, when driven mechanically by an engine or turbine, converts the mechanical power thus supplied into electrical power.

The essential parts of a generator are the magnetic field, the armature, and some device, such as slip rings or a commutator, for connecting the armature to the external circuit.

The magnetic field is produced by permanent magnets or by electromagnets. The armature consists of a rotating coil wound on a soft iron core to obtain an intense and uniform field. As the armature rotates, the flux through the coil changes because the perpendicular component of the magnetic field is changing, and a voltage is induced across the coil's terminals.

A simplified diagram of a generator is shown in Fig 34.1, in which a rectangular coil of wire *ABCD* is mounted on a longitudinal axis *OO'* between the magnetic poles N and S. The two ends of the coil are connected to two insulated rings mounted on the axis and rotating with the coil. Two blocks of carbon, called brushes, press against these rings as they rotate and provide electrical contact with the external circuit.

A cross section of the coil is represented in Fig. 34.2. The voltage induced in the coil will be zero at the position *AD*, or when the angle $\theta$ is zero, because at that instant the conductors are moving in a direction parallel to the magnetic field so that no change in flux occurs. The voltage will be at a maximum when $\theta$ is 90°, because at that instant the conductors are moving in a direction perpendicular to the magnetic field and the rate at which the flux is changing is at its maximum. The situation is sometimes expressed in terms of the conductors "cutting" the magnetic field lines. When the conductors move parallel to $B$, no lines are cut and no voltage is induced, whereas when they move perpendicular to $B$ they cut straight across the field lines and the induced voltage is maximum. Since these lines extend from the north pole to the south pole in Fig. 34.2, the component of the field $B$ perpendicular to the plane of the coil is $B \cos \theta$ and the flux through the coil is $BA \cos \theta$, where $A$ is the area that the coil bounds. The voltage induced in one loop is just the rate of change of this flux as $\theta$ changes with rotation of the coil, and if the coil consists of $N$ loops, the total voltage appearing at its terminals will be $N$ times this quantity. The rate of change of $\cos \theta$ with $\theta$ is $-\sin \theta$ times the rate of change of $\theta$, and if the coil in Fig. 34.2 is rotated clockwise at a constant angular velocity of $\omega$ radians per second, $\theta = \omega t$ and the rate of change of $\theta$ is $\omega$. Hence, the induced voltage at the coil terminals is given by

$$V = NBA\omega \sin \omega t \qquad (34.6)$$

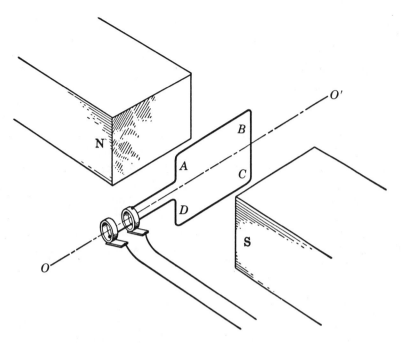

**Figure 34.1** *Principle of a simple generator*

The maximum value of $V$ is seen to be $NBA\omega$ and is usually called the amplitude $V_m$ of $V$. With this substitution, Equation 34.6 becomes

$$V = V_m \sin \omega t \qquad (34.7)$$

where $V$ is the instantaneous value of the induced voltage in volts, $V_m$ is its maximum value, $\omega$ is the angular velocity, and $t$ is the time in seconds. If instead of rotating at a constant angular velocity $\omega$, the coil is rotated through a small angle $\Delta\theta$ from an orientation $\theta$, the change of flux is $-BA\,\Delta\theta \sin \theta$; and if the rotation takes place in time $\Delta t$, the rate of change is $-BA(\Delta\theta/\Delta t)\sin\theta = -BA\omega \sin\theta$, where $\omega$ is the angular velocity of rotation through the small angle $\Delta\theta$. Thus, for rotation through a small angle, the induced voltage is given by

$$V = NBA\omega \sin\theta = V_m \sin\theta \qquad (34.8)$$

where $\theta$ is the angle through which the plane of the coil is displaced (from a position at right angles to the field) when the small rotation takes place.

Therefore, if the coil in this simple form of a generator is rotated at a constant speed, the induced voltage is at any instant proportional to $\sin\theta$. During the first half-revolution the voltage will be in one direction, or positive, while during the second half-revolution it will be in the opposite direction, or negative. This simple form of generator is an alternating current (or ac) generator, often called an alternator.

By using a model generator, it can easily be shown that the curve of the induced voltage has the form of a sine wave. The model generator used in this experiment is illustrated in Fig. 34.3. It consists of a flat, rectangular coil mounted so that it can rotate in a uniform magnetic field. The field is produced by two sets of permanent horseshoe magnets. The coil

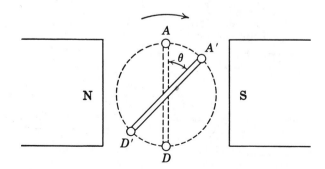

**Figure 34.2** *Armature coil rotating in a uniform magnetic field*

**Figure 34.3** *The model generator*

can be rotated in equal steps by means of a spring-and-ratchet mechanism. The spring provides a constant torque, while a ratchet wheel and a stop allow the coil to rotate through a 10° interval each time the ratchet is released. The terminals of the coil are connected by means of slip rings and brushes to a pair of binding posts.

If the coil is rotated through a series of small equal angles, the induced voltages can be measured by connecting it to a ballistic galvanometer. As discussed in Experiment 32, this instrument measures the quantity of charge passed through it. We can easily show that in the present application the maximum galvanometer deflection will be proportional to the voltage induced during each step $\Delta\theta$. Thus, if this voltage is given by Equation 34.8 and the total circuit resistance (coil and galvanometer) is $R$, then the current flowing during the step, that is, as the generator coil is rotating through $\Delta\theta$, will be

$$i = \frac{V}{R} = NBA\frac{\omega}{R}\sin\theta = NBA\frac{\Delta\theta\sin\theta}{\Delta tR}$$

The quantity of charge passing through the galvanometer is then $i\Delta t = NBA\,(\Delta\theta/R)\sin\theta$ and is thus proportional to the $V$ of Equation 34.8. It follows that the galvanometer deflection will be proportional to the voltage induced in the coil in each of the intervals. Thus, the model generator shows the relative values of the voltages that would be produced at any position of the coil if it were rotated at uniform speed.

## APPARATUS

1. Primary coil of about 100 turns of copper wire
2. Secondary coil of about 1000 turns of copper wire
3. U-shaped piece of iron
4. Straight piece of iron
5. Portable galvanometer
6. Ballistic wall galvanometer
7. Laser light source and scale as required by the galvanometer
8. dc ammeter (0–1 A)
9. Multimeter with ac voltage ranges to 150 V
10. Transformer delivering 6.3 V at 3 A
11. Model generator
12. Power supply capable of delivering at least 5 A at 10 V dc
13. Tubular rheostat (about 100 Ω)
14. Decade resistance box (1000 Ω)
15. A double-pole, double-throw reversing switch
16. Permanent bar magnet

## PROCEDURE

**1.** Connect the secondary coil to the portable galvanometer. Thrust the north pole of the bar magnet into the coil. Record the magnitude and direction of the deflection of the pointer for a fast speed of motion and for a slow speed. Observe whether the deflection is to the left or to the right; a deflection of the pointer to the right means that the current enters the galvanometer at its + terminal. Determine the direction of the current induced in the coil by noting the direction of the current through the galvanometer and the direction in which the coil is wound.

**2.** Repeat Procedure 1, withdrawing the north pole from the coil after having inserted it as in that procedure. Observe the direction of the current induced in the coil when the north pole is being withdrawn.

**3.** Repeat Procedures 1 and 2 using the south pole of the magnet.

**4.** Connect the primary coil in series with the rheostat, the ammeter, the double-pole, double-throw reversing switch, and the power supply, as shown in Fig. 34.4. Have the circuit approved by your instructor. Mount the two coils in a complete square of iron

**Figure 34.4** *Primary circuit used in studying electro-magnetic induction*

made up of the U-piece and the straight piece. Using a current of 0.2 A through the primary coil, observe and record the magnitude and direction of the deflection when the primary circuit is opened and when it is closed. Repeat the observations using primary currents of 0.4, 0.6, 0.8, and 1.0 A.

**5.** Using a current of 1 A, observe the effect of reversing the current in the primary coil. Record the direction of the induced current when the primary circuit is opened and when it is closed for each direction of the primary current.

**6.** Keep the current through the primary coil at some constant value, say 0.5 A. Observe the galvanometer deflection upon opening and closing the primary circuit, using the complete square of iron.

**7.** Repeat Procedure 6 using the straight piece only.

**8.** Repeat Procedure 6 using no iron at all.

**9.** Repeat Procedure 8 placing the two coils twice as far apart as in 8.

**10.** Test your two coils as a transformer by connecting the primary coil to the 6.3-V ac output of the commercial transformer and the secondary coil to the multimeter. CAUTION: Multimeters, or volt-ohm-milliammeters as they are often called, are easily destroyed by being connected in circuits while switched to an improper range. If you have not done Experiment 26, read the discussion of multimeters in the Procedure section of that experiment. Always check the position of the range switch on any multimeter you are using before connecting the meter leads. In the present case, be sure your multimeter is switched to its highest ac voltage range before connecting it in the circuit, and leave the line cord from the 6.3-V transformer unplugged until your setup has been checked by the instructor.

**11.** Begin by mounting the two coils in the complete square of iron made up of the U-piece and the straight piece. Plug in the 6.3-V transformer and adjust the multimeter range switch to the lowest ac voltage range for which the pointer remains on scale. Record the reading. Disconnect the multimeter and reconnect it to read the primary voltage. This should be close to 6.3 V, but the commercial transformer rating is nominal and obviously depends on the line voltage that happens to be present in the laboratory at the time you do the experiment. The primary voltage must therefore be checked as described. Record your result.

**12.** Repeat Procedure 11 using the straight piece only.

**13.** Repeat Procedure 11 using no iron at all.

**14.** Set up the ballistic galvanometer as described in Procedure 1 of Experiment 32. Connect the armature terminals of the model generator to the galvanometer, and then connect the resistance box in parallel with it as shown at $S$ in Fig. 34.5. Set the armature coil in the position for maximum induced voltage, that is, with the plane of the coil parallel to the direction of the magnetic field. Adjust the shunt resistance so that the maximum deflection of the galvanometer is about 20 divisions when the coil is rotated through a 10° interval.

**15.** Begin your observations with the armature coil set at right angles to the magnetic field. Rotate the coil through a 10° interval and observe the maximum deflection of the galvanometer. Take observations of the galvanometer deflection for each 10° movement of the coil over one complete revolution. Reset the spring after every second rotation of the coil to make sure that the spring tension remains the same and therefore that the speed of rotation of the coil remains the same. Record the galvanometer deflection observed for each movement of the coil. Notice that the deflection of the galvanometer corresponds to the position of the coil at the middle of its movement; that is, for the rotation 0° to 10°, $\theta = 5°$; for the rotation 10° to 20°, $\theta = 15°$; etc. The data table has been arranged so that the pertinent value of $\theta$ and the corresponding galvanometer deflection comes opposite the setting of the coil at the *beginning* of each 10° rotation increment.

**Figure 34.5** *Circuit for the model generator*

## DATA

| | Direction of current | Galvanometer deflection | |
|---|---|---|---|
| | | Slow motion | Fast motion |
| North pole into the coil | | | |
| North pole out | | | |
| South pole into the coil | | | |
| South pole out | | | |

| Primary current | Direction of secondary currents | | Galvanometer deflection | | Magnetic circuit |
|---|---|---|---|---|---|
| | Opening | Closing | Opening | Closing | |
| 0.2 A | | | | | Complete square of iron |
| 0.4 A | | | | | Complete square of iron |
| 0.6 A | | | | | Complete square of iron |
| 0.8 A | | | | | Complete square of iron |
| 1.0 A | | | | | Complete square of iron |
| 1.0 A | | | | | Complete square of iron |
| 0.5 A | | | | | Complete square of iron |
| 0.5 A | | | | | Straight piece only |
| 0.5 A | | | | | No iron |
| 0.5 A | | | | | No iron, double distance between coils |

| Core material | ac voltage | | Ratio primary/secondary |
|---|---|---|---|
| | Primary coil | Secondary coil | |
| Complete square of iron | | | |
| Straight piece only | | | |
| No iron | | | |

Number of turns on primary coil   ————   Ratio primary turns/secondary turns   ————

Number of turns on secondary coil   ————

| Setting of the coil | $\theta$ | Galvanometer deflection | Sin $\theta$ | Product of the maximum deflection and sin $\theta$ |
|---|---|---|---|---|
| 0° | 5° | | | |
| 10° | 15° | | | |
| 20° | 25° | | | |
| 30° | 35° | | | |
| 40° | 45° | | | |
| 50° | 55° | | | |
| 60° | 65° | | | |
| 70° | 75° | | | |
| 80° | 85° | | | |
| 90° | 95° | | | |
| 100° | 105° | | | |
| 110° | 115° | | | |
| 120° | 125° | | | |
| 130° | 135° | | | |
| 140° | 145° | | | |
| 150° | 155° | | | |
| 160° | 165° | | | |
| 170° | 175° | | | |
| 180° | 185° | | | |
| 190° | 195° | | | |
| 200° | 205° | | | |
| 210° | 215° | | | |
| 220° | 225° | | | |
| 230° | 235° | | | |
| 240° | 245° | | | |
| 250° | 255° | | | |
| 260° | 265° | | | |
| 270° | 275° | | | |
| 280° | 285° | | | |
| 290° | 295° | | | |
| 300° | 305° | | | |
| 310° | 315° | | | |
| 320° | 325° | | | |
| 330° | 335° | | | |
| 340° | 345° | | | |
| 350° | 355° | | | |
| 360° | | | | |

## CALCULATIONS

**1.** Plot the data of Procedure 4 using primary currents as abscissas and galvanometer deflections as ordinates. Note that the *magnitude* of these deflections for a given primary current should be the same when the switch is opened and when it is closed.

**2.** Calculate the ratios of the primary and corresponding secondary voltages measured in Procedures 11–13. Compare these voltage ratios with the turns ratio of your two coils, which you should calculate if the number of turns on each coil is known.

**3.** Plot a curve, putting the angular displacements of the model generator's armature coil along the x-axis and the corresponding ballistic galvanometer deflections obtained in Procedure 15 along the y-axis. Notice that some of the deflections are positive and some are negative; hence the x-axis should be chosen in the middle of the paper. The scales on the two axes should be so chosen as to utilize, as nearly as possible, the whole sheet of graph paper. Draw a smooth curve through the plotted points, using a solid line for this part of the graph.

**4.** Obtain the value of sin θ from your calculator for each angular displacement of the armature coil, and record the values. Multiply each of these by the maximum ballistic galvanometer deflection for the corresponding half cycle; that is, multiply the positive values of sin θ by the maximum positive deflection and the negative values by the maximum negative deflection (treated as a positive number so that these products will be negative).

**5.** Plot a second curve on the same sheet of graph paper, using the same axes. Put the angular displacements of the armature coil along the x-axis as in Calculation 3 and the product of the maximum deflection and sin θ along the y-axis. Draw a smooth curve through the plotted points, using a dotted line for this part of the graph. This one is a sine curve. Notice how it compares with the curve of Calculation 3.

## QUESTIONS _____

**1.** How does the experiment verify Lenz's law?

**2.** Show how the laws of electromagnetic induction were verified.

**3.** Discuss the essentials of a good magnetic circuit from the observations of Procedures 6–9 and 11–13.

**4.** Name three practical applications of electromagnetic induction.

**5.** By using Lenz's law, how can you predict the direction of flow of current in a secondary coil when the current in the primary is turned on? Assume a direction for the primary current and consider the two coils to be coaxial.

**6.** (a) Can an induced voltage be present in a circuit without an induced current flowing? (b) Can an induced current flow without an induced voltage?

**7.** Explain what happens when a transformer is plugged into a dc line by mistake.

**8.** A circular coil of 200 turns of fine wire has a diameter of 4 cm. It is held in a magnetic field of 0.5 T with its plane perpendicular to the field direction. If it is then pulled out of the field in 0.1 s, what is the induced voltage?

**9.** Suppose that the resistance of the coil of the model generator were doubled, while the size, shape, and number of turns remained the same. (a) How would this affect the voltage produced in the coil? (b) How would it affect the galvanometer deflection? Explain.

**10.** Referring to Fig. 34.1, assume that the coil *ABCD* is rotating clockwise around the axis *OO'*, looking at the coil in the direction *OO'*. Predict the direction of flow of the induced current at the instant when the coil has rotated through an angle of 90° from the position shown in the figure.

**11.** The current induced in the coils of an armature is always an alternating current. Sometimes it is desired to change the alternating current developed in the armature to current that always flows in the same direction in the outside circuit, that is, to make a direct current generator. Explain how this is accomplished by using a commutator.

**12.** A coil is composed of 50 turns of wire and has an area of 100 cm$^2$. It is rotated as the armature coil of a generator in a uniform magnetic field of 0.2 T at a rate of 180 revolutions per minute. Find the amplitude in volts and the frequency in cycles per second (hertz) of the induced voltage.

**13.** Compare the power required to drive a generator on open circuit (no current in the armature) with the power required to drive a loaded generator. Explain the difference.

**14.** We know that a battery provides direct current (dc) and that in older automobiles a dc generator using a commutator as mentioned in Question 11 was provided to charge the battery. Modern cars have alternators, that is, generators of ac. Find out and tell in your own words what is done to allow alternators to charge the batteries and power the dc circuits in today's automobiles.

# The Cathode Ray Oscilloscope 35

The cathode ray oscilloscope is one of the most important instruments used in electrical and electronic research, development, and testing, and is therefore encountered everywhere in laboratories and shops dealing with this kind of work. Basically it is a device for automatically and rapidly plotting one voltage, the so-called *horizontal voltage,* as the abscissa against another, the *vertical voltage,* as the ordinate. If the vertical voltage has a functional dependence on the horizontal voltage, that function is displayed electronically as a graph on the instrument screen. In addition, most oscilloscopes contain circuits which produce a voltage that increases linearly with time. When applied to the horizontal input, this voltage converts the screen's x-axis to a time axis in the resulting graph, and the time dependence of any applied vertical voltage is plotted. It is the purpose of this experiment to become familiar with a typical cathode ray oscilloscope by using it to observe some characteristics of varying voltages.

## THEORY

As noted above, the cathode ray oscilloscope is essentially a device for electronically plotting a graph of one variable against another, the two having a functional relationship and being represented by two voltages that can be applied to the oscilloscope's horizontal and vertical inputs. The complete instrument is built around a cathode ray tube—a long, evacuated glass bulb with an electron-emitting cathode in the small end followed by a series of electrodes used for accelerating the electrons and focusing them into a beam directed at the fluorescent screen which makes up the big end (see Fig. 35.1). This screen gives off light when struck by electrons, so that a bright spot appears at the point where the beam strikes. This spot may be moved around by deflecting the beam, and two sets of deflection plates mounted at right angles to each other are placed in the tube to do this. When a voltage is impressed across the pair of plates in one of these sets, the electrons in the beam are attracted to the positive plate and repelled by the negative one, and are accordingly deflected in their flight to the screen. A vertical displacement of the spot of light proportional to the voltage across the vertical deflection plates can thus be obtained, and similarly a voltage on the horizontal plates produces a proportional horizontal deflection. Thus if a varying voltage $v_x$ is connected to the horizontal plates and a voltage $v_y$, which depends on $v_x$ according to some function, is put on the vertical plates, the spot will trace out a graph of that function on the screen.

The oscilloscope instrument contains the power supplies necessary for the operation of the cathode ray tube and vertical and horizontal amplifiers whose outputs are connected to the vertical and horizontal deflection plates, respectively, and which serve to amplify small signal voltages up to the reasonably high levels required for appreciable deflection of the beam. Gain controls, some calibrated by the manufacturer and some requiring calibration by the user, are provided so that the input voltage required for a given deflection may be set. The calibration is conveniently given in centimeters per volt, and many oscilloscopes have a grid of 1-cm. squares, often called a *graticule,* superposed on the screen, so that the complete display looks like an actual graph. When a voltage that is a function of time is to be studied, the horizontal axis is made a time axis by applying a voltage proportional to $t$ to the horizontal input. Such a voltage is graphed in Fig. 35.2. It rises linearly with time until it reaches a value corresponding to full beam deflection and then drops quickly back to zero and starts over. For obvious reasons, it is often called a "sawtooth" voltage, and special circuits are provided in the oscilloscope to generate it. When applied to the horizontal deflection plates, it causes the spot to move across the screen at a constant rate and then snaps it back and "sweeps" it across again. During a sweep, the horizontal or x deflection is proportional to time, so that this axis may be calibrated in seconds. If the voltage to be studied is applied to the vertical plates while the spot is moving linearly across the screen, a plot of this voltage as a function of time will be traced out. Furthermore, if this voltage is periodic and the sawtooth is made to have the same period and to start at the same instant as a cycle does, a plot of one complete cycle will be traced over and over again. If the frequency is high

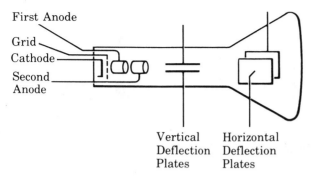

First Anode
Grid
Cathode
Second
Anode

Vertical
Deflection
Plates

Horizontal
Deflection
Plates

**Figure 35.1** *The cathode ray tube*

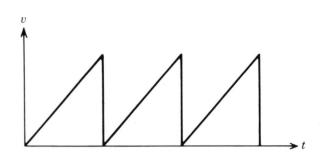

**Figure 35.2** *A sawtooth voltage*

enough so that one trace does not have a chance to fade from the screen before the next one starts (period short compared to the screen's *persistence time*), the plot of one cycle will appear to remain stationary and without flickering. To keep the sweep in step with the voltage under study, the oscilloscope includes *synchronizing* circuits, which deliver pulses to the sweep oscillator telling it when to start each sweep. The complexity of these circuits and the elaborateness of the associated controls vary greatly among different oscilloscopes; the details of the instruments provided in your laboratory will be explained by the instructor. In general, however, all oscilloscopes have the following basic controls:

*Intensity (or Brightness):* Controls the voltage on the cathode ray tube grid and hence the intensity of the electron beam and the brightness of the resulting spot or pattern on the screen.

*Focus:* Controls the voltage on the first anode and hence the potential difference between it and the second anode, which is at a fixed accelerating voltage. This potential difference has the effect of bringing the electron beam to a focus and may thus be adjusted to put the focal point at the screen.

*Vertical Centering:* Controls the dc component of the voltage between the vertical deflecting plates and therefore allows centering of the spot on the screen in the vertical direction. A complete pattern may also be centered to fill the screen evenly in this direction.

*Horizontal Centering:* Has the same function as vertical centering but in the horizontal direction.

*Vertical Gain:* Controls the gain of the vertical amplifier and hence the oscilloscope's vertical deflection sensitivity, that is, the number of centimeters of vertical deflection that will be produced by a given input voltage. On some instruments, this control is calibrated directly in centimeters per volt, but on the less expensive models you must make your own calibration.

*Horizontal Gain:* Same as vertical gain, but applies to the horizontal amplifier. Note that when a horizontal time axis is desired, the output of the sweep generator is switched to the horizontal amplifier input, and the horizontal gain then controls the amplitude of the sawtooth applied to the horizontal deflection plates. In this case the gain control sets the width of the pattern.

*Sweep Frequency:* Controls the sawtooth repetition rate so that it may be set equal to the repetition rate of a periodic voltage applied to the vertical channel for study. The sawtooth repetition rate may also be set equal to a submultiple of that of the voltage under study, in which case several cycles will appear on the screen. In high-quality oscilloscopes, this control is called *sweep speed* and is calibrated directly in seconds for the spot to travel 1 cm horizontally. Such an instrument provides a horizontal axis calibrated in time from which the period and hence the frequency of the vertical signal voltage may be read directly. Less expensive instruments do not have this feature and thus do not provide for measurement of the period of the observed alternating current.

*Synchronization Controls:* These vary in complexity with different oscilloscopes, but in general serve to control the introduction of pulses into the sweep oscillator that tell it when to start each sawtooth. Simple instruments merely have a gain control that sets the level of these pulses. It is advanced from zero until synchronization is obtained but should not be advanced too far, as "oversynchronization" can produce distortion in the pattern by interfering with the linear nature of the sawtooth sweep. There is also a switch to select the source of the pulses. Since the voltage under study and the sweep ordinarily are to have the same period, a very usual source is the voltage under study itself, and the *internal* position of the selector switch makes the necessary connection inside the oscilloscope. For studying voltages derived from the ac power line, a *60-Hz* or *line* position allows pulses taken from the power line to be introduced into the sweep generator. Finally, an *external* position is provided for cases when synchronization is to be effected by pulses from a separate source. Input terminals are provided on the oscilloscope panel for bringing in such pulses. For further details on the instruments provided in your laboratory, consult your instructor.

**Alternating Current**  Because generators produce voltages that vary sinusoidally with time, as shown by Equations 34.6 and 34.7 of the last experiment, such voltages occur repeatedly in practice, so that the term "alternating current" (or "ac") means not just any current that flows first in one direction and then in the other but rather more specifically a sinusoidally varying current such as would be obtained by applying a sinusoidal voltage to a resistor. Such a voltage (often called an ac voltage even though it's not a current) may be written as a function of time as

$$v = V_m \sin(\omega t + \phi_v) \qquad (35.1)$$

where $v$ is the value of the voltage at time $t$, $V_m$ is the maximum value or *amplitude* of this voltage, $\omega$ is the so-called *angular frequency* in radians per second, and $\phi_v$ is the *phase angle* in radians. Notice that the value of $v$ at the instant $t = 0$ is $V_m \sin \phi_v$, so that the purpose of the phase angle is to let Equation 35.1 give the correct value of $v$ at whatever instant the experimenter chooses to call $t = 0$. Fig. 35.3 shows $v$ plotted as a function of $t$ for the particular case of $v = 0$ when $t = 0$, so that $\phi_v = 0$ and $v = V_m \sin \omega t$. Notice that the voltage is positive until $t = \pi/\omega$, when it goes through zero and becomes negative, until returning to zero at $t = 2\pi/\omega$. Thereafter $v$ repeats the performance it has just gone through in the interval 0 to $2\pi/\omega$, and this complete oscillation, which is repeated over and over, is called a *cycle*. The time $2\pi/\omega$ seconds required for one cycle is called the *period* of the ac voltage, and its reciprocal $\omega/2\pi$,

which is the number of cycles completed in one second, is known as the *frequency f*. Equation 35.1 is therefore often written

$$v = V_m \sin(2\pi f t + \phi_v) \qquad (35.2)$$

where $f$ is in cycles per second (cps) or, according to currently accepted terminology, hertz (Hz).

If the voltage $v$ is inpressed across a resistance $R$, Ohm's law holds at every instant so that a current

$$i = \frac{v}{R} = \frac{V_m}{R} \sin 2\pi f t = I_m \sin 2\pi f t \quad (35.3)$$

flows through the resistance. Here we have arbitrarily chosen $v$ to be zero at $t = 0$ so that the phase angle is zero and have written $I_m$ for $V_m/R$, as this is clearly the amplitude of the resulting sinusoidal current. Notice that the power dissipated in the resistance is $iv = (V_m^2/R) \sin^2 2\pi f t$ and is a function of time. We therefore cannot simply multiply it by a time interval to get the energy delivered in that interval because, although power is the rate at which energy is delivered, this power is not constant. Hence an average value must be calculated for this purpose, and the average of $\sin^2 \pi f t$ over one cycle (and thus over any integral number of cycles) is $\frac{1}{2}$. Consequently

$$P_{av} = \frac{1}{2} \frac{V_m^2}{R} \qquad (35.4)$$

This average power can be multiplied by a time to give an energy delivered just as the power $V^2/R$ can in the dc case. For this reason a quantity $V_{rms}$ called the root mean square voltage (because it is the square root of the average of the square of $v$) is defined as $V_m/\sqrt{2}$. The root mean square (rms) voltage can then be used to calculate average power and energy deliv-

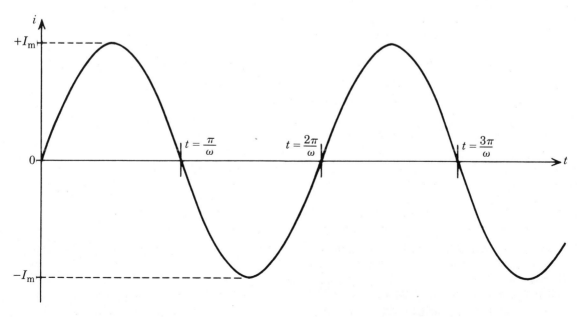

**Figure 35.3**  *A sinusoidal current as a function of time*

ered in the same way that a dc voltage can. This concept is so useful that it is the rms value of ac voltages and currents that is usually quoted. Thus when the ordinary household ac power line is said to be 120 V, it is the rms value, not the amplitude, that is meant.

**Lissajous Figures**   A very important application of the oscilloscope is the study of the pattern resulting from the application of sinusoidal voltages of the same frequency to both the vertical and horizontal inputs. Such a pattern is called a *lissajous figure*. In general, a lissajous figure is an ellipse and is a stationary pattern on the oscilloscope screen if the two frequencies are indeed equal. Thus, if one signal comes from a calibrated variable-frequency source and the other is unknown, tuning the calibrated source until a stationary ellipse is obtained allows the frequency of the unknown to be read from the calibrated source's dial. The nature of the ellipse can also reveal the phase difference and the ratio of the amplitudes of the two applied sinusoidal signals. Thus, if these signals are represented by

$$v_h = A \sin 2\pi f t \qquad (35.5)$$

and

$$v_v = B \sin (2\pi f t + \phi) \qquad (35.6)$$

where $v_h$ is the voltage applied to the horizontal input, $A$ is its amplitude, $v_v$ is the vertical input voltage, $B$ is its amplitude, and $\phi$ is the phase angle between $v_h$ and $v_v$, then the lissajous figure will be an ellipse whose major and minor axes make an angle $\theta$ with the horizontal and vertical axes given by

$$\tan 2\theta = \frac{2 \cos \phi}{(A/B) - (B/A)} \qquad (35.7)$$

Two interesting special cases deserve particular attention. The first is the very simple one obtained when $A = B$ and $\phi = 0$. Then $v_h = v_v$ and the pattern on the oscilloscope screen is a straight line through the origin at an angle $\theta = 45°$. The second is that of $A = B$ and $\phi = 90°$. In this case, $v_v = A \cos 2\pi f t$, and the relation between the vertical

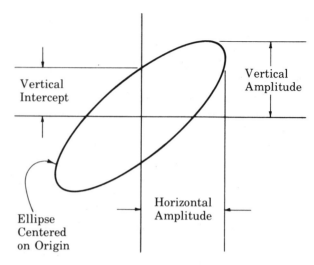

**Figure 35.4**   *Phase measurement with a lissajous figure*

and horizontal voltages is therefore

$$v_h{}^2 + v_v{}^2 = A^2 \qquad (35.8)$$

This is the equation of a circle of radius $A$. In the more general case of $A = B$ and an arbitrary phase angle, an ellipse is obtained from which the phase angle can be determined as follows (see Fig. 35.4): If the horizontal deflection is given by Equation 35.5, it will be zero at $t = 0$, $1/f$, $2/f$, etc., and at these times the vertical deflection, which is given by Equation 35.6, is $B \sin \phi$. The vertical deflection at zero horizontal deflection is the vertical intercept marked in Fig. 35.4 and can thus be read from the oscilloscope pattern. If $A$ is made equal to $B$, the vertical intercept is $A \sin \phi$ and thus gives $\sin \phi$ when divided by the vertical amplitude. We conclude that the phase angle can be obtained from the relation

$$\sin \phi = \frac{\text{vertical intercept}}{\text{vertical amplitude}} \qquad (35.9)$$

These characteristics of lissajous figures will be useful in our study of the performance of resistors, inductors, and capacitors in ac circuits.

## APPARATUS

1. Cathode ray oscilloscope
2. Transformer delivering 6.3 V at 3 A
3. Signal generator

4. 1.5 dry cell
5. Multimeter with rms ac voltage ranges
6. Assorted hookup wires and connectors

## PROCEDURE

**1.** Connect the output (secondary) of the transformer to the vertical input of the oscilloscope and set the horizontal circuits for a time sweep appropriate for viewing a 60-Hz voltage. Before plugging in the transformer, have the instructor check your ar-

rangement and explain the operation of your particular oscilloscope.

**2.** Plug in the transformer and adjust the oscilloscope to produce a stationary pattern of one cycle of a sine curve. Learn to use the sweep frequency and synchronizing controls on your instrument by readjusting them to get stationary patterns of two and three cycles. Note that since you are viewing a 60-Hz signal, you may use either "internal" or "line" synchronization.

**3.** Calibrate the vertical direction on the oscilloscope screen in centimeters per volt at the vertical input terminals. The method appropriate for your oscillocope will be described by the instructor. Then measure the transformer secondary voltage (which is also the vertical input voltage) with the multimeter. CAUTION: Multimeters, or volt-ohm-milliammeters as they are often called, are easily destroyed by being connected in circuits while switched to an improper range. If you have not done Experiment 26, read the discussion of multimeters in the Procedure section of that experiment. Always check the position of the range switch on any multimeter you are using before connecting the meter leads. In the present case, be sure your multimeter is switched to the range appropriate for measuring the 6-V ac output of the transformer before connecting it. Record the meter reading. Using the vertical calibration just obtained, measure and record the voltage amplitude by converting the amplitude of the sinusoidal pattern displayed on the oscilloscope to volts.

**4.** If your oscilloscope has a calibrated time base (horizontal sweep giving you a known time per centimeter along the screen's x-axis), use it to measure the period of one cycle of your transformer's secondary voltage. To do this with optimum precision, set the sweep speed to display two cycles of the ac voltage, adjust the horizontal centering control to bring one peak of the pattern exactly on a vertical graticule line, and measure the distance to the next peak. This will be in centimeters, which can then be converted to seconds using the chosen time base calibration (seconds per centimeter selected by the sweep speed switch on oscilloscopes so equipped).

**5.** If your oscilloscope has an "ac-dc" switch, notice that it does not set the instrument for operation from an ac or dc power line but controls the response of the vertical amplifier. In the dc position, any voltage, including a constant one, that is connected to the vertical input terminals will cause a vertical deflection of the trace on the screen. With this switch in the ac position, however, only ac signals will produce a vertical deflection, the oscilloscope's response cutting off when the frequency gets down to 3 or 4 Hz. Throw this switch back and forth while observing the two-cycle pattern displayed in Procedure 4 and note below whether you see any change.

**6.** Disconnect the transformer secondary from the vertical input, connect it to the synchronizing signal input, and set your oscilloscope for external synchronization. This will give you a horizontal line on the screen. Now connect the 1.5-V dry cell to the vertical input terminals. Throw the "ac-dc" switch back and forth as in Procedure 5 and record below what you see.

**7.** Use your observations in Procedure 6 and your vertical calibration to measure the voltage of the dry cell. Switch the multimeter to the proper scale and measure the dry-cell voltage with it. Record the two readings and compare them.

**8.** Disconnect the dry cell and the transformer secondary, unplug the transformer, and plug in the signal generator instead. Connect the latter's output to the vertical input of the oscilloscope. Set the signal generator frequency to about 400 Hz and obtain a two-cycle pattern of the generator output on the screen. Internal synchronization should be used in this case. Vary the generator output and also the generator frequency and describe your observations in the space below.

**9.** Switch off the oscilloscope's time base so that the instrument is set for an external horizontal input. CAUTION: Leave the vertical deflection due to the signal generator as it was in Procedure 8, so that a vertical line is displayed. Oscilloscopes should not be left with the beam on but no horizontal or vertical deflection, as the continuous electron bombardment at one point on the screen will soon burn away the fluorescent material there. Connect the transformer secondary to the *horizontal* input terminals, have the instructor check your setup, and then plug the transformer into the ac line. Set the signal generator frequency to about 30 Hz and adjust the horizontal and vertical gain controls to obtrain a lissajous figure filling most of the screen. Vary the signal generator frequency carefully in the neighborhood of 30 Hz until you obtain a stationary pattern having two loops, one above the other. Record in the Data section the exact reading of the generator dial when the pattern is stationary.

**10.** Shift the signal generator frequency to about 60 Hz and again move the dial slowly until a stationary lissajous figure is obtained. Notice that this may be a circle, an ellipse, or a straight line depending on the phase difference between the transformer secondary voltage and the generator output and on their relative amplitudes. Try to get a circle by making the generator's output amplitude equal to that of the transformer secondary voltage and shifting its freqency very carefully to bring about a 90° phase difference. Then try to obtain a straight line at 45° to the horizontal by bringing the phase difference to zero degrees. Finally, with the pattern stationary, observe and record the reading on the generator's frequency dial.

**11.** Increase the generator frequency to about 120 Hz, again adjust it carefully to obtain a stationary lissajous figure, and record the dial setting. Note that this time the pattern should have two loops, one beside the other.

**12.** Repeat Procedure 11 with the signal generator set for around 180 Hz. A three-loop stationary pattern should be obtained.

**13.** When you have finished the experiment, switch off the oscilloscope and signal generator and unplug the transformer.

## DATA

rms transformer secondary voltage
  (from multimeter) _____

Calculated amplitude of transformer
  secondary voltage _____

Distance between peaks of line voltage
  pattern _____

Frequency of line voltage _____

Dry-cell voltage from oscilloscope _____

Ampitude from oscilloscope _____

Percent error _____

Period of line voltage _____

Percent error _____

Dry-cell voltage from multimeter _____

| Frequency, Hz | 30 | 60 | 120 | 180 |
|---|---|---|---|---|
| Generator dial reading | | | | |

## CALCULATIONS

**1.** Compute the amplitude of the transformer secondary voltage from the rms value measured with the multimeter in Procedure 3. Compare this result with the amplitude as determined from the oscilloscope by calculating the percent error.

**2.** Calculate the period of the ac wave observed in Procedure 4. Then calculate its frequency and compare it with the "true" value of 60 Hz by finding the percent error. Note that the 60-Hz line frequency is not a nominal value but very precise (about 1 part in 10,000), as the accuracy of certain kinds of electric clocks depends on it.

**3.** Calculate the voltage of the dry cell from the displacement of the horizontal line observed in Procedure 7. Note the difference between the voltage so obtained and that measured with the multimeter.

**4.** Plot a calibration curve for your signal generator in the range 20 to 200 Hz using the data of Procedures 9–12. Draw the best smooth curve through your four plotted points, extrapolating it to 20 Hz at one end and 200 Hz at the other.

## QUESTIONS

**1.** Calculate the amplitude of the 120-V ac line voltage.

**2.** What is the time standard (analogous to the balance wheel in a windup watch or the pendulum in a pendulum clock) in an electric clock that does *not* use the ac power line for this purpose?

**3.** In high-quality oscilloscopes having the sweep frequency control calibrated directly in seconds per centimeter, the horizontal gain control is automatically switched out of the circuit when the sweep system is being used so that the pattern width is permanently preset rather than being adjustable. Why is this so?

**4.** Show that if there is no phase difference between the two sinusoidal voltages applied to an oscilloscope in a lissajous figure observation, the angle $\theta$ between the pattern axis and the horizontal axis is given by

$$\tan \theta = \frac{B}{A}$$

where $A$ is the horizontal and $B$ the vertical amplitude. What does the pattern look like under these circumstances? Does Equation 35.7 agree with the result when $\phi = 0$?

**5.** A certain cathode ray tube (CRT) has deflection plates of length $l = 2$ cm and spacing $d = 0.5$ cm located a distance $L = 30$ cm (measured from the center of the plates) from the fluorescent screen. Electrons are accelerated by a potential difference $V_{acc} = 1000$ V and focused in a beam directed down the axis of the tube between the deflection plates and onto the center of the screen (see Fig. 35.1). (a) Derive a formula for the deflection sensitivity of the CRT, that is, the distance $D$ through which the spot on the screen is deflected for each volt applied to the deflection plates. (b) Put the given values of $l$, $d$, $L$, and $V_{acc}$ into your formula to obtain a numerical result for the deflection sensitivity. (c) On the basis of your answer to Part (b), explain why vertical and horizontal amplifiers are needed.

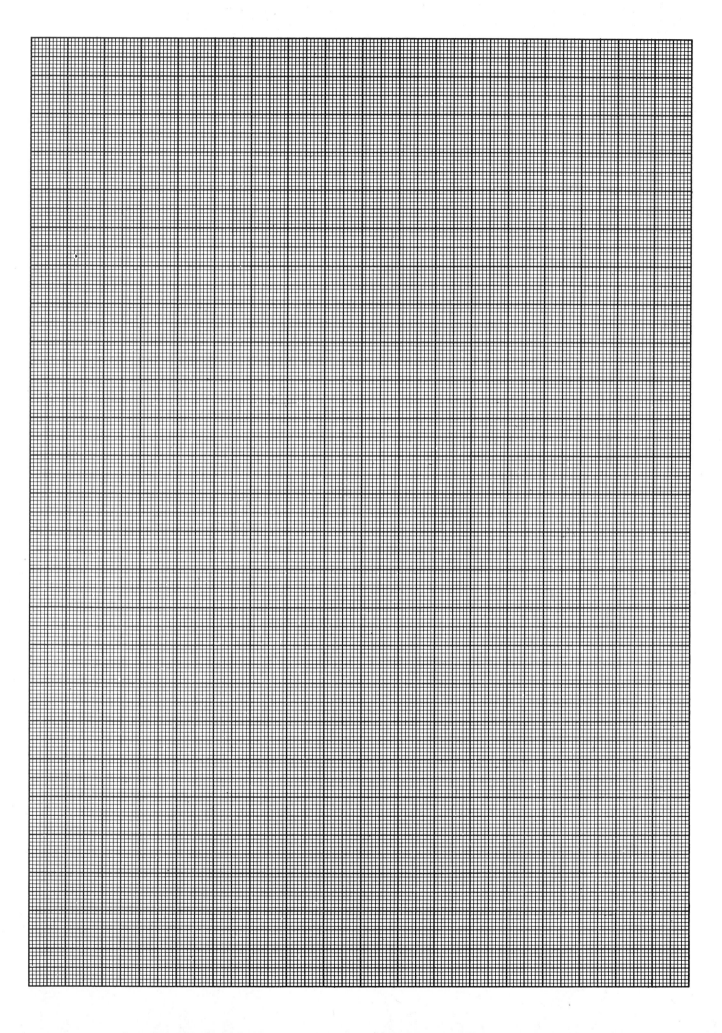

# Alternating Current 36

Because of the usefulness of the transformer, which does not work on direct current, alternating current has dominated the electrical industry for many years. Strictly speaking, the term "alternating current" should mean any current that varies with time so that it flows sometimes in one direction and sometimes in the other. However, as usually used, the term refers specifically to just one particularly simple form of such variation, namely a sinusoidal dependence on time. In this experiment, only currents and voltages that are sinusoidal functions of time will be dealt with, and the performance of circuit elements under these conditions will be studied.

## THEORY

An ac current that is a sinuosidal function of time may be written

$$i = I_m \sin(\omega t + \phi_i) \qquad (36.1)$$

where $i$ is the instantaneous value of the current in amperes, $I_m$ is the maximum value or *amplitude* of this current, $\omega$ is the so-called *angular frequency* in radians per second, $t$ is the time in seconds, and $\phi_i$ is the *phase angle* in radians. The purpose of the phase angle is to set the value of $i$ at $t = 0$. Thus, at this initial instant, $i = I_m \sin \phi_i$, which may have any value from 0 to $\pm I_m$ depending on the value chosen for $\phi_i$. Figure 36.1 shows $i$ plotted as a function of $t$ for the particular case of $i = 0$ at $t = 0$, so that $\phi_i = 0$ and $i = I_m \sin \omega t$. Notice that the current returns to zero after flowing in the positive direction when $t = \pi/\omega$ and again after its reverse (negative) flow when $t = 2\pi/\omega$. After $t = 2\pi/\omega$, the current repeats the performance it has just gone through in the interval 0 to $2\pi/\omega$, and this complete oscillation, which is repeated over and over, is called a *cycle*. The time $2\pi/\omega$ seconds required for one cycle is called the *period* of the alternating current, and its reciprocal $\omega/2\pi$ is the number of cycles per second and is known simply as the frequency $f$. Equation 35.1 is therefore often written

$$i = I_m \sin(2\pi ft + \phi_i) \qquad (36.2)$$

where $f$ is in cycles per second (cps) or, according to currently accepted terminology, hertz.

The question now arises as to what happens when a sinusoidal current passes through a circuit element such as a resistor. For alternating current, two other circuit elements, the inductor and the capacitor, are of interest. Notice that for direct current the ideal inductor, a coil of resistanceless wire, is simply a

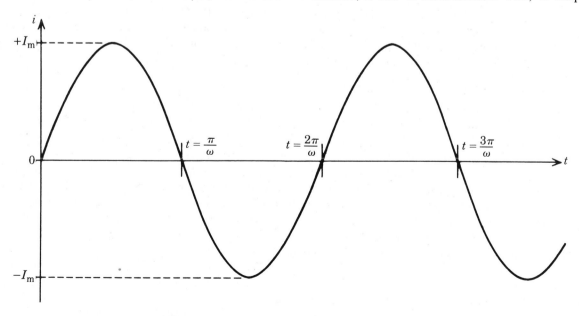

**Figure 36.1**  *A sinusoidal current as a function of time*

short circuit, whereas an ideal capacitor, two conducting plates separated by a perfect insulator, is an open circuit. With alternating current, however, the current through the inductor changes, hence the magnetic field changes and an induced voltage appears across the coil terminals. In the case of the capacitor, charge flows onto its plates with one polarity when the current is in one direction and the voltage across the capacitor builds up, but then the current reverses, discharging the capacitor and recharging it in the opposite direction. We will see that in each case there results a sinusoidal voltage across the circuit element that is proportional to the current (just as in the case of the resistance) so that a sort of Ohm's law holds for inductors and capacitors as well when the voltages and currents involved are sinusoidal. The proportionality constant, however, is called the reactance instead of the resistance in the case of inductors and capacitors. The reason for this will become apparent.

**Resistance in ac Circuits**  If the current $i$ flows through a resistance $R$, Ohm's law holds at every instant so that the voltage across the resistance is

$$v_R = iR = I_m R \sin 2\pi ft = V_{mR} \sin 2\pi ft \quad (36.3)$$

where we have taken zero phase angle for the current and have written $V_{mR} = I_m R$ since $V_{mR}$ is clearly the amplitude of the sinusoidal voltage that develops across the resistance. Notice that the power dissipated in the resistance is $iv_R = I_m^2 R \sin^2 2\pi ft$ and is a function of time but is never negative. This is because the current varies but energy is transformed into heat in the resistor no matter which way the current goes through it. Power is the rate at which energy is delivered, but in this case we cannot simply multiply $I_m^2 R \sin^2 2\pi ft$ by a time interval to get the total energy because this power is not constant. An average value of the time-varying power must be calculated for this purpose, and the average of $\sin^2 2\pi ft$ over one cycle (and hence over any integral number of cycles) is $\frac{1}{2}$. Thus,

$$P_{av} = \frac{1}{2} I_m^2 R \quad (36.4)$$

This average power can be multiplied by a time to give an energy delivered just as the power $I^2 R$ can in the dc case. For this reason, a quantity $I_{rms}$ called the root mean square current (because it is the square root of the average of the square of $i$) is defined as $I_m/\sqrt{2}$. The root mean square (rms) current can then be used to calculate average power and energy delivered in the same way that direct current can. This concept is so useful that it is the rms value of ac currents and voltages that is usually quoted. Thus, when the ordinary household ac power line is said to be 120 V, it is the rms value, not the amplitude, that is meant.

**Inductance in ac Circuits**  When an inductor (a coil of wire) carries a current, a magnetic field is produced. If the current and hence the field varies, there is a rate of change in the magnetic flux through the coil and an induced voltage appears across its terminals even though the resistance of the coil may be zero. If the current through an inductor is given by Equation 36.2 with $\phi_i = 0$, a flux proportional to this current will result, so that we can write

$$\Phi = lI_m \sin 2\pi ft \quad (36.5)$$

where $l$ is the proportionality constant and has the dimensions of webers per ampere. Now Faraday's induction law states that the induced voltage around a single loop is the negative of the rate of change of magnetic flux through that loop. The flux given by Equation 36.5 is certainly changing with time, and the rate of change of the sine of an angle is known to be the cosine of that angle multiplied by the angle's own rate of change. Thus the voltage induced around one turn of the coil is $-2\pi fl I_m \cos 2\pi ft$ and across the coil terminals is $-2\pi fNl I_m \cos 2\pi ft$, where $N$ is the number of turns. A voltage equal and opposite to this must therefore be applied to the coil terminals to keep the current flowing, and this will be

$$v_L = 2\pi fLI_m \cos 2\pi ft = V_{mL} \sin (2\pi ft + 90°) \quad (36.6)$$

where $L = Nl$ is called the *inductance* of the coil and is measured in henrys, in honor of the American physicist Joseph Henry (1797–1878), who was one of the early workers in electromagnetism. The inductance is a quantity that characterizes the coil and depends on the permeability of the core material (if any), the number of turns, and geometrical factors. Equation 36.6 suggests that $L$ is the proportionality constant between the rate of change of current through an inductor and the resulting voltage induced across the inductor's terminals, and it is often so defined.

The important thing about the result stated by Equation 36.6 is that the voltage across the inductor is sinusoidal just as the current through it is. The amplitude $V_{mL}$ of the voltage is equal to $2\pi fLI_m$ and is thus proportional to the amplitude $I_m$ of the current. The proportionality constant $X_L = 2\pi fL$ is called the *inductive reactance* of the coil and appears to play the role of resistance in relating $I_m$ and $V_{mL}$. Indeed, it has the dimensions of ohms. There are several differences, however, between this case and that of the resistor. Here, while the current goes as $\sin 2\pi ft$, the voltage goes as $\cos 2\pi ft$, that is, as the current with a phase shift of 90° or $\pi/2$ rad. Thus, while the current through and the voltage across a resistor are *in phase*, the current and voltage for an inductor are 90° *out of phase with the voltage leading the current*. Plots of $i$ and $v_L$ are shown in Fig. 36.2. Notice that when $i$ is

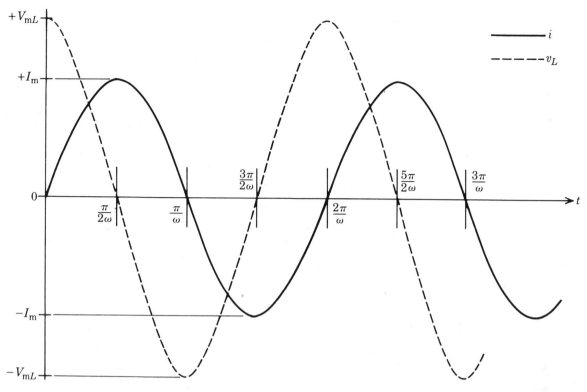

**Figure 36.2** *Voltage and current in an inductor*

at its maxima and minima so that for an instant it and the resulting magnetic flux are not changing, $v_L$ is zero, whereas $v_L$ has its maxima and minima at the instants when $i$ is zero, at the points where the sine curve changes most rapidly. Note also that, unlike $R$, $X_L$ is proportional to the frequency and becomes zero when $f = 0$. This is to be expected since a coil of resistanceless wire is a short circuit for direct current. Finally, the power delivered to the inductor cannot be obtained from the expression $i^2 X_L$ because $v_L$ is not equal to $iX_L$. Instead, the power delivered must be calculated from the original definition of electric power, namely the current times the voltage. But if $i$ is multiplied by $v_L$ we get $2\pi f L I_m^2 \sin 2\pi ft \cos 2\pi ft$, which is sometimes negative and sometimes positive. Moreover, the expression $\sin 2\pi ft \cos 2\pi ft$ averages to zero, showing that on the average no energy is delivered to the inductor. This is hardly surprising, since energy delivered to the inductor has to go somewhere, and the only place it can go is into the magnetic field. This is in fact what happens when the factor $\sin 2\pi ft \cos 2\pi ft$ is positive. When it is negative, the magnetic field is collapsing and the stored energy is being returned to the source. Thus, unlike the resistor, the inductor does not dissipate but rather temporarily stores energy.

**Capacitance in ac Circuits** When an electric current flows into a capacitor, the plates become charged

and the terminal voltage rises according to the relation

$$v_C = \frac{Q}{C} \tag{36.7}$$

where $Q$ is the charge on the capacitor in coulombs, $C$ is the capacity of the capacitor in farads, and $v_C$ is the capacitor voltage. If the charging current is dc, the capacitor voltage will simply build up to the source voltage, after which current flow will cease, for the capacitor is in fact an open circuit. However, if alternating current is used, the capacitor will charge up until the current reverses, after which it will discharge and then charge again with the opposite polarity. Thus current flows alternately into and out of the capacitor through the connecting wires, so that alternating current appears to pass through it, although no charges can actually travel through the dielectric from one plate to the other.

The situation is illustrated in Fig. 36.3. Looking at the first half-cycle of the current, we note that $i$ is positive throughout that half-period, and so the capacitor will be charging. It must reach its maximum voltage at the end of the half-cycle because thereafter $i$ reverses and the capacitor discharges. This process continues during the second half-cycle, the capacitor attaining maximum charge in the reverse direction and hence maximum negative voltage at the end of a full cycle of the ac current. A detailed study shows

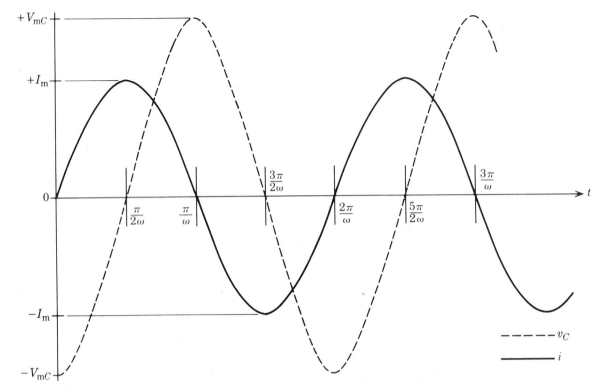

**Figure 36.3** *Voltage and current in a capacitor*

that the capacitor voltage, like the voltage across the inductor, is sinusoidal but is shifted by $-90°$ ($-\pi/2$ rad) with respect to the current. In other words, the phase shift is again 90°, but the current leads the voltage. Thus $v_C$ will be proportional to sin $(2\pi ft - 90°)$ or $-\cos 2\pi ft$.

The amplitude $V_{mC}$ of the capacitor voltage must surely be proportional to $I_m$, for this represents the maximum charging rate. Moreover, Equation 36.7 shows that $V_{mC}$ should be inversely proportional to $C$, since the greater the capacity the lower the voltage developed by a given amount of charge. We also expect the voltage amplitude to be proportional to the period of the ac current, for half of this period is just the time allowed for charging the capacitor in any one direction. In terms of the frequency, the reciprocal of the period, we should find $V_{mC} \sim 1/f$. Detailed calculation yields

$$v_C = -\frac{I_m}{2\pi fC}\cos 2\pi ft = V_{mC}\sin(2\pi ft - 90°) \quad (36.8)$$

The quantity $1/(2\pi fC)$ which multiplies $I_m$ to give $V_{mC}$ is called the *capacitative reactance* of the capacitor. Like the inductive reactance, it is the ratio of the voltage amplitude to the current amplitude and therefore has the dimensions of ohms. It is also dependent on the frequency but goes to infinity rather than zero as the frequency decreases, indicating that the capacitor is an open circuit rather than a short circuit for direct current. Finally, as in the case of the inductor, the 90° phase difference between the voltage and the current results in energy being alternately delivered to

and returned from the capacitor. In the capacitor, it is the electric field between the plates rather than the magnetic field of the coil that stores the energy; but as in the inductor, no energy is dissipated.

**The Series *RLC* Circuit**   Suppose a resistor, an inductor, and a capacitor are connected in series and a current $i = I_m \sin 2\pi ft$ is sent through them, as shown in Fig. 36.4. The total voltage $v$ across the combination must then be the sum of the voltages across the three series elements, and from Equations 36.3, 36.6, and 36.8 we have

$$v = v_R + v_L + v_C = I_m[R \sin 2\pi ft \quad (36.9)$$
$$+ (X_L - X_C)\cos 2\pi ft]$$

This expression is not in a convenient form because it does not display either the amplitude or the phase of $v$, and it does not really show that $v$ is sinusoidal. It can, however, be rewritten in the form

$$v = I_m Z \sin(2\pi ft + \phi_v) \quad (36.10)$$

That Equations 36.9 and 36.10 are equivalent is easily shown by expanding the latter with the usual formula for the sine of the sum of two angles. Doing this gives

$$v = I_m(Z\cos\phi_v \sin 2\pi ft + Z\sin\phi_v \cos 2\pi ft) \quad (36.11)$$

Equations 36.11 and 36.9 are identical if

$$R = Z\cos\phi_v \quad (36.12)$$

and $\qquad (X_L - X_C) \equiv X = Z\sin\phi_v \quad (36.13)$

**Figure 36.4** *The series* RLC *circuit*

where $X$ is simply defined as the reactance of the circuit and $\phi_v$ is the phase angle of the voltage with respect to the current.

The quantity $Z$ is called the *impedance* of the circuit. It multiplies the current amplitude $I_m$ to give the amplitude $I_m Z$ of the voltage across the complete circuit. Like the resistance and reactance, it is measured in ohms and is given by

$$Z = \sqrt{R^2 + X^2} \tag{36.14}$$

This solution for $Z$ is obtained by squaring and adding Equations 36.12 and 36.13. These equations can also be solved for $\phi_v$ by dividing the second by the first. This gives

$$\tan \phi_v = \frac{X}{R} \tag{36.15}$$

These last two results suggest that $Z$ can be represented by a vector drawn on a set of coordinate axes designated $X$ and $R$ as shown in Fig. 36.5. $X$ is positive on the left because $X_L > X_C$, and we say that the net reactance is inductive. On the right is a case of net capacitive reactance ($X$ negative). A most interesting possibility is that of $X = 0$ by reason of $X_L$ and $X_C$ having the same magnitude. In this case, which is called *series resonance*, the circuit acts as if only $R$ were present. The reason is that $v_L = X_L I_m \cos 2\pi ft$ and $v_C = -X_C I_m \cos 2\pi ft$, so that if $X_L = X_C$ these two voltages will be equal in magnitude but opposite in polarity at every instant and will thus cancel each other out. Note, however, that neither $v_C$ nor $v_L$ is zero. This illustrates the most important point that in a series ac circuit the amplitudes of the voltages across the individual elements may not be added, for these voltages will in general not have their maximum values at the same time due to their having different phase angles. It is therefore the instantaneous values that must be added, as was done in obtaining Equation 36.9. In particular, in the series *RLC* circuit, the voltage across the inductor is maximum and positive when the voltage across the capacitor is at its maximum negative value. Making the two reactances

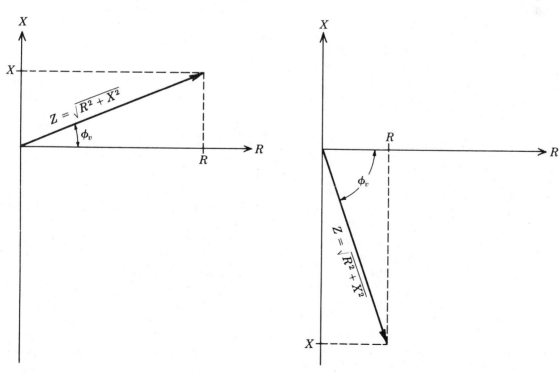

**Figure 36.5** *Impedance as a vector*

equal makes these two maxima equal, so that the total voltage across the inductor and capacitor in series is zero. Since as the frequency is increased $X_L$ gets larger while $X_C$ gets smaller, there will surely be a particular frequency $f_0$ for which they are equal. This is easily found to be

$$f_0 = \frac{1}{2\pi \sqrt{LC}} \qquad (36.16)$$

## APPARATUS

1. Cathode ray oscilloscope
2. Transformer delivering 6.3 V at 3 A
3. Iron-core inductor (about 1 H)
4. Capacitor decade with a range up to at least 8 $\mu$F
5. Carbon resistor, 1-W, 5% tolerance, 2200 $\Omega$
6. Decade resistance box (0–10,000 $\Omega$)
7. Multimeter with rms ac voltage ranges
8. Assorted hookup wires and connectors

## PROCEDURE

**1.** Wire the transformer, oscilloscope, carbon resistor, and decade box in the circuit diagrammed in Fig. 36.6. If you have not done Experiment 35, read the Theory section of that experiment to learn about the oscilloscope, and have your instructor explain the operation of the ones in your laboratory to you. The carbon resistor should be used in the position marked "Element Under Test." Notice that this element and the decade box are in series across the 6.3-V source (the oscilloscope inputs have very high impedance and thus draw negligible current) so that the same current flows in both and that the voltage across the element under test is applied to the oscilloscope's vertical input and the voltage across the decade resistance box to the horizontal input. These two voltages obviously have the same frequency, and so a stationary lissajous figure will be displayed on the oscilloscope screen. Note, however, that the "high" terminal of the horizontal input is connected to the "low" rather than the "high" terminal of the resistance box. This means that the polarity of the voltage applied to the horizontal input is reversed relative to that applied to the vertical input, so that the pattern will be reversed left-to-right as compared with that shown in Fig. 35.4 of the last experiment. This will have no effect on your measurements, however.

**2.** Set the oscilloscope to display the lissajous figure, that is, switch the horizontal channel to the input terminals rather than the sweep generator. CAUTION: Be sure you have a pattern rather than just a spot on the screen. Oscilloscopes should not be left with the beam on but no horizontal or vertical deflection, as the continuous electron bombardment at one point on the screen will soon burn away the fluorescent material there. Set the horizontal and vertical gain controls so that the horizontal and vertical sensitivities are equal and are such as to give a pattern filling most of the screen. To do this, set the resistance box to zero and temporarily shift the "high" horizontal input lead to the "high" vertical input terminal so that both oscilloscope inputs receive the same voltage, namely that across the resistor under test. Then adjust the gain controls so that a straight line extending almost all across the screen and at exactly 45° to the horizontal is obtained. The centering controls should be adjusted until this line passes through the origin, that is, the center of the screen. The angle of 45° is assured if the horizontal extent of the line (its horizontal component) is exactly equal to its vertical extent (the vertical component). Do not touch the centering or gain controls after these adjustments have been made. *Note:* With oscilloscopes having calibrated gain controls it should only be necessary to set both to the same sensitivity, choosing the value of this latter to fill the screen as described above. Check that the resulting straight line is indeed at 45° by seeing that it passes exactly diagonally through the squares of the *graticule*, that is, the grid of 1-cm squares usually found superposed on the screen of instruments of this quality.

**3.** Disconnect the horizontal input lead from the vertical input terminal and reconnect it to the "low" side of the resistance box so that your circuit is again as shown in Fig. 36.6. Adjust the resistance box until a line at exactly 45° to the horizontal is displayed. *Note:* this line will extend up to the left rather than up to the right like the line obtained in Procedure 2 because of the polarity reversal discussed in Procedure 1. Record the final setting of the decade box.

**4.** Remove the carbon resistor from your circuit and measure its resistance directly with the multimeter switched to be an ohmmeter. Have your instructor tell you how to do this if he or she has not already done so. Also note that this resistor has its nominal value marked on it by means of a color code. Your instructor will explain this code to you. The colors associated with the ten digits (0–9) are listed in Table XIV at the end of the book. Record the value of your carbon resistor as measured by the ohmmeter.

**5.** Connect the inductor in your circuit as the element under test. Because the inductor is not a pure inductance but presents an impedance consisting of a

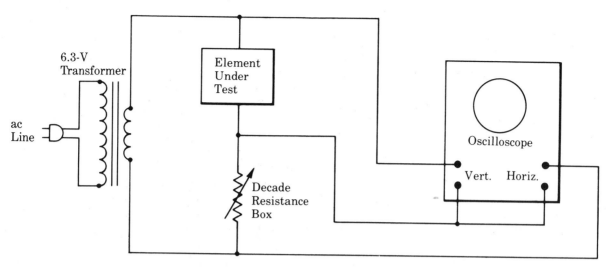

**Figure 36.6**  *Circuit for impedance measurement*

resistance and its inductive reactance in series, the pattern will not be a circle but an ellipse similar to that shown in Fig. 35.4 except for the left-right reversal. Adjust the resistance box so that the horizontal and vertical pattern amplitudes defined in that figure are exactly equal. Record the resistance box setting and the values of the vertical amplitude and vertical intercept.

**6.** Remove the inductor from your circuit and measure its dc resistance with the multimeter set up as an ohmmeter. Record your result.

**7.** Connect the capacitor decade in your circuit as the element under test and set it for a capacitance of 2 μF. Adjust the resistance box so that equal horizontal and vertical amplitudes are obtained. Practical capacitors approach the ideal rather closely, so that their impedance is usually very nearly a pure capacitative reactance. You should therefore get a good circular pattern. In this case, the vertical amplitude and vertical intercept are the same, and Equation 35.9 of Experiment 35 gives $\phi = 90°$ as it should for a pure capacitance. Observe this fact, and record the setting of the resistance box. *Note:* The transformer may disturb the sinusoidal nature of the voltage enough so

that some distortion may be apparent in the displayed circle, but if a reasonably good transformer is used, this effect will be negligible.

**8.** Repeat Procedure 7 with the capacitor decade switched to 4 μF.

**9.** Rewire your components in the circuit of Fig. 36.7 in order to observe series resonance. Use the decade capacitor at $C$, the iron core inductor at $L$, and the resistance box set to 5 Ω at $R$. Notice that $R$ serves as a current transducer, that is, the current through it produces a proportional voltage across it according to Ohm's law and this voltage is then displayed on the oscilloscope. The vertical oscilloscope deflection is thus proportional to the current flowing through the circuit. Set the oscilloscope to view the 60-Hz alternating current and look for series resonance by varying the decade capacitor until the sine curve appearing on the screen has maximum amplitude. This corresponds to resonance, for the circuit impedance is a minimum when $X = 0$; hence a fixed voltage source such as the transformer secondary coil will produce maximum current. Record the value to which $C$ is set when maximum current amplitude is observed.

**Figure 36.7**  *Circuit for observing resonance*

# DATA _____

### Resistance Measurement

| | | | |
|---|---|---|---|
| Resistance box setting | _____ | Percent error | _____ |
| Measured value of carbon resistor | _____ | | |

### Inductance Measurement

| | | | |
|---|---|---|---|
| Resistance box setting | _____ | Percent discrepancy | _____ |
| Pattern amplitude | _____ | Reactance of inductor | _____ |
| Pattern intercept | _____ | Inductance of inductor | _____ |
| Phase angle $\phi$ | _____ | Given value of the inductance | _____ |
| Resistance of inductor from oscilloscope | _____ | Percent discrepancy | _____ |
| Resistance of inductor from ohmmeter | _____ | | |

### Capacitance Measurement

| Value of capacitance | Resistance box setting | Capacitance | | Percent discrepancy |
|---|---|---|---|---|
| | | Reactance | Measured value | |
| 2 $\mu$F | | | | |
| 4 $\mu$F | | | | |

### Series Resonance

| | | | |
|---|---|---|---|
| Value of $C$ | _____ | Calculated value of the frequency | _____ |
| Value of $L$ | _____ | Percent error | _____ |

# CALCULATIONS _____

**1.** Refer to Fig. 35.4 of the last experiment and, by writing out an appropriate derivation in the space below, satisfy yourself that Equation 35.9 is correct.

**2.** The resistance box setting obtained in Procedure 3 should be equal to the resistance of the carbon resistor. Compare it with the value measured in Procedure 4 by calculating the percent error.

**3.** Calculate the phase angle between the current through and the voltage across the inductor from the data of Procedure 5. The resistance box setting obtained in this step is the inductor's impedance. Using Equations 36.12 and 36.13, compute the resistance and reactance of the inductor. Compare the resistance so obtained with that measured with the ohmmeter by calculating the percent error.

**4.** From the reactance of the inductor found in Calculation 3 and the known line frequency, compute the inductance of the inductor. Compare your result with the known value by calculating the percent discrepancy. *Note:* The value of the inductance in henrys marked on the inductor is nominal and will in general be smaller than your measured value because it is usually given as the effective value when a superposed dc current is flowing. Thus close agreement between the marked value and the result of your measurement cannot be expected. If a better value of the inductance is available, your instructor will give it to you when you have finished the experiment so that a more meaningful comparison can be made.

**5.** Calculate the capacitance from your data in Procedure 7 and compare it with the known value by calculating the percent discrepancy. The settings of the decade capacitor should be accurate to 3% or better.

**6.** Repeat Calculation 5 using the data of Procedure 8.

**7.** Using the resonant value of $C$ found in Procedure 9 and the inductance found in Calculation 4, compute the resonant frequency of your series *RLC* circuit. Compare your result with the known value of the ac line frequency, 60 Hz. *Note:* The value 60 Hz for this frequency is not nominal but very precise (about 1 part in 10,000), since the accuracy of certain kinds of electric clocks depends on it. Thus, if you have good values for $L$ and $C$, your calculated frequency should be very nearly equal to 60 Hz. Estimate the precision with which you could determine $C$ and the error to be attached to your value of the resonant frequency, and see if the "true" value of 60 Hz lies within this range.

## QUESTIONS _____

**1.** Discuss the precision with which you can read the amplitudes and intercepts on the oscilloscope screen in Procedures 5, 7, and 8 and estimate the resulting errors in your measurements of the inductance and resistance of the inductor and the capacitance of the capacitor. Then see if the known values of these parameters fall within the ranges of these errors.

**2.** Show that if the henry were written in terms of the dimensions of fundamental quantities in the SI system, it would be a newton-meter-second$^2$/coulomb$^2$.

**3.** The magnetic field of a very long solenoid (a single-layer cylindrical coil) is, to a good approximation, concentrated in the region inside the solenoid, where it is uniform, directed parallel to the solenoid axis, and given by

$$B = \frac{\mu N i}{h}$$

where $B$ is the field strength in teslas, $\mu$ is the permeability of the core material (the material on which the solenoid is wound), $N$ is the number of turns in the coil, $i$ is the current through the coil in amperes, and $h$ is the length of the solenoid in meters. Using the definitions of $\Phi$ and $L$, derive an expression for the inductance of a long solenoid.

**4.** Prove by any method you choose that the average of $(\sin 2\pi ft)(\cos 2\pi ft)$ is zero over one period and hence over any integral number of periods.

**5.** Derive Equations 36.14 and 36.15 as outlilned in the Theory section.

**6.** Derive Equation 36.16.

**7.** Explain why the decade box resistance obtained in Procedure 5 is equal to the impedance of the inductor being tested. Point out in particular why it is equal to the *impedance*, not the resistance or reactance of the inductor.

**8.** Explain why no vector representing $Z$ can ever be drawn in the left half-plane (the region to the left of the vertical axis) in diagrams such as those shown in Fig. 36.5 as long as only passive circuit elements (elements like resistors, capacitors, and inductors that do not include a source of power) are considered.

**9.** In high-quality oscilloscopes having the sweep frequency control calibrated directly in seconds per centimeter, the horizontal gain control is automatically switched out of the circuit when the sweep system is being used so that the pattern width is permanently preset rather than being adjustable. Why is this so?

**10.** What serves as the time standard in electric clocks that do *not* depend on the 60-Hz line for this purpose?

**11.** The A.M. broadcast band runs from 540 to 1600 kHz (thousands of cycles per second). One common tuning method used in older radio receivers was to have an inductor connected to a variable capacitor driven by the tuning dial. A station was selected by adjusting $C$ so that resonance was obtained with the fixed $L$ at the station frequency. (a) The variable capacitor used usually had a maximum capacity of 365 pF ($10^{-12}$F). Find the value of the inductance needed to tune in the low-frequency end of the broadcast band. (b) What must the variable capacitor's minimum value be in order that the high-frequency end of the broadcast band may be tuned in?

# Vacuum and Semiconductor Diodes

# 37

The diode or two-element electronic device allows conduction of an electric current in one direction only. Moreover, the volt-ampere characteristic (how the current through the diode depends on the voltage across it) is not linear even in the forward (current-carrying) direction and can in some cases be made to serve certain special purposes. These properties make the diode, whether in the form of a vacuum tube or a semiconductor device, a tremendously important circuit element in electronic systems. The purpose of this experiment is to study some characteristics of both thermionic (vacuum-tube) and semiconductor diodes.

## THEORY

**The Thermionic Diode** If a metal electrode is placed in a vacuum and heated, a certain number of electrons will break through the potential barrier at the surface and be emitted into the surrounding space. If a second electrode is located nearby, and is maintained at a positive potential with respect to the first, the negative electrons will be attracted to it and a current will apparently pass through the vacuum. If, on the other hand, the cold surface is made negative and the hot one positive, no current will flow. The field now wants to draw electrons from the cold to the hot electrode, but the cold one does not emit an appreciable number, and so there are few electrons available to cross the gap. This phenomenon was first observed by Edison and was called the Edison effect. It is the basis of operation of all modern thermionic diodes, which exist to perform the function just described—the conduction of current in one direction only—and which differ only in size and shape to suit various applications in which unidirectional conduction is desired.

The current per unit area carried by electrons emitted through a metal surface is given by

$$J = AT^2 e^{-e\phi/kT} \qquad (37.1)$$

where $J$ is the current density in amperes per square meter, $A$ is a constant having the dimensions of amperes per square meter per degree Kelvin squared, $e$ is the electronic charge in coulombs,* $\phi$ is a constant expressed in volts called the *work function* of the surface, $k$ is Boltzmann's constant in joules per degree Kelvin, and $T$ is the Kelvin or absolute temperature of the surface. Notice that $kT$ is a thermal energy, $\phi$ is the energy per unit charge required to get charge

(in the form of electrons) out of the surface, $e\phi$ is therefore the energy required to get one electron out, and the ratio $e\phi/kT$ is thus the ratio of energy required to extract an electron to the thermal energy available. Equation 37.1 is known both as Dushman's and as Richardson's equation, and the exponential dependence of emission current on temperature has been well established experimentally. However, the dependence of $J$ on the exponential is so predominant as to mask its dependence on $T^2$, which has never been decisively verified. Indeed, Richardson originally proposed a relation between $J$ and $T$ in which the $T^2$ factor was replaced by $\sqrt{T}$. Furthermore, although the Richardson-Dushman equation may be derived very nicely on the basis that the electrons in the metal constitute a perfect Fermi-Dirac gas, the constant $A$ appears as a combination of fundamental constants, and should thus not vary from metal to metal. Experimentally, $A$ is found to differ for different metals, and furthermore has the value $6.02 \times 10^5$ A/m$^2 \cdot$ K$^2$ for several materials, whereas the theoretical value is almost exactly twice this. These discrepancies have never been explained satisfactorily, but the Richardson-Dushman equation is nevertheless used, although in a rather empirical manner with experimentally determined values for $A$ and $\phi$.

If the positive electrode (usually called either the anode or the plate) is at a sufficiently high potential so that all electrons emitted by the hot electrode (the cathode, or in the case of a simple wire heated by the direct passage of a heating current, the filament) are immediately drawn over to it, the current is given by the Richardson-Dushman equation and is thus a strong function of the temperature. Note that the property of unidirectional conduction depends on this, for if the potential difference is reversed, the anode tries to play the cathode's role, but cannot simply

---

*This is the $e$ in the exponent. It should not be confused with the symbol used for the base of natural logarithms.

because it is relatively cold and therefore emits far fewer electrons. Under these conditions, the current in the forward direction (anode positive) is almost independent of the anode voltage, as all the emitted electrons are being collected. The anode (or plate) current is thus a function of the cathode temperature only, and is therefore said to be *temperature limited*.

It is possible (and is indeed the usual case in commercial vacuum tubes, which employ cathodes with highly efficient emitting surfaces) to have the plate voltage insufficiently great to draw over immediately all the emitted electrons. There then develops an electron cloud and hence a distribution of negative charge called a *space charge* in the region surrounding the cathode. This space charge forms a virtual cathode that supplies the electrons that flow to the plate and at the same time effectively presents a negative electrode to the cathode surface, thus shielding it from the plate's positive field. Electrons emitted by the cathode then tend to be repelled back to it. In the steady state, the space charge assumes a stable, equilibrium value at which the number of electrons leaving it for the anode is just balanced by the number that join it from the cathode. Under these conditions, the plate current depends on the plate voltage, which determines how many electrons will be drawn out of the space charge per unit time, and not on the cathode temperature, since excess electrons are merely repelled back to the cathode. Therefore an increase in temperature does not change the net emitted current. The diode's plate current is then said to be *space-charge limited*.

It may be shown that under space-charge-limited conditions, the plate current is proportional to the three-halves power of the plate voltage for all tube geometries, provided only that the electrons may be assumed to be emitted from the cathode with zero initial velocity. The three-halves power law was first derived for the case of plane parallel electrodes by Child and by Langmuir, who found that

$$J = B \frac{V_b^{3/2}}{d^2} \tag{37.2}$$

where $J$ is the current density as before, $V_b$ is the inter-electrode potential (plate voltage) in volts, $d$ is the electrode separation in meters, and $B$ is a proportionality constant composed of various fundamental quantities and having the dimensions of amperes per volt$^{3/2}$. This relation is thus often referred to as Child's law or Langmuir's equation. Except for effects due to the initial emission velocities of the electrons (which result in some dependence of $J$ on $T$), the three-halves power law is well established experimentally.

In rectifier service, that is, in applications where only the diode's property of conducting in one direction is made use of, the three-halves power law is of little interest. In fact, the current should preferably flow through the diode in the forward direction without any attendant voltage drop. Accordingly an *ideal diode* is defined as a circuit element that is an open circuit for one polarity and a short circuit for the other. The vacuum diode is indeed an open circuit when its plate is negative, but the three-halves power law means that in the forward direction it is not ideal but has a certain voltage drop across it. Therefore it also dissipates some power (namely the voltage across it times the current through it) and thus gives rise to a loss of energy in the system. We shall see below that semiconductor diodes, besides being smaller and requiring no cathode heating, are somewhat better approaches to the ideal and are therefore preferred in almost all present-day rectifier applications. Thus the volt-ampere characteristic for a typical vacuum diode is shown in Fig. 37.1 (a). By contrast, the volt-ampere characteristic for a semiconductor diode, whose features will now be discussed, appears in Fig. 37.1 (b).

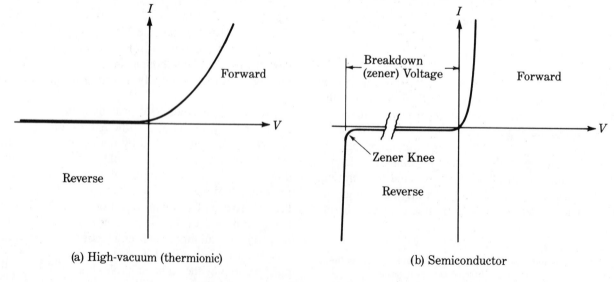

(a) High-vacuum (thermionic)

(b) Semiconductor

**Figure 37.1** *Diode volt-ampere characteristics*

**The Semiconductor Diode**  It is possible to obtain an essentially unidirectionally conducting device by forming a junction between two types of semiconductor material called P-type and N-type. Both types are made by "doping" pure semiconductor materials such as silicon or germanium with appropriate impurity elements. Pure silicon and germanium are called *intrinsic semiconductors* and behave in a rather simple way. In both materials, the electrons are bound in the crystal lattice and require a certain small amount of energy to break free. This is in contrast to metals, in which large numbers of electrons are free to move. Quantum mechanically speaking, we note that in metals there is a continuum of energy states extending upward (on an energy scale) from the level normally occupied by the electrons, so that it is possible to give an electron any energy, however small, and it will immediately accept it and start to move through the lattice with the kinetic energy it has thus acquired. In the semiconductor, on the other hand, there is a range of energies that the electrons in the material cannot have, extending from the levels normally occupied up to a certain level above which is again found a continuum of allowed states. This continuum is called the *conduction band,* while the array of normally occupied states is called the *valence band.* Thus the difference between a metal and a semiconductor is that there is a gap between the valence and conduction bands in the semiconductor whereas there is none in the metal. Hence, in a semiconductor, an electron must first be excited with sufficient energy to jump the gap from the valence to the conduction band before it finds a continuum of available states that will allow it to move about freely, that is, to take part in electrical conduction as a charge carrier. Since few electrons have this energy under ordinary conditions, few are available for conduction; hence the conductivity of an intrinsic semiconductor is relatively low. Note that the conductivity increases as the temperature is raised, for the higher the temperature, the more electrons will have sufficient thermal energy to cross the gap into the conduction band.

A P-type semiconductor is made by doping an intrinsic semiconductor with impurity atoms having a valence one less than the intrinsic semiconductor material. The presence of the impurity atoms in the crystal lattice raises the conductivity of the material because there are some energy states available to electrons that are unoccupied, for the impurity atoms bring insufficient electrons with them. Other electrons can thus move into these sites just as in a row of parked cars with one space open the car behind can move into the space, leaving a space where it formerly stood into which the next car can move, and so forth. Note that if this situation were observed from high in the air, it would look as if a space were moving in the direction opposite from that of the cars.

Similarly, the unoccupied sites obtained when an intrinsic semiconductor is doped with a lower valence impurity move in an applied field as if they were positive particles, and indeed these "holes" may be regarded as positive electrons and treated as positive charge carriers. Semiconductors that have been doped in this way are thus called P-type because positive charge carriers appear to be responsible for conduction.

Clearly, an N-type semiconductor is made by doping the intrinsic material with impurity atoms having a larger valence. These atoms bring in extra electrons, so that again the conductivity of the intrinsic semiconductor is increased, although this time by reason of the presence of negative charge carriers. Conduction therefore proceeds in a manner similar to that in metals.

The modern solid-state rectifier diode is made by forming a junction between pieces of N-type and P-type semiconductor material. At this so-called PN junction the holes in the P-type material are brought together with the electrons in the N-type so that they combine and neutralize to form a *depletion layer* in the neighborhood of the junction in which there are few free charge carriers. The flow of holes into the depletion layer from the P-type material leaves a negative charge on that side of the junction, whereas the N-type material becomes positively charged. In this way, a potential barrier is set up across the depletion layer that prevents further current flow. Placing across the junction an external potential of such a polarity that the N-type side is made positive may be thought of as drawing the electrons back from the junction in the N-type material and the holes back in the P-type, which is made negative in this situation. The thickness of the depletion layer is thus increased. The external potential may be considered as adding to the internal potential across the depletion layer and thus helping to stop all motion of the *majority* charge carriers (electrons in the N-type material and holes in the P-type). Note, however, that even in N-type material there are a few holes and in P-type there are a few free electrons that had enough thermal energy to reach the conduction band. These are called *minority* charge carriers, and under the conditions just described a small current will flow across the junction because of them. This is called a *reverse current,* and when the N side of the junction is positive the junction is said to be *reverse biased.*

If the N-type material is made negative, however, electrons are driven into the depletion layer, and at the same time holes are driven in from the P-type side. There is a large amount of recombination, the depletion layer is reduced in thickness, and a large current flows. One may consider that the applied potential, by supplying positive charge to the P-type material and negative charge to the N-type, neutralizes the internal potential barrier so that the large current flow can take place. Under these conditions, the junction is said to be *forward biased*.

A PN-junction diode is thus an excellent rectifying (unidirectionally conducting) device. Besides being small and requiring no cathode heating, it has the additional advantage over vacuum diodes that in the forward direction the current rises steeply after a volt or so has been applied—much more steeply than the three-halves power function that is found to hold for vacuum diodes. The semiconductor rectifier is thus a closer approximation to the ideal and has accordingly become more and more popular. When reverse-biased, however, a PN junction passes a small current and is thus inferior in this respect to the vacuum diode, which passes no current in the reverse direction no matter what the applied reverse voltage right up to the point where the insulation between the anode and cathode breaks down. The semiconductor diode's reverse current is usually negligible, however, if the temperature is not allowed to rise appreciably. Since the reverse current is due to minority carriers originating from electrons in the crystal occasionally having enough thermal energy to break their bonds and create electron-hole pairs, an increase in temperature results in an increased reverse current. Because of the high reverse voltages often encountered, considerable energy dissipation may be associated with this reverse current, and under some conditions the temperature may continue to rise and a runaway effect occur, ending in the destruction of the diode. Keeping the junction cool is thus an important consideration in semiconductor applications, hence the sometimes odd-looking designs of cases for semiconductor devices, the widespread use of "heat sinks" to carry off the heat generated, and the attention paid to temperature in semiconductor rating charts. Moreover, the *peak inverse voltage* (P.I.V.) rating is particularly critical for a semiconductor diode, and good circuit design dictates that a large safety factor be employed. This derating of the diode is especially necessary when large temperature variations may be encountered, since elevated temperatures, because they result in increased numbers of mi-

nority charge carriers, tend to reduce the value of the breakdown voltage.

Semiconductor diodes do, however, exhibit an interesting and useful characteristic at breakdown. At the breakdown or peak inverse voltage, minority charge carriers being drawn across the depletion layer acquire sufficient energy to knock other electrons out of the atomic lattice, and a so-called avalanche develops. Under these conditions, the diode passes a large reverse current and has the additional interesting and important property that the voltage across it is very nearly constant and independent of the current being passed over a considerable current range. Thus, in contrast to a resistor, whose volt-ampere characteristic is Ohm's law, the reverse-biased semiconductor diode has a volt-ampere characteristic in which the voltage is fixed at some value determined in the diode's manufacture, regardless of the current. Figure 37.1 (b) gives a typical volt-ampere characteristic for a semiconductor diode and should be contrasted, as already mentioned, with the high-vacuum diode characteristic of Fig. 37.1 (a).

Because of the constant-voltage characteristic exhibited by a semiconductor diode when operated in the reverse polarity beyond breakdown, many diodes are made specifically for operation in this mode and are manufactured with breakdown voltages ranging from 2.4 to 200 V. A diode intended for such operation is called a zener diode after Clarence Zener (b. 1905), who first suggested an explanation for its operation, and the breakdown voltage is sometimes called the zener voltage. In addition, the sharp break in the reverse volt-ampere characteristic is referred to as the *zener knee*. A zener diode can be used as a voltage regulator in the circuit of Fig. 37.2. Here an unregulated voltage higher than the diode's zener voltage is applied at the input. The diode breaks down and current flows through it and $R_t$ in series. Since the voltage across the diode is fixed and independent of the current, this latter must automatically take the value that makes the drop across $R_t$ equal to the difference between the zener voltage and the input voltage. Notice that the input voltage must be

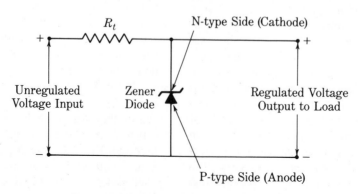

**Figure 37.2** *Zener diode regulator circuit*

higher than the desired regulated output. This is characteristic of all voltage regulators, which operate by automatically adjusting the input-output difference to maintain the output voltage constant despite variations in the input. In the present case, the automatic adjustment results from the constant-voltage characteristic of the zener diode. If the input voltage goes up or down, the current through $R_t$ goes up or down as needed to make the drop across this resistor equal to the difference between whatever value the input voltage may assume and the fixed voltage across the diode. If the load draws more current, the diode draws less to compensate, and *vice versa*. Thus, as long as the diode is not required to pass a current that would exceed its power rating (remember that it must dissipate a power equal to the product of the current

passed and the zener voltage), the regulated output voltage will remain essentially constant at the zener or breakdown value. This voltage is characteristic of the diode and cannot be varied except by changing diodes. For this reason, zener diodes are made with a wide choice of zener voltages, so that a readily available commercial unit can usually be found to deliver any voltage that might be desired.

In the present experiment, we will investigate the properties of thermionic electron emission in a special high-vacuum diode and will also determine the volt-ampere characteristics of two typical commercial semiconductor diodes, one intended for rectification, that is, simply unidirectional conduction, and the other a zener diode intended for voltage regulator service.

## APPARATUS

1. Special high-vacuum diode (see Fig. 37.3)
2. Type 1N5391 semiconductor diode
3. Type 1N5342 zener diode (6.8 V)
4. Filament transformer, 6.3-V, 3-A, centertapped secondary
5. Variable line transformer ("Variac" or "Powerstat")
6. Multimeter having a very low-voltage dc range

7. dc milliammeter (0–100 mA)
8. ac voltmeter (0–10 V)
9. 15-V regulated dc power supply
10. 0–300-V dc power supply
11. 6-V dry battery (lantern battery or equivalent)
12. Tubular rheostat (about 100 $\Omega$)
13. Carbon resistor, 68 $\Omega$, 2 W, 10% tolerance
14. Assorted hookup wires and connectors

## PROCEDURE

**1.** The special high-vacuum diode used in this experiment is pictured in Fig. 37.3. It consists of a cylindrical anode and a concentric filamentary cathode, that is, a filament wire to be heated by the passage of a heating current but serving also as the electron emitter. Connections to this filament are brought out to pins in the tube's base that plug into jacks on the top of a mounting box supplied with the tube; the jacks are connected inside the box to terminals to which connection to the external circuit can be made. This circuit is shown in Fig. 37.4. Notice that the cathode connection is made to the *centertap* (a connection to the middle turn) of the filament transformer secondary winding. This is a frequent arrangement when a vacuum tube has a filament that serves both as the heating element and the cathode. Such a tube is said to be directly heated as distinguished from an indirectly heated tube in which the cathode is a separate element insulated from an electric heater that serves only to do the heating and plays no part in the electronic circuit. In a directly heated tube there is of course a voltage across the filament, which means that the anode-to-cathode voltage is different at one end of the filamentary cathode than at the other. Moreover, in the case of an ac heater supply such as that shown in Fig. 37.4, this voltage varies at the ac supply frequency. This effect can be

more or less balanced out and thus minimized by the symmetrical connection to the center of the heating power source.

Wire the circuit of Fig. 37.4 but do not plug in or turn on either the power supply or the variable transformer until your setup has been approved by the instructor. Make sure that all switches are off, that the variable transformer and the power supply are turned all the way down, and that the multimeter is switched to the proper range (250 V or more, dc). Then plug in the variable transformer and power supply, turn their switches on, and prepare to take data.

**2.** Turn the power supply voltage up to 250 V. Then *slowly* turn up the variable transformer. Watch the tube filament, which should begin to glow red as it heats up, and be alert for the first sign of current on the milliammeter. As soon as anode current is observed, be especially careful about advancing the variable transformer, for (as Equation 37.1 shows) this current varies rapidly with cathode temperature and hence with the variable transformer's setting when the diode is temperature-limited. Adjust this transformer with great care to give an anode current $I_b$ of exactly 40 mA, readjusting the power supply to keep the anode voltage $V_b$ at exactly 250 V if necessary. Record the filament voltage $V_a$ as read on the ac voltmeter.

**Figure 37.3** *High-vacuum diode for temperature- and space-charge-limited volt-ampere measurements*

**Figure 37.4** *Circuit for measuring the vacuum diode's volt-ampere characteristic*

**3.** Reduce $V_b$ in 25-V steps, recording $I_b$ as read on the milliammeter at each step.

**4.** When $V_b$ has been reduced to zero, turn down the variable transformer and repeat Procedures 2 and 3, this time setting $I_b$ at 20 mA when $V_b$ is 250 V.

**5.** To investigate the tube's space-charge-limited characteristic, turn the power supply voltage ($V_b$) down to zero and set the filament voltage at 6 V ac as read on the filament voltmeter $V_a$. Be careful about this setting. The filament will be white hot, and when you start increasing $V_b$ the dissipated power $I_b V_b$ will contribute to heating the tube. Work as fast as you can and turn the power supply down to zero as soon as you have finished. Take data by increasing $V_b$ until $I_b$ is 10 mA and noting the $V_b$ so determined. Proceed in this manner, recording $V_b$ for anode currents of 20, 40, 60, 80, and 100 mA. *Do not leave the apparatus operating with $I_b = 100$ mA* but reduce $V_b$ to zero as soon as the data has been taken. Turn off both the power supply and the filament power when you are finished with the vacuum diode.

**6.** Remove the multimeter and milliammeter from your vacuum-tube circuit and wire the circuit of Fig. 37.5. Note the standard schematic symbol for the type 1N5391 semiconductor diode. The arrowhead points in the direction of conventional current flow and thus corresponds to the diode's anode or the P side of the PN junction. Also note the 68-$\Omega$ protective resistor inserted in series with the anode to limit the current when the forward characteristic is being investigated. Have your circuit approved by the instructor.

**7.** Make sure the rheostat is set up for zero voltage before turning on the power supply. Note that if you have a variable-voltage power supply you have two ways of changing the voltage applied to the diode. There are two reasons for this. First, variable power supplies often cannot be turned down to exactly zero output voltage, and you want to be able to

do this in the present work. Secondly, the current through a semiconductor diode rises steeply with voltage in the forward direction so that a fine adjustment of the voltage is needed, especially as you start up from zero. Initially setting the power supply at some voltage considerably less than its 15-V maximum allows you to do this by spreading the lower voltage thus selected over the full range of the rheostat. The forward voltage across the diode will probably not exceed 1 V, and so the multimeter should be switched to its lowest dc voltage range. CAUTION: Be very careful when the multimeter is set to this range. Do not make any changes in your circuit without either disconnecting this meter or switching it to a higher range. If this is your first experience with a multimeter, have your instructor explain its operation to you and/or read the discussion of such meters in the Procedure section of Experiment 26.

When you are sure everything is in order, turn on the power supply and set its output at 6–10 V. Slowly advance the rheostat until you just observe some current flow through the diode. Record the voltage at which current flow just begins as that corresponding to zero current. Actually, it is the maximum forward voltage before the onset of current through the diode, all lower voltages corresponding to zero current.

**8.** Advance the rheostat setting until the diode current is 10 mA, turning up the power supply output as required. Record the diode voltage observed with this current. Repeat for currents of 20, 40, 60, 80, and 100 mA.

**9.** Reset the rheostat to zero, turn off the power supply, and interchange the connections to the 1N5391. *Switch the multimeter to a higher voltage range* (10 or 15 V). Turn the power supply back on, set it for maximum output (15 V), and advance the rheostat to see if a reverse current through the diode

**Figure 37.5** *Circuit for measuring the semiconductor diode's volt-ampere characteristic*

is observed for any voltage up to 15 V. Report your finding in the space provided in the data sheet.

**10.** Turn off the power supply and replace the 1N5391 with the 1N5342 zener diode. For this step, connect the diode to forward bias it as shown in Fig. 37.5. Take data for the zener diode's forward characteristic by repeating Procedures 7 and 8.

**11.** Repeat Procedure 9 with your zener diode, but note that in this case you expect a reverse current to develop suddenly when the reverse voltage is increased to about 6.8 V. Determine the point at which reverse current starts to flow as accurately as you can from the multimeter scale you are using and record this voltage in the space provided in the data sheet.

**12.** To investigate the reverse characteristic of the zener diode more precisely, change the circuit of Fig. 37.5 to that of Fig. 37.6, making sure the power supply is switched off and the rheostat set for zero voltage before you do any rewiring. Before connecting the battery in the circuit, measure its voltage as accurately as you can with the multimeter and record your result. Note that the zener diode is connected so as to be reverse-biased and that its schematic symbol is the same as that for an ordinary diode except for the two little flags added to the cathode side. The battery produces a so-called bucking voltage, the multimeter now reading the difference between this voltage (which you know from the measurement just made) and the zener voltage. Thus, if the battery voltage were exactly 6 V and the zener voltage were

6.8 V, the multimeter would read 0.8 V and could be set to a very low voltage scale. In this case, small variations in the zener voltage that would hardly be apparent on the 10-V scale needed to measure the zener voltage directly are easily observed. Bucking circuits like this are often used when it is desired to study small changes in a relatively large voltage, and the example presented here will enable you to study the small variations that occur in the zener voltage as the diode current is varied.

To begin measurements, however, first switch the multimeter to its 10-V or higher scale. Then turn on the power supply, set its output near maximum, and advance the rheostat until reverse current just begins to flow, as you did in Procedure 11. Then reset the multimeter to as low a voltage range as possible and record its reading as the voltage corresponding to zero reverse current. Actually, this is the maximum reverse voltage that can be applied before breakdown (it is in fact the breakdown or zener voltage); all lower reverse voltages correspond to zero current.

**13.** Advance the rheostat until the diode reverse current is 10 mA. Record the multimeter reading with this current. Repeat for currents of 20, 40, 60, 80, and 100 mA.

**14.** Return the multimeter to its 10-V or higher scale, reduce the rheostat and power supply controls to zero, and switch off the supply when you have finished your work.

**Figure 37.6** *Circuit for studying the zener diode's reverse characteristic*

# DATA _____

| Anode voltage $V_b$, V | Anode current $I_b$, mA | |
|---|---|---|
| | $V_a =$ | $V_a =$ |
| 250 | 40.0 | 20.0 |
| 225 | | |
| 200 | | |
| 175 | | |
| 150 | | |
| 125 | | |
| 100 | | |
| 75 | | |
| 50 | | |
| 25 | | |
| 0 | | |

Space-charge-limited characteristic:
   $V_a = 6.0$ V

| Anode current $I_b$, mA | 0 | 10 | 20 | 40 | 60 | 80 | 100 |
|---|---|---|---|---|---|---|---|
| Anode voltage $V_b$, V | | | | | | | |

Measured exponent in three-halves     Percent error            _____
   power law          _____

| Diode current $I_b$, mA | 0 | 10 | 20 | 40 | 60 | 80 | 100 |
|---|---|---|---|---|---|---|---|
| 1N5391, forward $V_b$, V | | | | | | | |
| 1N5342, forward $V_b$, V | | | | | | | |
| 1N5342 reverse characteristic, multimeter reading, V | | | | | | | |
| 1N5342 reverse $V_b$, V | | | | | | | |

Reverse current observed with 1N5391 _____

Zener voltage measured in
   Procedure 11          _____

Battery voltage          _____

# CALCULATIONS _____

   **1.** Using the data of Procedures 2 and 3, plot a graph of $I_b$ against $V_b$ on a sheet of linear graph paper. Plot $V_b$ as the abscissa and $I_b$ as the ordinate and choose scales so that your curve covers most of the sheet. Indicate the temperature-limited region on your graph.

**2.** On the same sheet of graph paper, plot $I_b$ against $V_b$ using the data of Procedure 4. Again mark the temperature-limited region on this graph.

**3.** Using the data obtained in Procedure 5, plot a graph of $I_b$ against $V_b$ on the 1-cycle by 2-cycle logarithmic graph paper. Use the 2-cycle scale for $V_b$ and the 1-cycle scale for $I_b$. Refer to the instructions on plotting graphs in the Introduction. Draw the best straight line through your points.

**4.** From the slope of the graph plotted in Calculation 3, find the exponent to which $V_b$ should be raised to make the result proportional to $I_b$. According to the three-halves power law this exponent should be $\frac{3}{2}$. Compare your result with this theoretical one by finding the percent error.

**5.** Using the data of Procedures 7, 8, and 9, plot a graph of $I_b$ against $V_b$ on a second sheet of linear graph paper. Plot the voltage values as abscissas and the current values as ordinates. Arrange your scales to cover as much of the sheet of graph paper as possible, but choose an origin so you can plot negative (reverse) voltages and currents and thus indicate your findings with regard to reverse biasing the 1N5391 in Procedure 9.

**6.** Add the battery voltage measured in Procedure 12 to the multimeter readings obtained in Procedures 12 and 13, recording these sums as the reverse voltage measurements for the 1N5342 when it is operating as a zener diode.

**7.** Using the data of Procedures 10, 12, and 13 and the results of Calculation 6, plot a graph of $I_b$ against $V_b$ for your zener diode on the third sheet of linear graph paper. Note that you will be plotting positive (forward) and negative (reverse) voltages and currents. Plot your values of $I_b$ as ordinates, choosing an origin so that you can plot current from $-100$ mA to $+100$ mA with the zero in the middle and a scale that lets you cover most of the graph paper in the vertical direction. Plot your voltage values as abscissas and note that you will need to choose a scale that will show the small values of forward voltage and the small variation in the relatively large zener voltage. You can do this by placing your origin somewhat to the right of center horizontally, thus leaving room on the left to plot the zener voltage values, not much space being needed on the right to accommodate the forward voltages. Refer to the illustrative graph of Fig. 37.1(b).

## QUESTIONS _____

**1.** (a) Compare the forward volt-ampere characteristics of the vacuum-tube diode and the semiconductor diode. (b) Which approaches the ideal diode most closely? Explain your answer.

**2.** (a) Show that the power dissipated in a vacuum diode is proportional to $I_b^{5/3}$. (b) What power is dissipated in an ideal diode? (c) Using your graph from Calculation 5, explain why you would expect the power dissipated in a semiconductor diode to be proportional to the current $I_b$.

**3.** What is the purpose of the 68-$\Omega$ resistor in the circuit of Fig. 37.5? Is its value critical?

**4.** A thin ribbon of tungsten 2 mm wide and 5 cm long is to be used as the filamentary cathode of a high-power vacuum tube. For tungsten the constant $A$ in Equation 37.1 has the value of $6.02 \times 10^5$ $A/M^2 \cdot K^2$, and the work function is 4.5 V. The tube is to have a maximum current rating of 200 mA, but because this is to be under space-charge-limited conditions, the cathode must be capable of emitting five times that, or 1 A. (a) Remembering that both sides of the tungsten ribbon may be used and that the temperature in Equation 37.1 is in degrees *Kelvin*, find the temperature in degrees Celsius to which the filament must be heated. (b) Look up the melting point of tungsten and see if the filament can in fact be heated to this temperature. (c) Tungsten filaments are often treated with thorium to reduce their work function. Suppose the present filament had its work function reduced to 3.0 V in this way. To what temperature would it then have to be heated to yield the same emission?

**5.** When data was taken for the volt-ampere characteristics of the high-vacuum diode in Procedures 2–4, the voltage $V_b$ was used as the independent variable, whereas the current was used in the case of the 1N5391. Why?

**6.** A 15-V zener diode capable of dissipating 1 W is to be used with a 28-V unregulated power supply to deliver a regulated (that is, constant) 15-V supply to a load that may draw anything from 0 to 50 mA. The circuit is shown in Fig. 37.2. Calculate the appropriate resistance for $R_t$. What power must this resistor be capable of dissipating?

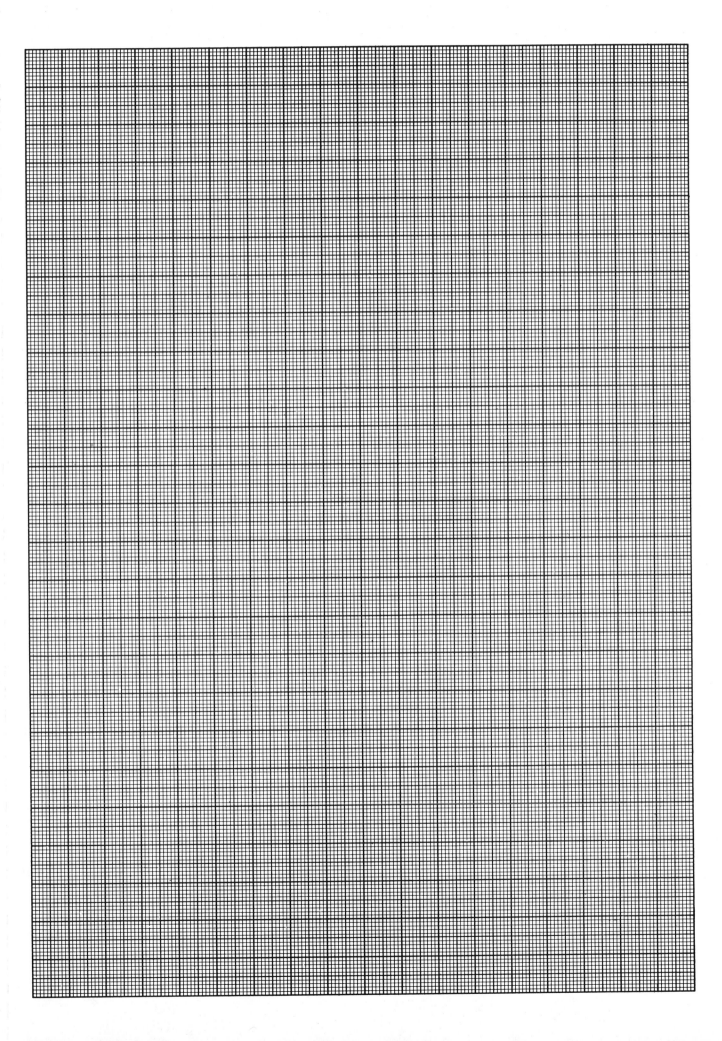

# Rectifier Circuits and Power Supplies 38

Most electrical equipment containing electronic circuitry requires a supply of dc power, yet household and industrial power lines deliver ac. We have seen (Experiment 34) that the adoption of ac power distribution is largely due to the fact that the transformer, which provides a simple, efficient, trouble-free method of changing any given voltage to any desired value, works on ac only. Indeed, many household appliances are marked "ac only," not because they really work on alternating current but because they require a voltage other than that on the power line, so that the first thing the power cord connects to inside the appliance is a transformer designed to change the line voltage to whatever the appliance's components need.

If these components also operate on dc, some means must be provided to convert to dc the ac that has been brought to the proper voltage by the trans-

former. Clearly, a device that can conduct electric current in one direction only can accomplish this purpose, and we saw in Experiment 37 that high-vacuum or semiconductor diodes have this property. If you have not done that experiment, a review of its Theory section is recommended as an introduction to the present work. Here we will study some circuits in which diodes are used as *rectifiers* to convert ac to dc. A combination of a transformer designed to change the power-line voltage to some specified value and a diode rectifier circuit that converts the transformer's ac output to dc is called a *power supply*. Power supplies may also include *filters* that smooth out the variations or *ripple* left in the rectifier output and *regulator circuits* that maintain the output voltage at a fixed value regardless of variations in the line voltage or the load. Simple examples of a filter and a regulator are included in this experiment.

## THEORY

The principal use of diodes is for rectification, which means the application of their property of unidirectional conduction to the conversion of alternating current (current that flows first in one direction and then in the other) to direct current (which flows in one direction only). An obvious circuit for doing this is shown in Fig. 38.1(a). Notice the symbols used. On the left, the circle with the tilde in it represents a voltage source and is marked with the voltage's polarity. In this case of an alternating voltage, the

actual sinusoidal dependence of voltage on time is given and the + and − signs mean that if the plus end is treated as the "high" side, the observed voltage will be as given, whereas if the polarity is reversed, the observed voltage will be $-V_m \sin 2\pi ft$, corresponding to a 180° phase reversal. The symbol ▸|  is universally used to represent a diode; the direction of conduction ("forward" direction) is indicated by the arrow. Note that if this symbol stands for a vacuum diode, the arrow represents the anode or plate.

(a) The Circuit

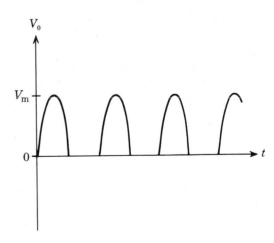

(b) The Output Waveform

**Figure 38.1** *Half-wave rectification*

323

The other end is then the cathode or filament, and the circuit for heating the cathode is understood to be there without being actually shown on the drawing. We will, however, deal only with semiconductor diodes in the present experiment because high-vacuum diode rectifiers are rarely seen today. The small size, absence of a cathode heating requirement, and low forward voltage drop make semiconductor diodes much to be preferred, except in a few specialized applications.

An ideal diode is one that is a short circuit in the forward (conducting) direction (zero forward voltage drop) and an open circuit in the reverse or nonconducting direction. If $C_r$ in Fig. 38.1(a) is ideal, the output voltage $V_0$ obtained across the load resistor $R$ is as shown in Fig. 38.1(b), where it is plotted as a function of time. It is direct in the sense that it never goes negative, but it is hardly constant at a particular value as is the dc voltage from a battery. In fact, the diode (or *rectifier*, as a diode used in this service is often called) looks like an open circuit during the intervals in which the voltage $v$ from the voltage source is negative, so that $V_0$ is zero during these intervals. The rectifier has thus "thrown away" all the negative half-cycles and passed only the positive ones. The arrangement of Fig. 38.1 is consequently called a *half-wave rectifier* circuit.

Despite the variations plotted in Fig. 38.1(b), the rectified output voltage can be made almost pure direct current (almost constant) by special circuits called *filters*. The purpose of a filter is to distinguish between the ac and dc components making up the time-dependent voltage of Fig. 38.1(b), eliminating the ac components and retaining the dc or constant one. The dc component refers to the average value of $V_0$, which can be shown to be $V_m/\pi$ or approximately one third of the amplitude of the ac input voltage in the case of the half-wave rectifier. There are many different kinds of filter circuits, but one of the simplest and most effective consists of a large capacitor connected in parallel with the load resistance $R$, as shown in Fig. 38.2. Assuming that the diode and ac voltage source can supply whatever current is needed, we note that during the positive half-cycles a large charging current will flow into the capacitor in addition to the current delivered to the load. In fact,

the capacitor will ideally receive charge $Q = CV_m$ where $C$ is its capacity in farads and $V_m$ is the amplitude of the ac input voltage in volts. During the negative half-cycles, when the diode is an open circuit, the capacitor supplies current to the load, discharging in the process. In fact, as soon as the ac voltage $v$ decreases from its maximum value $V_m$ to a value less than the voltage on the capacitor, the diode becomes reverse-biased and hence an open circuit. Thereafter the output voltage is simply the exponentially decreasing voltage obtained when a capacitor discharges through a resistor. The exponential decrease continues until the next positive half-cycle, when $v$ rises above the voltage across $C$, the diode becomes forward-biased once more, and the capacitor is again charged up to $V_m$. If the $RC$ time constant is large, the voltage decrease does not get a chance to proceed very far in one period of the ac wave, and so the output voltage not only never falls to zero but indeed stays near the value $V_m$ of the ac voltage amplitude. Hence this filter does more than just get rid of the ac components in the wave of Fig. 38.1(b). It also raises the average to a value close to $V_m$. In fact, one form of ac voltmeter called a peak-reading voltmeter uses the circuit of Fig. 38.2 with $v$ as the voltage to be measured and $R$ the meter resistance. Note, however, that both good filtering and making the average approach $V_m$ depend on the time constant $RC$ being very long compared with a period of the alternating current. In the peak-reading voltmeter, this is easy to do with reasonable values of $C$ because $R$, the resistance of the dc voltmeter used as the load, can be made very large. In power supplies, however, $R$ represents the load, and in the case of a heavy load (one drawing a lot of current), $R$ is small. Good filtering then requires that $C$ be very large.

The waveform (graph of voltage as a function of time) shown in Fig. 38.1(b) has a low average and large ac component principally because the negative half-cycles of the ac input are not used, voltage $V_0$ being simply zero during those time intervals. This situation would be greatly improved by a rectifier circuit that made use of the negative half-cycles, somehow changing their polarity so that a positive output would be obtained. Such a circuit appears in Fig. 38.3. Notice that there are two identical voltage sources, each producing $V_m \sin 2\pi ft$, on the left. The upper one is in a circuit with $C_{r1}$ and $R$ identical to the circuit of Fig. 38.1(a) and gives rise to a voltage across $R$ identical to that plotted in Fig. 38.1(b). In addition, the circuit consisting of the lower voltage source, $C_{r2}$, and $R$ is also identical to that of Fig. 38.1(a) except that the polarity of the source is reversed, the minus end rather than the plus end going to the diode. As a result of this phase reversal, the negative half-cycles appear as positive ones to $C_{r2}$ and are thus the ones passed by this diode to the upper end of $R$ to fill in the spaces between the positive

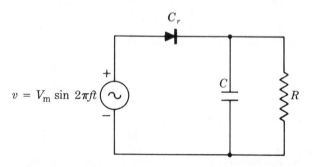

**Figure 38.2**   *The half-wave rectifier with capacitor filter*

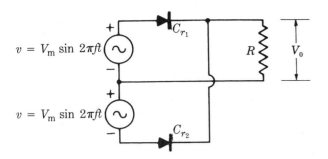

**Figure 38.3** *The full-wave rectifier*

half-cycles passed by $C_{r1}$, as illustrated in Fig. 38.4. Because both positive and negative half-cycles are used, the present scheme is called a *full-wave rectifier*, and because another half-cycle is inserted in each empty space between the half-cycles delivered by the half-wave rectifier, the average output of the full-wave rectifier is twice that of the half-wave circuit, or $2V_m/\pi$. Moreover, positive peaks come at twice the repetition rate of those obtained in the half-wave case, making filtering correspondingly easier.

The disadvantage of the full-wave rectifier circuit of Fig. 38.3 is the need for two voltage sources. This is less of a problem than it might at first appear because of their series connection. The usual ac voltage source supplying a rectifier is the secondary coil of a transformer, and this winding is easily made to give $2V_m \sin 2\pi ft$ (an ac voltage having the amplitude of the two sources of Fig. 38.3 in series) and to have a centertap (a connection to the middle turn of the winding). Then $V_m \sin 2\pi ft$ appears between the centertap and either end, these two voltages are in series with the proper polarity, and in fact the centertapped

secondary coil is in every way identical to the arrangement of the two separate sources of Fig. 38.3. The resulting circuit is shown in Fig. 38.9. Because of the centertap, it is called the *full-wave center-tapped rectifier* system. Note, however, that the transformer secondary coil has to be wound to give a total amplitude of $2V_m$ in order to produce a peak voltage of only $V_m$ at the rectifier output. In other words, a factor of two in voltage has been sacrificed. This is somewhat compensated for by the fact that the average current delivered to the load can be twice that for which the transformer is rated, since each half of the secondary coil supplies load current only half the time. However, the fact that the secondary coil has to be wound for twice the load voltage and centertapped occasionally dictates the use of the different arrangement described below.

The *bridge rectifier* system features full-wave rectification requiring four diodes but without sacrificing a factor of two in output voltage. To understand this system, look first at Fig. 38.5, which shows two full-wave centertapped rectifier circuits driven from a single pair of voltage sources. Note that in the circuit employing diodes $C_{r3}$ and $C_{r4}$, these diodes are reversed in polarity so that a negative output voltage is obtained at terminal $N$ relative to the common terminal $G$. Such an arrangement is sometimes used when both a negative and a positive voltage relative to a common ground point are needed. Thus point $P$ is at a positive voltage with respect to $G$, and this voltage has all the characteristics of the output of the full-wave centertapped rectifier as illustrated in Fig. 38.4. An identical voltage appears between $N$ and $G$ except that $N$ is negative with respect to $G$.

**Figure 38.4** *Full-wave rectifier waveforms*

**Figure 38.5** *Development of the bridge rectifier circuit from two full-wave centertapped systems*

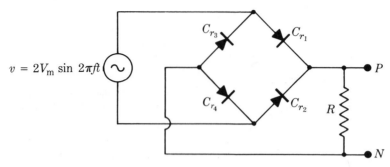

$$v = 2V_m \sin 2\pi ft$$

**Figure 38.6**  *The bridge rectifier circuit*

Now, if $R_1 = R_2$, the two currents returned through the common ground line to the centertap (the connection between the two voltage sources in Fig. 38.5) are equal and opposite, so that the net current in this line is zero. Hence it and the centertap connection can be omitted without changing anything. Point $G$ is then also omitted, and the voltage between $P$ and $N$ is seen to be twice the full-wave centertapped rectifier output, since in effect we have two such systems in series. Thus between $P$ and $N$ the peak voltage is $2V_m$ and hence is equal to the amplitude of the voltage across the two sources in series. The complete circuit is usually drawn as shown in Fig. 38.6, where the two sources of Fig. 38.5 (which now have no connection to the line between them) have been combined into a single source giving twice the amplitude, and the load resistors $R_1$ and $R_2$ (which are now simply in series) have been combined into an equivalent resistance $R = R_1 + R_2$. You should compare Figs. 38.5 and 38.6 to convince yourself that they are in fact identical except for the omission of the terminal $G$ and the associated common ground line. Further comparison of Fig. 38.6 with the circuit of the Wheatstone bridge (Fig. 27.1) should reveal why the bridge rectifier circuit is so called.

In Experiment 37 it was pointed out that a reverse-biased semiconductor diode has a critical peak inverse voltage (P.I.V.) rating beyond which it breaks down and conducts freely in the reverse direction. This breakdown does not damage the diode unless the resulting power dissipation overheats it, and it exhibits the interesting property of maintaining a voltage across it that remains nearly constant at the

breakdown value regardless of current. It was further explained that a diode operated in this manner could be used as a voltage regulator in the circuit of Fig. 37.2, repeated here as Fig. 38.7, and that this application has led to the development of diodes called *zener diodes*, which are specifically designed for this service. The regulated output voltage is fixed at the breakdown or zener voltage characteristic of the diode, and resistor $R_t$ is chosen so that the drop across it will equal the difference between the zener voltage and the higher unregulated input. The choice of $R_t$ will depend on the current that is to flow, which in turn depends on the current to be drawn by the load. Clearly, if a high load current is required, $R_t$ will be small, but if the load is disconnected, all this current must pass through the diode, which will then need to dissipate a power given by the product of this current and the zener voltage. Although high-power zener diodes are available so that this can be done, such a procedure wastes energy, and so *shunt regulators*, that is, regulators that use a constant-voltage element in parallel with the load, are seldom used with high-current loads. They are, however, simple and useful when the current demand is small.

Another interesting application of the zener diode is the generation of a square wave. Thus suppose that in the circuit of Fig. 38.7 the input voltage, instead of being dc, was an ac wave whose amplitude was considerably higher than the diode's breakdown voltage. On the positive half-cycle the voltage across the diode would build up until the breakdown (zener) value was reached, after which a current would flow through $R_t$ such that the voltage across the diode

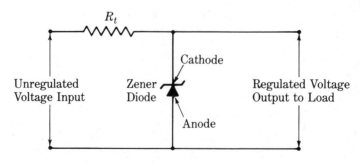

**Figure 38.7**  *Zener diode regulator circuit*

would remain constant at the zener value until the sine wave fell below it. On the negative half-cycle the diode is, of course, forward biased and behaves like any semiconductor diode, being essentially a short circuit (except for a small forward voltage drop). Hence for this half-cycle, $R_t$ is in effect directly across the input and the output voltage is nearly zero. The output therefore appears as a positive-going square wave whose height is approximately equal to the zener voltage. True, the sides of the pulse, being parts of a sine wave, are not really straight, but if the amplitude of the input ac is much

higher than the zener voltage so that these sides are only the lower parts of the curve, their curvature will hardly be noticed. We will demonstrate this application of the zener diode in the present experiment after first studying some rectifier circuits and a simple filter. Note that a regulator circuit is itself an effective filter by reason of the very fact that it preserves a constant output voltage. Thus at the end of the experimental work we will have produced a complete power supply delivering a fixed, pure dc output voltage from the ac power-line input.

## APPARATUS

1. Four type 1N5391 semiconductor diodes
2. Type 1N5342 zener diode (6.8 V)
3. Filament transformer, 6.3 V, 3 A, centertapped secondary
4. Oscilloscope
5. Volt-ohm-milliammeter (multimeter)
6. Four carbon 2-W, 10%-tolerance resistors (220; 1000; and two 10,000 Ω)

7. 6-V dry battery (lantern battery or equivalent)
8. Decade capacitor with a range up to at least 8 μF
9. Electrolytic capacitor, 10 μF, 50-V rating
10. Electrolytic capacitor, 100 μF, 50-V rating
11. Assorted hookup wires and connectors

## PROCEDURE

**1.** Assemble and wire the half-wave rectifier circuit of Fig. 38.8. Be sure the multimeter is switched to its 10- or 15-V dc range. CAUTION: Multimeters (volt-ohm-milliammeters) are easily damaged by being connected in a circuit while switched to the wrong range. If you have not used a multimeter before, have your instructor explain its operation to you and/or read the discussion of such instruments in the Procedure section of Experiment 26. In any case, have your circuit approved by the instructor before you plug in the transformer.

**2.** When your circuit has been checked, plug in the transformer and adjust the oscilloscope to display two cycles of the ac voltage wave. Use line synchronization to stabilize the pattern. Note that a dc voltmeter reads the *average* of a rapidly varying voltage applied to it. Record the multimeter reading as the average output voltage, and read and record the peak output voltage displayed by the oscilloscope. *Note:* To read the peak voltage from the oscilloscope pattern, the vertical axis must be calibrated. You will

have to carry out this calibration if your instrument does not have it provided for you. See Experiment 35.

**3.** Some oscilloscopes have an "ac-dc" switch on the front panel. This switch does not set the instrument for operation from an ac or dc power line but controls the response of the vertical amplifier. In the dc position, any voltage, including a constant one, that is connected to the vertical input terminals will cause a vertical deflection of the trace on the screen. Thus if the oscilloscope is set for a vertical deflection sensitivity of 1 V/cm and you connect a 1.5-V battery to the vertical input, the trace will shift vertically by 1.5 cm and remain in the new position. With the input switch at ac, on the other hand, only ac signals will produce a vertical deflection, the oscilloscope's response cutting off when the frequency gets down to 3 or 4 Hz. Connecting the battery as before will produce only a momentary deflection, the trace returning to its zero vertical position at once.

The rectifier output voltage is time-varying but does not have a zero average and may therefore be

**Figure 38.8**  *Half-wave rectifier test circuit*

**Figure 38.9** *Full-wave rectifier test circuit*

thought of as the sum of a dc component equal to its average value and a "true ac" component having zero average. With the oscilloscope set for direct current, both components are displayed on the screen. When the oscilloscope is set for alternating current, the dc component is blocked. Hence, *switching the vertical input from alternating to direct current displaces the pattern vertically by an amount equal to its average value.* If your instrument has such a switch, make and record a second measurement of the rectifier output's average value in this way.

**4.** Investigate the effect of a capacitor filter by connecting the capacitor decade across the output of your rectifier (in parallel with the 10,000-Ω load resistor). Observe the oscilloscope pattern and read and record the average output voltage indicated by the multimeter for capacitances of 0.5, 1, 2, 4, 8, and 18 μF. Connect the 10-μF electrolytic capacitor in parallel with the decade terminals to obtain the 18-μF value. *Note:* If possible, have the vertical amplifier set for dc when making these observations. If your oscilloscope permits and the instructor suggests it, measure the average value of the rectifier output voltage by using the ac-dc switch as described in Procedure 3 for each value of capacitance tried. Note also that the electrolytic capacitor must be connected in accordance with the polarity markings on its jacket.

**5.** Unplug the transformer and rewire your circuit using two 1N5391s to form the full-wave center-tapped rectifier system diagrammed in Fig. 38.9. Have your instructor check the wiring, plug in the transformer, and observe the pattern displayed on the oscilloscope screen. Record the average value of the rectifier output voltage as read on the multimeter and determine the peak value from the oscilloscope display. If possible and if the instructor so suggests, make a second measurement of the average output voltage using the oscilloscope's ac-dc switch as described in Procedure 3.

**6.** Unplug the transformer, reverse the connections to both diodes, plug the transformer in, and observe the pattern now displayed by the oscilloscope. Notice that the polarity of the multimeter must be reversed. Report your observations in the space below.

**7.** Unplug the transformer and rewire your circuit using four 1N5391s to form the full-wave bridge rectifier of Fig. 38.10. Have your circuit checked by the instructor and then plug in the transformer. Mea-

**Figure 38.10** *Bridge rectifier test circuit*

**Figure 38.11**  *Zener diode regulated power supply*

sure and record the average and peak values of the output voltage as in Procedure 5.

**8.** Unplug the transformer and rewire the circuit of Fig. 38.10 to produce the regulated power supply of Fig. 38.11. Note the use of the large (100-$\mu$F) filter capacitor in this circuit. Have your wiring approved by the instructor and then plug in the transformer. Measure the output voltage $V_0$ with the multimeter and record your result.

**9.** Test your regulator's operation by "bucking" the output voltage $V_0$ with a 6-V battery as described in Procedure 12 of Experiment 37. Fig. 37.6 of that experiment shows how the battery and multimeter are to be connected across the 1N5342. Leave the multimeter set to the range used in Procedure 8 until you are sure your bucking circuit is working correctly. Then switch it to its lowest dc voltage range so that you can observe small variations in $V_0$. Begin by observing and recording the multimeter reading (which is the difference between the battery voltage and $V_0$) for the no-load condition, that is, with nothing that draws any current connected to your power supply's output (across the 1N5342). Then apply a load by connecting the 1000-$\Omega$ resistor across the output and again observe and record the meter reading. The difference between these two readings is a measure of the ability of the regulator to maintain $V_0$ constant as the applied load changes.

**10.** Disconnect the 1000-$\Omega$ resistor from the power supply output and connect it across the output of the rectifiers, that is, across the 100-$\mu$F capacitor. Observe and record the multimeter reading under

these conditions. The difference between this reading and that obtained in Procedure 9 with no load indicates the ability of the regulator to maintain $V_0$ constant despite variations in the voltage supplied to it by the rectifiers and filter capacitor. To find out what variation you produced by connecting the 1000-$\Omega$ resistor across the rectifier output, remove the multimeter from the bucking circuit, switch it to the range you use for measuring voltages near 10 V dc, and connect it across the capacitor. Record its reading, remove the 1000-$\Omega$ resistor, and again read the meter. The difference between these readings is a measure of how the unregulated rectifier-filter circuit responds when a load is applied and is also the variation that the regulator is supposed to combat. You should find that the change in rectifier output voltage made only a very small change in $V_0$.

**11.** Unplug the filament transformer and remove the 1N5391 diodes and the filter capacitor from your circuit. Rewire the 220-$\Omega$ resistor and the zener diode in the circuit of Fig. 38.12 to produce the square-wave generator described in the Theory section. When your wiring has been checked, plug in the filament transformer and, using line synchronization as in Procedure 2, adjust the oscilloscope to display two or three cycles of the square wave. Read and record the height of your square wave in volts by applying your oscilloscope's vertical calibration to the displayed pattern. This height will be somewhat greater than the 6.8-V rating of the 1N5342 because, like all diodes, zener diodes have a small forward voltage drop. If your oscilloscope has the ac-dc

**Figure 38.12**  *Zener diode square-wave generator*

switch described in Procedure 3, set it to dc and establish your pattern's zero line by temporarily disconnecting the oscilloscope input from your circuit and adjusting the vertical centering control to bring the zero-input trace to a convenient position. Then reconnect the oscilloscope and observe and record the pattern's extent both above and below the zero line. Note that the extent above the zero line should be very nearly equal to 6.8 V while the extent below gives you your diode's forward voltage drop. The sum of the two is of course the full height of the square wave as already observed.

## DATA

| Rectifier circuit | Filter capacitor, $\mu$F | Peak value of output voltage | Average value of output voltage | | |
|---|---|---|---|---|---|
| | | | Meter | Oscilloscope | Calculated |
| Half-wave | 0 | | | | |
| | 0.5 | | | | |
| | 1 | | | | |
| | 2 | | | | |
| | 4 | | | | |
| | 8 | | | | |
| | 18 | | | | |
| Full-wave centertapped | 0 | | | | |
| Full-wave bridge | 0 | | | | |

Regulated power supply output voltage $V_0$ _____

Differences between $V_0$ and the battery voltage:

No load _____

1000-$\Omega$ load across output _____

Resulting change in $V_0$ _____

No load _____

1000-$\Omega$ load across capacitor _____

Resulting change in $V_0$ _____

Rectifier output voltage without 1000-$\Omega$ resistor _____

Rectifier output voltage with 1000-$\Omega$ resistor across capacitor _____

Resulting change in rectifier output voltage _____

Measured height of square wave _____

Extent of square wave above zero _____

Extent of square wave below zero _____

Sum _____

## CALCULATIONS _____

**1.** From the peak output voltage values observed in Procedures 2, 5, and 7, calculate the average output voltage expected for each rectifier circuit. Compare the three (two if your oscilloscope has no ac-dc switch) average values obtained in each case.

**2.** Find the difference between the two meter readings recorded in Procedure 9 to get the change in your regulated power supply's output voltage $V_0$ when the 1000-$\Omega$ load was applied.

**3.** Find the difference between the meter reading recorded at the start of Procedure 10 and the no-load reading, which you should copy from Procedure 9. This is the variation in $V_0$ due to the change made in the rectifier output.

**4.** To find the rectifier output voltage change mentioned in Calculation 3, take the difference between the rectifier output voltages measured with and without the 1000-$\Omega$ resistor in the latter part of Procedure 10. This difference is not only the change giving rise to the variation in $V_0$ found in Calculation 3 but also an indication of how an unregulated power supply performs.

**5.** Compare the extent of your square wave above zero recorded in Procedure 11 with the value of $V_0$ (which is your 1N5342's zener voltage) recorded in Procedure 8. Add the extent of your square wave below zero (which is your 1N5342's forward drop) to obtain the square wave's full height. Make sure your sum agrees with the direct measurement of the full height made at the start of Procedure 11.

## QUESTIONS

**1.** Suppose a sinusoidal ac voltage were displayed on an oscilloscope whose vertical amplifier was equipped with an ac-dc switch. What would you observe when this switch was thrown? Explain your answer.

**2.** How do the patterns obtained in Procedures 5 and 7 compare? What is the principal difference between them?

**3.** A peak-reading ac voltmeter is made by measuring the output of a half-wave rectifier with an electronic voltmeter (EVM) having an input resistance of 10 MΩ ($10^7$ Ω). What capacitance must be used in parallel with the EVM input so that the time constant will be 10 times the period of the 60-Hz power line voltage?

**4.** A *full-wave voltage doubler* is made by substituting large capacitors for the two lower diodes ($C_{r2}$ and $C_{r4}$) in the circuit of Fig. 38.6. If $R$ is very large, the resulting circuit delivers a dc voltage equal to twice the amplitude of the ac input. Explain the action of this circuit.

**5.** Despite the fact that the bridge rectifier circuit is full-wave and hence easier to filter than the half-wave arrangement of Procedure 4, a much larger ($100\text{-}\mu\text{F}$) filter capacitor was considered necessary in the circuit of Fig. 38.11. Explain why this is true.

**6.** (a) From your data in Procedures 8 and 10, determine the current flowing through the zener diode under no-load conditions (nothing connected to your power supply output) in the circuit of Fig. 38.11. (b) What is the maximum load (minimum load resistance) you would dare connect to your power supply's output terminals and still be sure that $V_0$ remained properly regulated by the zener diode?

**7.** Suppose you considered the regulated power supply in Fig. 38.11 to be an almost but not quite ideal voltage source that could be represented by the equivalent circuit of Fig. 30.1. Use your data from Procedures 8 and 9 to find the appropriate values of the open-circuit voltage $V_{oc}$ and the internal resistance $R_{int}$ for this case. See the Theory section of Experiment 30 for a discussion of how to do this.

**8.** If you have not already done so, answer Question 6 in Experiment 37. What should be the power rating of the zener diode chosen in this question?

# Amplification with Triode and Pentode Vacuum Tubes 39

Until relatively recently, the high-vacuum electronic tube has been the mainstay of electronic communications, computing, and control systems of all kinds, and even now it has not been completely superseded by semiconductor devices such as transistors and integrated circuits. Its importance as an amplifier, especially in systems where the voltage is higher than can be readily handled by transistors, entitles it to special study, and the purpose of this experiment is to see how a tube amplifies and to observe some of the characteristics of vacuum-tube amplifier circuits.

## THEORY

The construction and operation of a diode (two-element) vacuum tube has been described in Experiment 37. It was pointed out that when the cathode is arranged to provide copious electron emission, a negative electron cloud called a space charge forms in its vicinity, and the plate current is controlled by the plate voltage according to the three-halves power law. Now, if a third electrode—a grid of fine wire mesh—is interposed between the cathode and the plate, the potential applied to it will have a great effect on the space charge and will thus provide a control over the plate current. In particular, if the grid is made negative with respect to the cathode, it will help the space charge to repel electrons back towards the cathode and will thus cause a decrease in plate current. In fact, if the grid is made very negative, the plate current will be reduced to zero despite the plate's being at a high positive voltage. The tube is then said to be *cut off*, and the grid voltage required to just reduce the plate current to zero with a particular plate voltage is called the *cutoff grid bias* for that plate voltage. Less negative grid voltages allow some plate current to flow, the amount depending on the grid voltage value. Thus, the *triode*, or three-electrode (cathode, grid, and plate) vacuum tube is a control device in which the grid voltage controls the plate current. The plate current is affected much more strongly by changes in grid voltage than by changes in plate voltage; hence amplification is possible. Indeed, a quantity called the *amplification factor* $\mu$ is defined for any particular triode as the ratio of a small change in plate voltage to the change in grid voltage that would have the same effect on the plate current. Thus, in a triode with an amplification factor of 20, a change of 20 V in the plate voltage would have the same effect on the plate current as a change of 1 V in grid voltage. In addition, you should note that as long as the grid is negative (the normal case in small amplifiers, for plate current flows over a wide range of negative grid voltages between zero and the cutoff value) it will not collect any electrons and will therefore draw no current. Thus a negative grid just sits there in the tube looking like an open circuit. Nevertheless, the grid voltage, despite the fact that it has negligible current and negligible power associated with it, controls an appreciable plate current associated with an appreciable plate voltage. Vacuum tubes are therefore capable of large power gains.

In the triode, the plate current depends on both the plate voltage and the grid voltage and is thus a function of two variables. To plot this function completely would therefore require a three-dimensional graph with three mutually perpendicular axes for the plate voltage, plate current, and grid voltage, respectively. Since this is impractical, plots of one variable against another for various constant values of the third are published for most commonly available tubes. Of the three possibilities, the most useful and hence the most often given are the so-called plate characteristic curves obtained by plotting plate current $I_b$ as a function of plate voltage $V_b$ for various values of grid voltage $V_c$. Those for a triode-connected type 6AU6 are shown in Fig. 39.1. Note how each curve goes down to zero plate current at a plate voltage for which the pertinent grid voltage is the cutoff bias.

The type 6AU6 is actually a *pentode* or five-element tube, there being two extra grids placed between the grid already described and the plate. This latter grid, which is nearest the cathode and is the one used to control the plate current, is accordingly called the *no. 1* or *control* grid. When the tube is to be used as a triode, the other two are simply connected to the plate so that the combination of these three electrodes functions as a single electron collector or anode. When the tube is used as a pentode, on the other hand, the second and third grids have special functions that will now be described.

A less-than-ideal feature of the triode is the fact

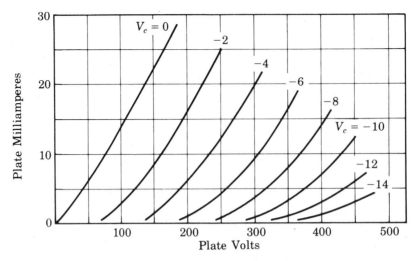

**Figure 39.1** *Plate characteristic curves for the type 6AU6 connected as a triode*

that the plate current, which is supposed to be controlled by the grid voltage, also depends on the plate voltage. This dependence can be much reduced by the insertion of a second grid between the control grid and the plate, this second grid being maintained at a fixed positive potential. It therefore looks like the plate as far as the cathode and space charge are concerned; and since its voltage is held constant, the current through the tube is controlled by the control grid voltage only. The plate voltage has little effect on this current because the plate is effectively shielded or screened from the cathode region by the second grid. This grid is therefore called either the *no. 2* or the *screen* grid. A set of plate characteristic curves for the type 6AU6 operated as a pentode with the

screen grid at $+100$ V is shown in Fig. 39.2. Notice that in contrast to the curves of Fig. 39.1, these are very flat, indicating that there is little change in plate current with plate voltage once the latter is made large enough to collect the electrons. The pentode is said to have a high *dynamic plate resistance* as compared to the triode. The dynamic plate resistance ($r_p$) is the ratio of a small change in plate voltage to the corresponding change in plate current. It is called "dynamic" because the curves of Figs. 39.1 and 39.2 are not linear, so that the tube does not obey Ohm's law and a resistance obtained by dividing plate voltage by plate current has no meaning. Note, however, that the *slopes* of the curves in both Fig. 39.1 and Fig. 39.2 give the ratio of a small change in plate

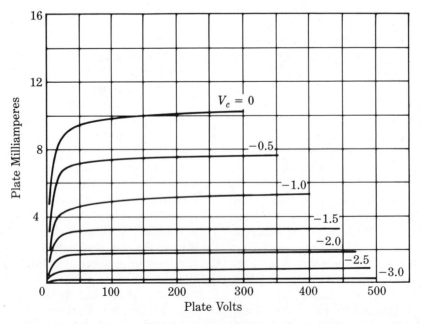

**Figure 39.2** *Plate characteristic curves for the type 6AU6 pentode (screen grid voltage = 100 V)*

current to the corresponding small change in plate voltage and thus represent a *dynamic plate conductance* $g_p$. The reciprocal of $g_p$ is clearly $r_p$, and the fact that the pentode characteristics have almost zero slope means that for the pentode, $g_p$ is very small and $r_p$ correspondingly large.

Because the screen grid is operated at a positive potential, the possibility exists of its attracting electrons and drawing current—an undesirable effect since this current simply passes through the screen grid power supply and represents a power dissipated for no purpose. Screen grid current can be minimized by aligning the screen and control grid wires so that those in the screen are in the shadow of those in grid no. 1. However, secondary electrons coming from the plate can be attracted to the screen and result in a current that both detracts from the plate current and is wasted, as noted above. Secondary emission from the plate resulting from electrons in the surface being knocked out by those arriving from the cathode occurs in both the triode and the pentode, but in the triode the secondaries see the negative grid and are immediately repelled back into the plate. In the pentode the secondaries are confronted with the positive screen and can easily be attracted over to it, especially when the plate voltage is low. To prevent this, a third grid is placed between the screen and the plate and operated at a low potential, usually the same as the cathode. Electrons accelerated by the screen have enough energy to get through the third grid and proceed to the plate, but secondary electrons are shielded from the screen's field by this third grid and therefore return to the plate as in the triode. Because the function of the third grid is to suppress secondary emission, it is called the *suppressor grid*.

When a triode is to be used as a *voltage amplifier*, that is, a device for increasing the amplitude of an ac signal voltage, it is connected in a circuit such as that drawn schematically in Fig. 39.3. A power supply supplies voltage $V_{bb}$ to the circuit con-

sisting of the tube (plate and cathode) and the so-called load resistor $R_L$ in series, as a result of which current $I_b$ flows around this circuit. Notice the direction of $I_b$ indicated in Fig. 39.3: it is opposite to the electron flow direction because electrons are negative, whereas the conventional direction for current is the direction of flow of positive charge.

The purpose of $R_L$ is to convert the varying current $I_b$ (which varies as a result of the control of this current by a signal applied to the grid) into a varying voltage, since a voltage output is wanted. Ohm's law tells us that the potential across $R_L$ is $I_b R_L$, and since this is the difference between the power supply voltage $V_{bb}$ and the plate voltage $V_b$, we have

$$V_{bb} - V_b = I_b R_L$$

or

$$I_b = \frac{V_{bb}}{R_L} - \frac{V_b}{R_L} \tag{39.1}$$

Notice that Equation 39.1 is a relation between the triode's plate current $I_b$ and its plate voltage $V_b$ and can thus be plotted on the tube's plate characteristics as shown for a typical triode in Fig. 39.4. The plot is seen to be a straight line with slope $-1/R_L$ and intercept $V_{bb}/R_L$. Because it is determined by the load resistor, this plot is called the *load line*.

Once the load resistor $R_L$ and the power supply voltage $V_{bb}$ have been established, the relationship between $I_b$ and $V_b$ is fixed by Equation 39.1, and changes in plate current due to control by the grid voltage will result in the changes in plate voltage dictated by this equation. These changes will be made from some resting or *quiescent* value of plate voltage and current that the tube will assume in the absence of any input signal and that determine a point called the *operating point*, as shown in Fig. 39.4. The operating point is located at the intersection of the load-line and the plate characteristic drawn for the grid voltage applied to the tube in the absence of an input

**Figure 39.3** *Triode voltage amplifier circuit*

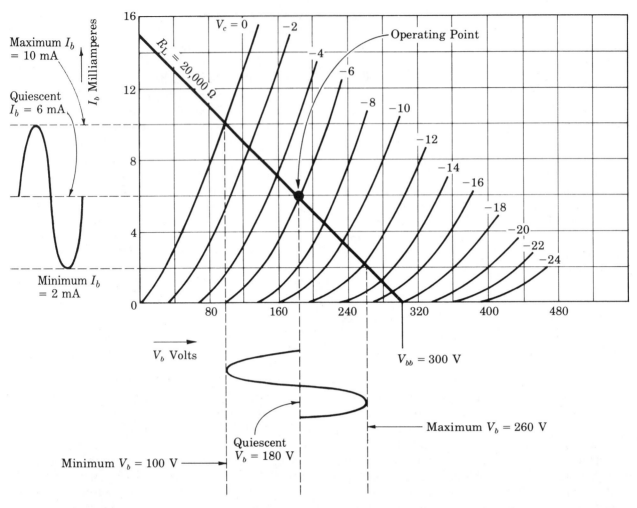

**Figure 39.4**   *Plate characteristics and a sample load line for a typical triode*

signal, and in fact constitutes the graphic solution of two simultaneous equations, namely Equation 39.1 and the plate characteristic in question. The operating point is thus chosen by choosing the voltage applied to the grid under quiescent or no-signal conditions. This is the voltage $V_{cc}$ delivered by the power supply on the left in Fig. 39.3. Note that $V_{cc}$ and $V_c$ are the same with no input signal, since with no grid current there is no voltage drop across resistor $R_g$.

The choice of the operating point is made on the basis of allowing the greatest possible linear swing of the plate voltage and current in either direction. Tubes are nonlinear devices, as the characteristic curves of Figs. 39.1 and 39.2 show, and so if the amplified output voltage is to be a faithful reproduction of the signal voltage at the input, operation must be limited to ranges of plate current and voltage over which the tube's characteristics are reasonably linear. In particular, note that the characteristics become particularly nonlinear at low plate currents, so that operation of the triode of Fig. 39.4, for example, should be limited to currents above, say, 2 mA. This

sets a somewhat arbitrary lower limit on $I_b$. The upper limit is set by the intersection of the load line with the plate characteristic drawn for zero grid voltage, for in this kind of amplifier the grid must not be made to swing positive.* This is because no provision has been made for the flow of grid current, which would result in a voltage drop across the grid resistor $R_g$. In fact, large values of $R_g$ are chosen on the assumption that there will be no current through it. The tube's operation with signal can thus be considered to be along the load line from the point where it intersects the low plate current limit to the point where it intersects the characteristic curve for $V_c = 0$. Since the input signal is likely to swing negative as much as positive (this will be exactly true if the input is sinusoidal), the operating point is chosen in the middle of this range, as shown in Fig. 39.4, and the corresponding grid voltage value is selected for $V_{cc}$. If

---

*This kind is designated "class A." There are class B and class C amplifiers in which the grid is driven positive and grid current is drawn, but they will not be considered here.

the location so chosen does not happen to fall on one of the given plate characteristic curves, interpolation between the neighboring curves is used to determine the grid bias.

With the above design parameters fixed, the amplifier is ready to accept an input signal. This is applied to the input terminals marked $V_{in}$ in Fig. 39.3 and passes through capacitor $C_g$. The purpose of $C_g$ is to pass the ac signal but to block the dc grid bias voltage from getting back to the signal source. The signal voltage develops across $R_g$ so that the grid voltage $V_c$ is the sum of the dc component $V_{cc}$ (the bias), which sets the operating point, and the ac signal voltage $V_{in}$, which makes $V_c$ swing above and below its quiescent value. The tube's plate current varies with these swings, and the corresponding plate voltage variations shown in Fig. 39.4 appear across $R_L$. $C_p$ stops the dc component of plate voltage while passing the variations on to the output terminals. Because $C_g$ and $C_p$ block dc voltages from appearing where they are not wanted, these capacitors are often called *blocking capacitors*. They are also called *coupling capacitors* because they couple the signal voltages from one circuit to the next. However, it should be recalled from Experiment 36 that while a capacitor appears to pass an alternating current, it exhibits a reactance $X_C = 1/2\pi fC$, which is measured in ohms and acts in some ways as a resistance to the passage of ac. Thus the combination of $C_g$ and $R_g$ behaves as a sort of voltage divider (see Experiment 30) with the element represented by $C_g$ introducing a reactance that gets larger and larger as the frequency is reduced. Consequently, at high frequencies for which $X_{C_g}$ is small compared to $R_g$, the capacitor $C_g$ is essentially a short circuit for the signal and the effect of its reactance may be neglected. But as the frequency is reduced, $X_{C_g}$ becomes significant in decreasing the signal voltage applied to the grid below the input value $V_{in}$, so that the net gain of the amplifier as a whole is reduced below its so-called midband value, which is the value it has for frequencies high enough to make $X_{C_g} \ll R_g$. Eventually a frequency called the *lower half-power point* is reached at which the gain is reduced to $1/\sqrt{2}$ times its value at midband. Notice that as it is voltage gain we are talking about, reducing the frequency to the lower half-power point with a fixed input signal amplitude reduces the output voltage by a factor of $1/\sqrt{2}$. This means that the square of the output voltage, to which output power would be proportional, is cut in half, hence the designation half-*power* point. There is also an *upper half-power point,* which occurs when stray capacities between the tube's plate with its associated wiring and the common ground line exhibit a reactance sufficiently small to load the amplifier output down to where again the overall gain is $1/\sqrt{2}$ times its midband value. The amplifier's frequency response or

*bandwidth* is often defined as the range of frequencies between the two half-power points. If the load being driven by the amplifier has an impedance so large that we may neglect the effect of $C_p$, the lower half-power point may be shown to occur when the reactance of $C_g$ becomes equal to $R_g$. The frequency of the upper half-power point may also be calculated if the stray capacities mentioned above are known, but we will not attempt to do this in the present work both because we have no way of measuring these capacities and because the upper half-power frequencies obtained with the amplifier circuits discussed herein are high enough to require special precautions in their measurement.

The gain of a voltage amplifier may be determined graphically by a construction such as that shown in Fig. 39.4. Here the maximum allowed swing in grid voltage (and hence the maximum acceptable input signal amplitude) is 6 V above and below the chosen grid bias $V_{cc} = -6$ V. The corresponding plate voltage swing may be read off the voltage axis. The quiescent plate voltage is seen to be 180 V, this being the dc component that must be blocked from the output by $C_p$, and the swing is up to about 260 V and down to about 100 V for an output swing of 80 V in either direction. This represents the maximum signal-handling capability of the amplifier, and under these conditions the gain is 80/6 or 13.3. Note that an input signal with an amplitude larger than 6 V is not permissible. Indeed, for minimum distortion, input signal levels should be kept appreciably less than this. For this reason, circuits like that of Fig. 39.3 are often called *small-signal amplifiers*. Note also that because an increase in plate current causes a decrease in plate voltage in consequence of the action of $R_L$, a rising grid voltage produces a falling plate voltage. This type of amplifier therefore reverses the phase of the input signal so that the output voltage, in addition to being larger than the input, is the negative of it. This feature of voltage amplifiers of the sort diagrammed in Fig. 39.3 must often be taken into consideration, and such circuits are sometimes put into a system for the express purpose of reversing the phase of the signal. An amplifier so used is called a *phase inverter*.

The gain of a voltage amplifier of the sort considered here is always less than the amplification factor $\mu$ of the tube, but it can be made to approach $\mu$ by making the value of $R_L$ large. This can be seen from the fact that the larger $R_L$ is, the smaller the variations in plate current needed to produce a given output voltage. Thus, as $I_b$ comes closer to being constant, the change in plate voltage becomes more nearly equal to $\mu$ times the change in grid voltage. One method of measuring the amplification factor of a triode is therefore to see how the gain it delivers in the circuit of Fig. 39.3 varies as $R_L$ is increased. The amplification factor will be the limit that this gain approaches.

Pentodes can also be used as voltage amplifiers in circuits similar to that of Fig. 39.3, except that a source must be provided for the screen grid voltage. This may be the same as the plate voltage supply, but since the screen is usually operated at a potential considerably less than $V_{bb}$, means must be provided for dropping the voltage to a suitable level while at the same time maintaining that level constant. A simple series resistor chosen so that the current drawn by the screen will drop the voltage to the proper value is often used, with a capacitor connected from this grid to the cathode to remove voltage variations that would result from variations in screen current. However, a better method is to provide a screen voltage regulated at the proper value by a regulator such as the zener diode circuit described in Experiment 37. If this experiment has not been done, the part of its Theory section describing the zener diode and the regulator circuit of Fig. 37.2 should be studied. Because the screen grid of a well-designed pentode draws very little current, a shunt regulator of this type works quite well, and the circuit of Fig. 37.2 is often used to provide a fixed screen voltage. The zener diode is of course chosen to have a zener voltage equal to the value recommended for the screen grid of the pentode in question. Thus, in the present case, Fig. 39.2. shows the 6AU6 plate characteristics for a screen voltage of +100 V. Hence if we are to use these curves in the design of a 6AU6 pentode voltage amplifier, we would choose a 100-V zener as the screen voltage regulator.

Because there is very little dependence of plate current on plate voltage (very high plate resistance) in a pentode, almost the entire swing in grid voltage is effective in producing a plate current swing and hence an output voltage across $R_L$. If the effect of changing plate voltage can be disregarded, the change in plate current is equal to the change in grid voltage multiplied by a quantity $g_m$ called the *transconductance*. This parameter is the real measure of how good the tube is at controlling the current through its plate-cathode circuit by means of the voltage on its grid. Being the ratio of a change in plate *current* to a corresponding change in grid *voltage*, it is a *conductance* and the prefix *trans-* expresses the fact that it is the ratio of a current in one circuit (the plate) to the voltage in another (the grid). Transconductance was formerly called *mutual conductance*, which is the reason for the subscript m in $g_m$, the standard symbol for this parameter.

In view of the above considerations, the variations in the plate current of a pentode will be $g_m$ times the variations in grid voltage (the input signal), and the output voltage will be the plate current changes times $R_L$, as in the triode. Hence the voltage gain, which is the output signal divided by the input signal, should be equal to $g_m R_L$. This will be approximately true in a pentode amplifier provided $R_L$ is not made too large. However, although $\mu$ for a triode remains almost constant regardless of the choice of operating point, $g_m$ for a pentode does not remain constant and will therefore change when different load resistors and hence load lines and operating points are chosen. Therefore, $R_L$ and the amplifier gain are not truly proportional. Nevertheless, for any given value of $R_L$ (provided it is small compared to $r_p$), a value of $g_m$ can be found that is valid at the operating point associated with that $R_L$. Several such values will be determined in this experiment and can be compared with those listed in tube manuals for the type 6AU6 operated under similar conditions.

Because of their large $r_p$, pentodes offer considerably higher gain than do triodes. They are, however, less linear, so that signal voltage levels must be strictly limited if excessive distortion is to be avoided. Pentodes are therefore used to best advantage in applications where a small signal is to be amplified as much as possible by a single tube.

## APPARATUS

1. Type 6AU6 tube
2. Type 1N5378 zener diode (100-V)
3. 7-pin miniature tube socket, mounted
4. Power supply, 300 V dc and 6.3 V ac
5. Oscilloscope
6. Audio signal generator
7. Electronic voltmeter
8. Decade resistance box (10,000 Ω)
9. Decade resistance box (1000 Ω)
10. 6-V dry battery (lantern battery or equivalent)
11. Five carbon, 1-W, 5%-tolerance resistors (220,000; 100,000; 47,000; 27,000; and 15,000 Ω)
12. 20,000-Ω, 10-W wirewound resistor
13. Volume-control-type potentiometer (5000 Ω, wirewound)
14. Three paper capacitors (0.01-μF, 400-V rating)
15. Assorted hookup wires and connectors

## PROCEDURE

**1.** On the triode plate characteristics for the type 6AU6 (Fig. 39.1), draw load lines for a plate supply voltage of 300 V and load resistances of 15,000; 27,000; 47,000; and 100,000 Ω. Assuming a mini-

**Figure 39.5**  *Triode amplifier circuit*

mum permissible plate current of 0.5 mA, choose a proper operating point on each load line. Record the values of plate current, plate voltage, and grid voltage at each of your operating points.

**2.** Assemble the triode voltage amplifier circuit using the type 6AU6 as shown schematically in Fig. 39.5. Use the 10,000-$\Omega$ decade resistance box for $R_1$, the 1000-$\Omega$ box for $R_2$, and the 15,000-$\Omega$ resistor for $R_L$. Set $R_1$ to 1000 $\Omega$ and $R_2$ to zero. Notice that the tube's heater is operated from a source of 6.3 V ac included in the power supply. The electronic voltmeter's common ("low" side) lead should be connected to the tube's cathode (the common ground line). The EVM's probe is then available to test voltages at various points in the circuit. The "low" side of the oscilloscope's vertical input is also connected to the common ground line, and the "high" side is provided with a lead that may be connected either to the amplifier output terminal or to point $A$ to observe the output from the signal generator. In addition, the oscilloscope's External Synchronization input is connected to point $A$, as the oscilloscope will be synchronized from this signal rather than internally from the signal at the vertical input.

The type 6AU6 is a so-called miniature tube and has a standard 7-pin miniature base. This base has seven wire connecting pins arranged in a circle and spaced as if there were supposed to be eight. The eighth pin position is blank and serves to determine the one way in which the tube may be plugged into its socket. The pins on miniature tubes are easily bent and care must be taken when inserting the 6AU6 to avoid damaging them. Note that the pins are numbered *clockwise when viewed from the base end,*

starting with pin no. 1 on one side of the blank space and proceeding in order until pin no. 7 is reached on the other. Remember that if the tube is viewed from above (base down), the pin numbering appears to be counterclockwise. The base diagram for the type 6AU6 is given in Fig. 39.6, and for convenience the pin numbers are marked on the connections to the respective tube elements in the schematic diagram of Fig. 39.5.

When your circuit is completely wired, call the instructor. Do not plug in or turn on the power supply

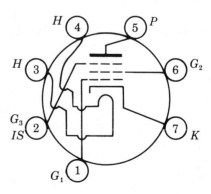

**6AU6**

| | |
|---|---|
| $P$ = Plate | $K$ = Cathode |
| $G_1$ = Control Grid | $H$ = Heater |
| $G_2$ = Screen Grid | $IS$ = Internal Shield |
| $G_3$ = Suppressor Grid | |

**Figure 39.6**  *Tube base connection diagram for the type 6AU6*

until the instructor has checked your work. CAU-TION: The power supply output voltage is high enough to give you a painful shock. Be very careful when working around your circuit and do not make any changes with the power on. When measuring voltages with the EVM, be sure you have the meter switched to the right range and, when touching the probe to a particular circuit point, use care not to short-circuit it to any adjacent circuit points.

**3.** Adjust the 5000-$\Omega$ potentiometer to give the grid bias voltage corresponding to the operating point you chose on the 15,000-$\Omega$ load line in Procedure 1. *Note:* This so-called potentiometer is really a rheostat with connections made to both ends of its resistance element as well as to the movable contact arm, but such a device is commonly called a potentiometer in electronic work because of its resemblance to the slide-wire arrangement in Experiment 30. Measure $V_{cc}$ at the movable contact arm (middle terminal) of the potentiometer. Do not attempt to measure the voltage at the tube grid, for even though the EVM has an input resistance of about 10 M$\Omega$, it will draw enough current when connected to the grid to cause a noticable voltage drop in the 220,000-$\Omega$ grid resistor $R_g$.

**4.** Plug in and turn on the power supply. Allow a minute or two for the tube to warm up, and then adjust $V_{bb}$ to exactly 300 V as measured by the EVM. Measure and record the 6AU6 plate voltage $V_b$.

**5.** Check that $R_1$ is set to 1000 $\Omega$ and $R_2$ to zero. Then turn on the signal generator and set it for a frequency of 1000 Hz. Connect the oscilloscope vertical input to point A (Fig. 39.5) and adjust the signal generator output to give a sine wave with an amplitude of 1 to 2 V. Set the oscilloscope to display two or three cycles of this wave. Use external synchronization to stabilize the pattern. With the oscilloscope's vertical and external synchronization inputs both connected to point $A$, there is no difference between external and internal synchronization, but there will be when the oscilloscope is used to observe the amplifier output. Adjust the vertical gain so that the pattern has a convenient amplitude. Note that a vertical calibration is not necessary for this experiment, but an easily reproducible amplitude should be chosen. This means that the vertical gain should be set to bring the sine curve peaks to easily remembered marks on the oscilloscope screen so that this amplitude can be duplicated later. Do not change the vertical gain setting after this has been done.

**6.** Disconnect the oscilloscope vertical input from point $A$ and connect it to the amplifier output. Increase the value of $R_2$ until the displayed sine curve has the same amplitude as that established in Procedure 5. Record the value of $R_2$ needed to do this.

**7.** Compare the sine curves displayed on the oscilloscope in Procedures 5 and 6 and report your ob-servation in the space below.

**8.** Determine your amplifier's lower half-power point as follows: Resistances $R_1$ and $R_2$ form a voltage divider as discussed in Experiment 30. Equation 30.2 in that experiment shows that the actual input signal $V_{in}$ for the amplifier is cut down from the signal generator output by a factor $R_2/(R_1 + R_2)$. In Procedure 6 you set $R_2$ so that the voltage divider reduced the signal generator output by an amount equal to the gain of the amplifier, thus making the amplifier and signal generator outputs identical. Therefore, the amplifier gain *at 1000 Hz* is $(R_1 + R_2)/R_2$. You now want to find a frequency for which the gain is $1/\sqrt{2}$ times this value. In this case $R_2$ will have a new value $r$ such that

$$\frac{1}{\sqrt{2}} \frac{R_1 + R_2}{R_2} = \frac{R_1 + r}{r} \qquad (39.2)$$

Solving for $r$ and noting that $R_1 = 1000 \ \Omega$ yields

$$r = \frac{1414R_2}{1000 - 0.414R_2} \qquad (39.3)$$

Using your value of $R_2$ from Procedure 6, find $r$, record it, and reset decade box $R_2$ to this new value. Then reduce the signal generator frequency until the amplifier output returns to its original level. The signal generator is now set at the frequency of the lower half-power point. Read this frequency from the dial and record it.

**9.** Turn off the power supply and substitute the 27,000-$\Omega$ resistor for the 15,000-$\Omega$ resistor at $R_L$. Then repeat Procedures 3–6 for $R_L = 27,000$ ohms. Also repeat Procedure 8 to determine the lower half-power point for this case.

**10.** Repeat Procedure 9 with $R_L = 47,000 \ \Omega$ and again with $R_L = 100,000 \ \Omega$.

**11.** Turn off the power supply and unplug it. Rewire your circuit to use the 6AU6 as a pentode as shown in Fig. 39.7 with $R_L = 15,000 \ \Omega$. Notice how the screen grid voltage is regulated by the 100-V type 1N5378 zener diode. Compare this part of the circuit with Fig. 37.2. Capacitor $C_s$ bypasses to the common ground line any ac voltages that might develop across the 1N5378 and thus assures that the 6AU6 screen voltage will be pure direct current. For this reason, $C_s$ is called a *bypass capacitor*. Other parts of the pentode circuit are very similar to those for the triode. Do not plug in the power supply until your circuit has been checked by the instructor.

**Figure 39.7** *Pentode amplifier circuit*

**12.** Before turning on your pentode amplifier, repeat Procedure 1 using the pentode plate characteristics of Fig. 39.2. In choosing operating points on your four load lines, again assume a minimum $I_b$ of 0.5 mA, but this time notice that before some of these lines intersect the characteristic curve for $V_c = 0$, they go through a region where the pentode plate characteristics become very nonlinear. Hence, for these values of $R_L$, a minimum plate voltage rather than intersection with the curve for $V_c = 0$ must be specified to determine the high plate current operating limit. Use 50 V for the 6AU6 and take the high plate current limit on any given load line as the point where that line intersects either the ordinate at $V_{bb} = 50$ V or the curve for $V_c = 0$, whichever oc-

curs first. Choose your operating points halfway between the minimum plate voltage and minimum plate current points on each load line and record the corresponding values of plate current, plate voltage, and grid voltage.

**13.** Repeat Procedures 3–10 for your pentode amplifier. After turning on the power, check the 6AU6 screen voltage. The rated operating voltage for the 1N5378 is 100 V, but this varies a little either way with different samples of the type. Any voltage between 98 and 102 V is satisfactory. Report the observations made when repeating Procedure 7 in the space below. Be sure to shut off all power when you have finished the experiment.

# DATA _____

## Triode

| Load resistor $R_L$, Ω | Calculated from curves | | | Measured | | | | | |
|---|---|---|---|---|---|---|---|---|---|
| | Grid bias $V_{cc}$ | Plate voltage $V_b$ | Plate current $I_b$ | Plate voltage $V_b$ | Plate current $I_b$ | Value of $R_2$ | Gain | Value of $r$ | Lower half-power point |
| 15,000 | | | | | | | | | |
| 27,000 | | | | | | | | | |
| 47,000 | | | | | | | | | |
| 100,000 | | | | | | | | | |

Estimated value of the amplification factor $\mu$ _____

Value of $\mu$ listed by the manufacturer _____

## Pentode

| Load resistor $R_L$, Ω | Calculated from curves | | | Measured | | | | | | |
|---|---|---|---|---|---|---|---|---|---|---|
| | Grid bias $V_{cc}$ | Plate voltage $V_b$ | Plate current $I_b$ | Plate voltage $V_b$ | Plate current $I_b$ | Value of $R_2$ | Gain | Value of $g_m$ | Value of $r$ | Lower half-power point |
| 15,000 | | | | | | | | | | |
| 27,000 | | | | | | | | | | |
| 47,000 | | | | | | | | | | |
| 100,000 | | | | | | | | | | |

Calculated value of the lower half-power frequency _____

# CALCULATIONS _____

**1.** From Equation 39.1, calculate and record $I_b$ for each value of $R_L$ used in both the triode and pentode amplifiers. For the triode, $V_{bb} = 300$ V and the necessary values of $V_b$ were measured in Procedures 4, 9, and 10. For the pentode, $V_{bb}$ is again 300 V and $V_b$ was measured for each load resistance in Procedure 13. Compare each measured value of $I_b$ and $V_b$ with those predicted from the plate characteristics.

**2.** As already mentioned in Procedure 8, resistances $R_1$ and $R_2$ form a voltage divider whose purpose is to produce a voltage $V_{in}$ across $R_2$ which is the fraction $R_2/(R_1 + R_2)$ of the voltage impressed across $R_1$ and $R_2$ in series by the signal generator. Notice that $V_{in}$ is the amplifier input signal and that $R_2$ was adjusted in Procedures 6, 9, 10, and 13 until the output voltage $V_0$ was equal to the voltage at Point $A$ in each case. Hence the amplifier was amplifying the input signal back up to the same level from which the voltage divider cut it down. The gain is therefore $(R_1 + R_2)/R_2$. Use this conclusion to compute the gain obtained with each load resistor in both the triode and pentode amplifiers.

**3.** Plot your values of the triode amplifier gain against the corresponding load resistances using the values of $R_L$ as abscissas and the gain values as ordinates. Choose your scales so as to cover as much of the sheet of graph paper as possible but allow for some extrapolation of your curve to higher $R_L$. *Note:* If $R_L$ were zero there would be no output voltage and therefore no voltage gain, so that the origin is a point on your graph. Draw a smooth curve through the five points you have plotted.

**4.** Extrapolate the graph of Calculation 3 to higher $R_L$ and estimate the limit that the gain approaches as $R_L$ gets very large. This should be the amplification factor of the type 6AU6 operated as a triode. Look up the rated value in a tube manual and compare it with your estimate.

**5.** Assuming that the gain of your pentode amplifier is given to a good approximation by $g_m R_L$ for the load resistors you used, calculate and record the value of the transconductance displayed by your 6AU6 under each of your four sets of operating conditions. Look up the values listed by the manufacturer in a tube manual and compare them with your results. *Note:* The manual may list several values for different operating conditions. Observe which listed conditions most nearly approach those in effect with each of your load resistors and compare appropriate pairs of $g_m$ values.

**6.** From the fact that the lower half-power point should occur at the frequency for which $X_{C_g} = R_g$, compute a theoretical value for this frequency and enter it in the space provided in the data sheet. *Note:* It is assumed that the oscilloscope input resistance is so high that the output blocking capacitor $C_p$ has no effect on this frequency.

## QUESTIONS

**1.** Define plate resistance and explain why the gain of a triode amplifier should approach the tube's amplification factor when the load resistance is made larger than the plate resistance.

**2.** The relation $\mu = g_m r_p$ holds for all triodes. Give a qualitative derivation of this relation.

**3.** External synchronization of the oscilloscope pattern was used in Procedure 5 so that an important difference between the patterns displayed in Procedures 5 and 6 could be observed in Procedure 7. What was this difference and why did external rather than internal synchronization have to be used to observe it?

**4.** Using the plate characteristics given in Fig. 39.2, find by any method you wish the transconductance you would expect a type 6AU6 pentode to display at each of the four operating points chosen in this experiment. Describe your method and compare your results with the values of $g_m$ you determined in Calculation 5.

**5.** Why does no one talk about the amplification factor of a pentode?

**6.** Under what conditions does the approximation $g_m R_L$ for the gain of a pentode amplifier fail? Why?

**7.** Carry out the derivation of Equation 39.3.

**8.** How much current could the screen grid of your pentode amplifier draw from the zener diode regulator before the regulation would fail?

**9.** Prove that when $X_{C_g} = R_g$ the signal voltage at the tube grid and hence the gain of the amplifier as a whole is reduced by a factor of $1/\sqrt{2}$ below the value in effect if $C_g$ were a short circuit, that is, at midband frequencies. *Hint:* Remember that a reactance, in addition to having a magnitude measured in ohms, introduces a phase shift. Thus $C_g$ shifts the phase of the amplifier input voltage relative to that delivered by the signal generator.

**10.** It was mentioned in the Theory section that an *upper half-power point* existed at a frequency for which the stray capacity between the plate with its associated wiring and ground decreased the effective load impedance (which now consists of this capacity and $R_L$ in parallel) sufficiently to reduce the amplifier gain to $1/\sqrt{2}$ times its midband value. Consider a pentode amplifier having stray plate-to-ground capacity $C_w$ and operated with a load resistance $R_L$ small enough for the midband gain to be given by $g_m R_L$. (a) Show that the upper half-power point will occur at a frequency for which $X_{C_w} = R_L$, where $X_{C_w}$ is the reactance of $C_w$. (b) If $C_w = 10$ pF ($10^{-11}$ F), what upper half-power frequencies would you expect with each of the load resistors used in your 6AU6 pentode amplifier?

**11.** In Calculation 6 you found a lower half-power frequency determined by the grid coupling capacitor $C_g$ and the grid resistor $R_g$. Should all the lower half-power points measured for both your triode and pentode amplifiers be equal to each other and to this theoretical value? Are they?

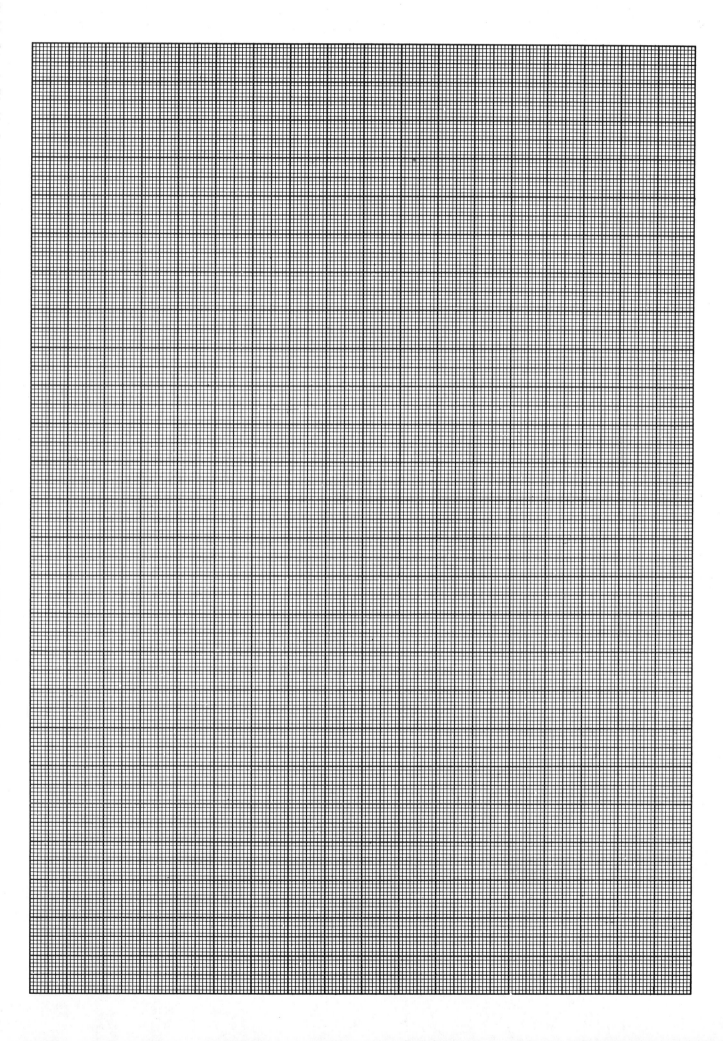

# Amplification with Transistors 40

In the last three decades, the solid-state control device called the transistor has developed to the point where it has supplanted vacuum tubes in almost all low-power applications. Its advantages include smallness and the absence of a hot cathode requiring heating power and possibly burning out. In operation, the transistor differs from the vacuum triode (or, more correctly, the vacuum pentode, since transistor and pentode characteristic curves look very similar) in that it operates at lower voltage and higher current levels and that the controlling signal is basically a current rather than a voltage. It is, however, a control and amplifying device similar to a vacuum tube, and many of the considerations discussed in Experiment 39 apply to it. A review of that experiment (or reading the Theory section if Experiment 39 has not been done) is suggested in preparation for the present work, whose purpose will be to study the characteristics of a typical NPN transistor. A review of the section on semiconductor diodes and the PN junction in Experiment 37 may also prove helpful.

## THEORY

In Experiment 37, we saw how a PN junction could behave as a unidirectional conductor. It is also possible to make a semiconductor triode with amplifying properties, that is, control of current in an output circuit by a relatively small amount of power in an input circuit. Such a device is called a *transistor*. In its modern form, it consists of two PN junctions back-to-back and may be thought of as two diodes connected in series opposing (cathode-to-cathode or anode-to-anode), although if two semiconductor diodes were actually connected in this way, they would not work as a transistor because junction transistor action depends on actual physical proximity of the two junctions.

A PNP transistor is made by sandwiching a piece of N-type semiconductor material between two pieces of P-type, in which case two PN junctions with their N sides connected together are formed. A piece of P-type material can also be sandwiched between two N-type pieces, thus forming two PN junctions with their P sides connected together. Such a transistor is designated NPN. The two kinds have very similar properties; the principal difference is the reversal of the polarity of all voltages for one relative to the other. Figure 40.1 illustrates the NPN case. The material in the middle is always called the *base,* that on one end the *emitter,* and that on the other the *collector.* The terms "emitter" and "collector" have some meaning in the light of present-day transistor operation since they respectively emit and collect charge carriers, but the name "base" stems from the first point-contact transistors, in which the emitter and collector were point contacts on a "base" of semiconductor material and which are now obsolete. The term has persisted, however, and is now defined as referring to the material in the middle of the sandwich.

The diode formed by the emitter-base junction is shown forward-biased in Fig. 40.1. It therefore offers little resistance to the passage of current. On the other hand, the base-collector junction is reverse-biased and would pass negligible current if it were isolated. However, the emitter-base junction conducts by reason of the electrons in the N-type emitter being driven across the junction into the base. There they take some time to recombine with the holes, so that if the base is made very thin, the probability of these electrons drifting across the base-collector junction can be quite high. It is in fact possible to "collect" from the base most of the electrons "emitted" into it by the emitter, so that most of the emitter current goes to the collector and the base current is comparatively small. Nevertheless, the presence of the emitter-base bias and the resulting base current is essential to the existence of any appreciable collector current at all. A large collector current is thus controlled by a small base current, so that the transistor exhibits amplifying properties. Note, however, that

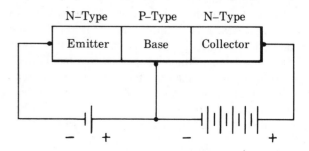

**Figure 40.1** *Transistor arrangement and biasing*

the collector (output) current is controlled by the base (input) *current*, whereas in the vacuum tube the plate (output) current is controlled by the grid (input) *voltage*. A fundamental difference between the tube and the transistor is that the tube is a voltage-controlled device whereas the transistor is current-controlled.

Nevertheless, the transistor is similar to the tube of Experiment 39 in that the collector is comparable to the plate, the base to the grid, and the emitter to the cathode: the emitter and cathode both emit charge carriers, the collector and plate both collect them, and the base and grid both control how many get through. The NPN transistor is particularly close to the tube since the charge carriers are electrons in both cases. While there is a PNP transistor, there is no practical high-vacuum tube through which the current is carried by positive charge carriers. In this experiment, the operation of the type 2N2219 NPN transistor will be studied, and the similarity to much of the work in Experiment 39 should be noted.

To begin with, a collector family of characteristic curves may be drawn for a transistor just as a plate family can for a tube. Figure 40.2 shows such a set for the type 2N2219. Here it is the collector current $I_C$ that is plotted as a function of collector-to-emitter voltage $V_{CE}$ for *various values of base current*. Note carefully that these curves correspond to different values of base *current*, not voltage. As already noted, it is a current rather than a voltage that is the control variable in a transistor. Thus the curves are similar in shape to those for a pentode vacuum tube (see Fig. 39.2) except for the voltage scale and the fact that they are labeled with values of base current rather than grid voltage.

When a transistor is to be used as a voltage amplifier, it is connected in a circuit such as that of Fig. 40.3. Notice the standard symbol for a transistor shown in this figure. The arrowhead on the emitter lead indicates the direction of current flow and is drawn in this figure for an NPN transistor. The same diagram could be used for a PNP transistor if the polarity of $V_{CC}$, the direction of $I_C$, and the direction of the arrowhead on the transistor symbol were all reversed. A battery or other power supply supplies voltage $V_{CC}$ to the circuit consisting of the collector-emitter path through the transistor and the load resistor $R_L$ in series, as a result of which current $I_C$ flows around this circuit. $R_L$ converts variations in $I_C$ into voltage variations, which are passed by blocking capacitor $C_C$ to the output terminals, where they appear as the output signal voltage $V_0$. As in the case of vacuum tubes, the drop across $R_L$ is the difference between the supply and collector voltages, so that

$$V_{CC} - V_{CE} = I_C R_L$$

or

$$I_C = \frac{V_{CC}}{R_L} - \frac{V_{CE}}{R_L} \qquad (40.1)$$

in analogy with Equation 39.1 for the vacuum-tube plate current. Thus a load line can be drawn on the collector characteristics of a transistor just as it can on the plate characteristics of a tube, and an operating point can be chosen in a similar manner. Since the transistor curves resemble those for a pentode, a minimum collector voltage and a minimum collector current are chosen to limit the operating range as was done in Procedure 12 of Experiment 39. The choice of operating points determines the input bias as in the vacuum-tube case, but now it is a *base bias current* rather than a grid bias voltage that must be provided. Figure 40.3 shows that this is done by supplying a current $I_B$ to the base through resistor $R_B$ from the $V_{CC}$ supply. If the base-emitter junction were a perfect short circuit, this base current would simply be given by

$$I_B = \frac{V_{CC}}{R_B} \qquad (40.2)$$

In fact, this approximation is extremely good for the 2N2219 in the circuit of Fig. 40.3, where $V_{CC}$ is always large compared to the base-emitter voltage $V_{BE}$ and the input resistance (resistance of the base-emitter junction) is small compared to the values required for $R_B$. Equation 40.2 can therefore be used to calculate $R_B$ so that the base current needed to establish the chosen operating point will be obtained.

Variations in $I_B$ result in variations in $I_C$, and since (like the dynamic plate resistance of a pentode) the dynamic collector resistance of a transistor is high, the attendant collector voltage variations due to the action of $R_L$ have little effect. The ratio of the change in $I_C$ due to a change in $I_B$ is called the transistor's *forward current transfer ratio* and is designated either $\beta$ or $h_{fe}$. It is analogous to the transconductance $g_m$ of a pentode. However, note that because the controlling signal in a tube is a voltage whereas that in a transistor is a current, $g_m$ is a conductance, being the ratio of a current change to a voltage change, but $\beta$ is dimensionless, being the ratio of two current changes. For the pentode, variations in the grid voltage constitute the input signal $V_{in}$, and the voltage gain $V_0/V_{in}$ can be approximated by $g_m R_L$ as described in Experiment 39. For the transistor, however, we can only say that $V_0 \approx \beta R_L \Delta I_B$, where $\Delta I_B$ represents variations in base current due to the input signal. If this signal is a voltage (as it will be if we are trying to make a voltage amplifier), it can cause variations in $I_B$ by being applied to the base, as Fig. 40.3 indicates. In this case, $V_{in}$ develops across the transistor's *dynamic input resistance* $h_{ie}$, so that $\Delta I_B = V_{in}/h_{ie}$. The voltage gain of the transistor amplifier is then given approximately by

$$A_v \equiv \frac{V_0}{V_{in}} \approx \frac{\beta R_L}{h_{ie}} \qquad (40.3)$$

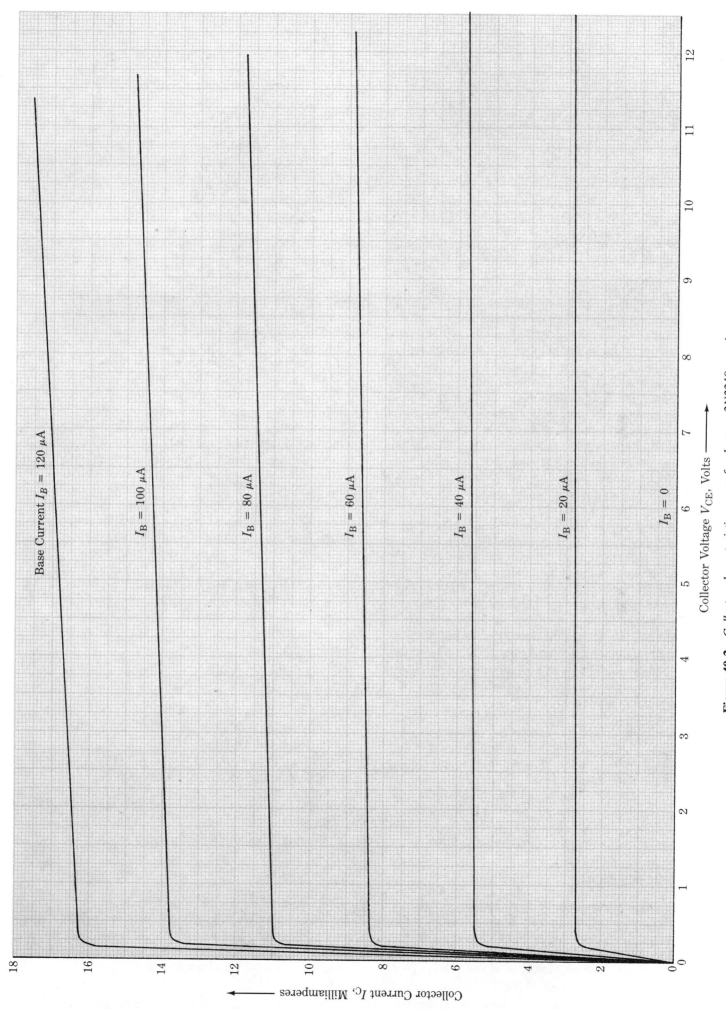

**Figure 40.2** *Collector characteristic curves for the type 2N2219 transistor*

Collector Voltage $V_{CE}$, Volts ——→

Collector Current $I_C$, Milliamperes ——→

Base Current $I_B = 120\ \mu A$

$I_B = 100\ \mu A$

$I_B = 80\ \mu A$

$I_B = 60\ \mu A$

$I_B = 40\ \mu A$

$I_B = 20\ \mu A$

$I_B = 0$

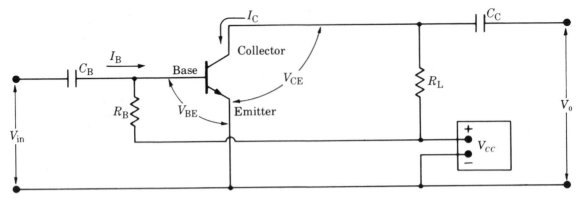

**Figure 40.3** *Transistor amplifier circuit*

where $A_v$ is the gain factor (output signal divided by input signal) and $V_{in}$ is the input signal voltage applied as shown in Fig. 40.3. Equation 40.3 assumes that variations in the collector-to-emitter voltage $V_{CE}$ have no effect on $I_C$ owing to the high dynamic collector resistance of the transistor, and this assumption is fully justified for most transistors as long as $R_L$ is not made too large. Thus Equation 40.3 predicts that $A_v$ will be proportional to $R_L$ over at least some range of load resistor values provided $h_{ie}$ and $\beta$ remain constant as the operating point is changed. This, however, is not the case. Although $\beta$ remains relatively fixed, $h_{ie}$ does not and must be measured for each operating point. We have, however, a convenient way of doing this. In the Theory section of Experiment 39 (which you should review, especially if you have not done the experiment) the concept of a lower half-power point was introduced, and the fact that the gain of the amplifier as a whole was reduced at low frequencies because of the reactance of the input coupling capacitor was discussed. In that case the lower half-power point was the frequency at which the reactance of input capacitor $C_g$ (see Fig. 39.3) became equal to $R_g$. By analogy in the present work, the lower half-power point is determined by the reactance of the input coupling capacitor $C_B$ (Fig. 40.3) becoming equal to the transistor input resistance $h_{ie}$. Knowledge of $C_B$ and a measurement of the lower half-power frequency therefore allows $h_{ie}$ to be calculated for each set of operating conditions.

The forward current transfer ratio $\beta$ may be found from the characteristic curves by noting the collector voltage at the operating point chosen for a particular load resistance and reading off the collector currents for this voltage on the curves immediately above and below that point. The difference between these currents is the collector current change produced by the change in base current from the value for one of the two curves to the value for the other, and the ratio of these changes is by definition $\beta$. Thus a value of $\beta$ can be determined for each operating point, although it will be seen not to vary much from one operating point to another. In the present experiment, we will measure $h_{ie}$ and $\beta$ as just described, compute a gain from Equation 40.3, and compare this result with the gain actually displayed by the transistor amplifier.

## APPARATUS

1. Type 2N2219 transistor
2. Audio signal generator
3. 15-V regulated, adjustable power supply
4. Oscilloscope
5. Electronic voltmeter (EVM)
6. Decade resistance box (0–10,000 Ω)
7. Decade resistance box (0–1000 Ω)
8. Eight carbon 1-W, 5%-tolerance resistors

(470,000; 220,000; 100,000; 4700; 2200; 1500; 1000; and 680 Ω)
9. Volume-control-type rheostat (200,000 Ω)
10. Paper or electrolytic capacitor (10 μF, 35-V rating)
11. Electrolytic capacitor (about 100 μF, 35-V rating)
12. Assorted hookup wires and connectors

## PROCEDURE

**1.** On the collector characteristics for the type 2N2219 transistor (Fig. 40.2), draw load lines for a supply voltage $V_{CC}$ of 12 V and load resistances of 680; 1000; 1500; 2200; and 4700 Ω. Then choose operating points on each load line as described in Proce-

dure 12 of Experiment 39. Note that the curves for the 2N2219 have the same general shape as those for the 6AU6 pentode and so are very nonlinear at low collector voltage. Moreover, there is no zero-grid-voltage characteristic to limit the operating range at

2N2219

**Figure 40.4**   *Amplifier test circuit for the type 2N2219*

the low-voltage end; hence a minimum voltage must be specified for all the load resistances in the present case rather than only some as was the case with the pentode. Use a minimum permissible collector current of 0.5 mA and a minimum permissible collector-to-emitter voltage of 1 V, and choose each operating point midway between the points where the pertinent load line intersects these limits. Note and record the base current corresponding to each operating point. Interpolate between neighboring characteristic curves if necessary to find the base current required for an operating point that doesn't fall on one of the given ones. Also record the collector current and collector voltage determined by each of your operating points.

**2.** Wire the transistor voltage amplifier circuit given in Fig. 40.4. Use the 10,000-$\Omega$ decade resistance box for $R_1$, the 1000-$\Omega$ box for $R_2$, and the 680-$\Omega$ resistor for $R_L$. Set $R_1$ to 1000 $\Omega$ and $R_2$ to zero. The electronic voltmeter's common ("low" side) lead is connected to the 2N2219's emitter (the common ground line), and the EVM probe is then available to test voltages at various points in the circuit. The "low" side of the oscilloscope's vertical input is also connected to the common ground line, and the "high" side is provided with a lead that may be connected either to the amplifier output terminal or to point $A$ to observe the output from the signal generator. In addition, the oscilloscope's External Synchronization input is connected to point $A$, as the oscilloscope will be synchronized from this signal rather than internally from the signal at the vertical input. The connection diagram for the type 2N2219 is given in Fig. 40.5, and the base resistance $R_B$ is made up of a suitable combination of the 100,000-; 220,000-; and 470,000-$\Omega$ resistors in series with the 200,000-$\Omega$ rheostat. From Equation 40.2, calculate the value of

$R_B$ needed to establish the base bias current found in Procedure 1 for the 680-$\Omega$ load line, and pick a resistor combination that will bring this value within the range of the rheostat. Wire this arrangement into your circuit as $R_B$, setting the rheostat at about the middle of its range. Have your circuit approved by the instructor before applying power to it.

**3.** Plug in the power supply, turn it on, and set its output voltage to exactly 12 V as read by the EVM. Then measure the collector voltage $V_{CE}$. Set the rheostat in your circuit so that $V_{CE}$ is exactly equal to the value found at the end of Procedure 1 for the 680-$\Omega$ load.

**4.** Check that $R_1$ is set to 1000 $\Omega$ and $R_2$ to zero. Then turn on the signal generator and set it for a frequency of 1000 Hz. Connect the oscilloscope vertical input to point $A$ (Fig. 40.4) and adjust the signal generator output to give a sine wave with an amplitude of about 0.3 V. Set the oscilloscope to display two or three cycles of this wave. Use external synchronization to stabilize the pattern. Note that with the oscilloscope's vertical and external synchronization inputs both connected to point $A$, there is no difference between internal and external synchronization, but there will be when the oscilloscope is used to observe the amplifier output. Adjust the vertical gain so that the

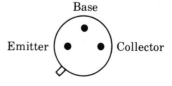

Bottom (Lead End) View

**Figure 40.5**   *Connection diagram for the type 2N2219*

pattern has a convenient amplitude. If your oscilloscope has a vertical calibration, you can use it to check the voltage at point $A$ and set the signal generator output accordingly, but this is not essential for the experiment. However, the final setting of the vertical gain should in any case be such that the displayed sine curve covers most of the oscilloscope screen and falls between easily remembered marks so that its amplitude may be duplicated later. Do not change the vertical gain once this setting has been made.

**5.** Disconnect the oscilloscope vertical input from point $A$ and connect it to the amplifier output. Increase the value of $R_2$ until the displayed sine curve has the same amplitude as that established in Procedure 4. Record the value of $R_2$ needed to do this.

**6.** Compare the sine curves displayed in Procedures 4 and 5 and report your observation in the space below.

**7.** Determine your amplifier's lower half-power point by applying Procedure 8 of Experiment 39 to the present case. In the present experiment, $R_2$ has been found in Procedure 5.

**8.** Turn off the power supply and remove the combination of resistors forming $R_B$ without disturbing the setting of the rheostat. Connect one end of this combination to the common ground line, set the EVM to its Ohms range, zero it, and measure the actual resistance you were using as $R_B$. Record this value in the space provided in the data sheet.

**9.** Substitute the 1000-$\Omega$ resistor for the 680-$\Omega$ one at $R_L$. Then repeat Procedures 2–5, 7, and 8 for $R_L = 1000\ \Omega$.

**10.** Repeat Procedure 9 with $R_L = 1500$; 2200; and 4700 $\Omega$ in turn. Be sure to turn off your equipment when you have finished the experiment.

# DATA

## Values Calculated from Collector Characteristics

| Load resistor $R_L$, $\Omega$ | Collector voltage $V_{CE}$, V | Collector current $I_C$, mA | Base bias current $I_B$, $\mu$A | $\beta$ | Gain $A_v$ |
|---|---|---|---|---|---|
| 680 | | | | | |
| 1000 | | | | | |
| 1500 | | | | | |
| 2200 | | | | | |
| 4700 | | | | | |

## Measured Values

| Load resistor $R_L$, $\Omega$ | Base resistor $R_B$, $\Omega$ | Value of $R_2$, $\Omega$ | Lower half-power point, Hz | Input resistance $h_{ie}$, $\Omega$ | Gain $A_v$ |
|---|---|---|---|---|---|
| 680 | | | | | |
| 1000 | | | | | |
| 1500 | | | | | |
| 2200 | | | | | |
| 4700 | | | | | |

## CALCULATIONS _____

**1.** Calculate and record the forward current transfer ratio $\beta$ in effect at each of your chosen operating points by the method outlined in the Theory section.

**2.** As already mentioned in Procedure 7 (Procedure 8 of Experiment 39), resistances $R_1$ and $R_2$ form a voltage divider whose purpose is to produce a voltage $V_{in}$ across $R_2$ that is the fraction $R_2/(R_1 + R_2)$ of the voltage impressed across $R_1$ and $R_2$ in series by the signal generator. Notice that $V_{in}$ is the amplifier input signal voltage and that $R_2$ was adjusted in Procedures 5, 9, and 10 until the output voltage $V_0$ was equal to the voltage at point $A$. Hence the amplifier was amplifying $V_{in}$ back up to the same level from which the voltage divider cut it down. In other words, the amplifier was simply undoing what the voltage divider had done, so that the gain is $(R_1 + R_2)/R_2$. Use this conclusion to compute the measured gain $A_V$ obtained with each load resistor.

**3.** From the frequencies of the lower half-power points measured for each load resistor, calculate and record the input resistance $h_{ie}$ exhibited by the 2N2219 at each operating point. Use the fact that at a lower half-power point $h_{ie}$ must equal the reactance of $C_B$. *Note:* $C_C$ (Fig. 40.4) was made quite large and the oscilloscope input resistance is assumed to be so high that this capacitor has no effect on these frequencies.

**4.** Use the values of $h_{ie}$ found in Calculation 3 and the values of $\beta$ recorded in Calculation 1 to compute a theoretical gain $A_V$ from Equation 40.3 for each value of $R_L$. Compare these results with the gains you determined in Calculation 2.

## QUESTIONS

**1.** Why was it necessary to limit the signal generator output to an amplitude of about 0.3 V? What would have happened if a much larger amplitude had been used?

**2.** External synchronization of the oscilloscope pattern was used in Procedure 4 so that an important difference between the patterns displayed in Procedures 4 and 5 could be observed in Procedure 6. What was this difference and why did external rather than internal synchronization have to be used to observe it?

**3.** The collector conductance of a transistor (analogous to the plate conductance $g_p$ of a vacuum tube) is the ratio of a small change in collector current to the corresponding small change in collector-to-emitter voltage with no change in base current. This parameter, usually designated $h_{oe}$, is thus the slope of the collector characteristic passing through the point where $h_{oe}$ is to be evaluated. Find $h_{oe}$ for your transistor at $I_B = 60$ $\mu$A and $V_{CE} = 6$ V. How does its reciprocal (analogous to the tube's $r_p$) compare with the values of $R_L$ used in this experiment?

**4.** How does the input resistance of a transistor amplifier compare with that of an amplifier using a vacuum tube? How will this affect its operation?

**5.** Did your theoretical and experimentally measured values of the gain of your transistor amplifier agree closely? If not, discuss features of the experiment that might have contributed to the disagreement.

**6.** In the derivation of Equation 40.3, the signal voltage $V_{in}$ was assumed to be applied across only the transistor input (base-emitter) resistance $h_{ie}$ so that $\Delta I_B$ was equal to $V_{in}/h_{ie}$. Resistor $R_B$ is also connected to the transistor base, so that $V_{in}$ is actually applied across $h_{ie}$ and $R_B$ in parallel. Why could $R_B$ be neglected in the determination of $\Delta I_B$?

**7.** A type 2N3394 transistor has $\beta = 100$ and $h_{ie} = 2750\ \Omega$ when $I_C = 1$ mA and $V_{CE} = 10$ V. If the operating point fixed by these values of $I_C$ and $V_{CE}$ is to be used and a 15-V power supply is available for $V_{CC}$, find (a) the necessary load and base resistances $R_L$ and $R_B$ and (b) the gain of the resulting amplifier. *Note:* To find the base current required at the operating point, assume $\beta$ to be a constant right down to zero collector current, that is, assume $I_C$ and $I_B$ to be strictly proportional over their entire range.

**8.** Find out what a transistor parameter known as the *reverse amplification factor* $\mu_{re}$ is. Were you justified in neglecting it in your work with the 2N2219?

# Operational Amplifiers 41

An operational amplifier is one characterized by high gain (20,000 to a million), a reasonably low output impedance, and a frequency response that extends from some fairly high value all the way down to zero (that is, it is a so-called dc amplifier). The name comes from the fact that operational amplifiers are often used to perform the electrical analogue of various mathematical operations such as addition, multiplication, integration, and their inverses. Today operational amplifiers are all solid-state units built with integrated-circuit techniques and completely encapsulated in small, sealed packages of which there are several standard styles. Connections for input, output, and supply power are brought out as "leads," that is, metal tabs suitable for connection in printed circuits. Operational amplifiers can also be built using vacuum tubes, and many such were produced before the advent of transistors. However, because tubes are relatively large, fragile, and characterized by much higher power requirements than the solid-state units, they are now considered obsolete. The purpose of this experiment is to see how a typical modern operational amplifier works in some standard circuits and to observe its performance.

## THEORY

Operational amplifiers may include several stages of amplification and much complex circuitry, but the details of their construction will not concern us. We will be interested only in the fact that we have a unit with an input and an output characterized by a large voltage gain $A_v$, a reasonably low output impedance so that the output may supply a certain amount of current and thus power to a load, a reasonably high input impedance to avoid loading the source of the input signal, and a frequency response extending from zero (dc) up to a value sufficiently high to allow needed response speeds to be realized. The schematic symbol for an operational amplifier is an isosceles triangle pointing in the direction of amplification, that is, with the output lead shown coming out of the apex and the input lead or leads going into the base. Fig. 41.1(a) is the schematic diagram for a simple operational amplifier, defined as one having a single input. Input voltage $V_{in}$ is applied between this input and common ground, and output voltage $V_0$ appears between the output terminal and common ground. The amplifier exhibits voltage gain $A_v$, which means that as long as the input signal has a frequency within the amplifier's response range and the output is lightly loaded, $V_0$ is $A_v$ times $V_{in}$. Because operational amplifiers are, as we will see, usually used in negative-feedback circuits, the amplifier is designed to invert, that is, to have an increase in $V_{in}$ result in a decrease in $V_0$. Thus,

$$V_0 = -A_v V_{in} \qquad (41.1)$$

Modern operational amplifiers are almost all of the differential type, which can be represented schematically as in Fig. 41.1(b). Here there are two inputs, and the input voltage to be amplified is applied between them. They are labeled + and − to indicate that if the + input is made increasingly positive with respect to the − input, the output voltage $V_0$ will rise (become more positive) with respect to ground. Con-

(a) Simple               (b) Differential

**Figure 41.1** *Operational amplifier schematic diagrams*

sequently, the + input is often called the *noninverting input*, whereas the − input is the *inverting* one. A differential operational amplifier may be turned into a simple one by grounding one input, in which case it inverts if the + input is grounded and does not if we ground the − input. In addition, $V_0$ is supposed to be affected only by a voltage applied *between* the two inputs and not at all by a voltage applied between the paralleled inputs and ground. Thus if the two inputs are connected together and a voltage is applied between ground and the two of them in common, the output voltage should remain zero and the amplifier's operation should be unaffected no matter what this applied voltage may be (provided it stays in a range quoted in the amplifier's specification sheet). Such a voltage is called a *common mode voltage*, and the amplifier's ability to remain unaffected by it is called its *common mode rejection* (CMR). Actually, $V_0$ will, in practice, be slightly affected by the application of a common mode voltage, and a common mode gain (that is, the ratio of the change in output voltage to the change in common mode voltage that produced it) can be quoted. Usually the ratio of $A_v$ to the common mode gain, a quantity called the *common mode rejection ratio* (CMRR), is what is listed in operational amplifier data sheets. Because the common mode gain is likely to be less than one and the amplifier's voltage gain $A_v$ likely to be on the order of 100,000, the CMRR is often very large indeed. It may therefore be given in decibels, in which case the quoted number is 20 times the common logarithm of the actual gain ratio. Differential operational amplifiers have the advantages of (1) not requiring that the input voltage be referred to ground and (2) being either inverting or non-inverting. Because the differential or "push-pull" amplifier circuit is better from the point of view of circuit balance and is no harder to build than the simple or "single-ended" type, almost all modern operational amplifiers have differential inputs.

Figure 41.1 shows the power supply connections explicitly, although in many schematic diagrams the fact that power must be supplied to the amplifiers is taken for granted, and these connections are consequently not shown. Notice that because the am-

plifier's output voltage may swing both above and below ground, both a positive and a negative power supply are required. Most solid-state operational amplifiers are designed for a standard power supply voltage $V_{cc}$ of 15 V but will work perfectly well with a $V_{cc}$ of 12 V, so that standard storage batteries may be used. The junction of the positive and negative power supplies is the common ground point.

Figure 41.2 shows the basic circuit for an operational amplifier employing negative feedback. On the assumption that the amplifier's input impedance is so high that negligible current flows into the input, a current

$$i = \frac{V_1 - V_0}{R_1 + R_f}$$

flows around the circuit composed of $V_1$, $R_1$, $R_f$, and $V_0$ in series. The amplifier's input voltage $V_{in}$, which because the + input is grounded is the voltage between the − input and ground, is then seen to be given by

$$V_{in} = V_1 - iR_1 = V_1 - \frac{V_1 - V_0}{R_1 + R_f} R_1$$

But the output voltage must be given by Equation 41.1, so that substitution for $V_{in}$ yields

$$-\frac{V_0}{A_v} = V_1 - \frac{R_1}{R_1 + R_f} (V_1 - V_0)$$

which, after an appropriate rearrangement of terms, becomes

$$V_0 \left( \frac{R_1}{R_1 + R_f} + \frac{1}{A_v} \right) = -V_1 \left( \frac{R_f}{R_1 + R_f} \right) \quad (41.2)$$

Now the voltage gain $A_v$ is very large for all frequencies in which we are interested; thus, $1/A_v$ is very small. In particular, $1/A_v \ll R_1/(R_1 + R_f)$ and may therefore be neglected in comparison. This can be appreciated by noting that $A_v$ will be at least 20,000 whereas $R_f$ will be at most ten times $R_1$, so that we are comparing 1/20,000 or less with 1/11 or more. Hence to a very good approximation, Equation 41.2 may be reduced to

$$V_0 \left( \frac{R_1}{R_1 + R_f} \right) = -V_1 \left( \frac{R_f}{R_1 + R_f} \right)$$

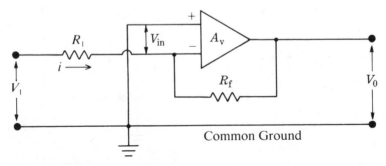

**Figure 41.2** *The basic inverting multiplier*

and finally to

$$V_0 = -V_1 \frac{R_f}{R_1} \qquad (41.3)$$

This most important result tells us that the output voltage $V_0$ is the negative of the input voltage $V_1$ multiplied by the fixed factor $R_f/R_1$. We can say that we have a device that multiplies $V_1$ by $-R_f/R_1$ or that our complete circuit is an inverting amplifier having a gain $G$ with feedback (as distinguished from the operational amplifier's unfeedback gain $A_v$) of $R_f/R_1$. The important point is that $G$ is fixed by the values of a couple of resistors (which can be made very precise and stable) and *does not depend on the gain of the amplifier at all*. In fact, *as long as $A_v$ is large enough so that the approximation made in deriving Equation 41.3 remains valid, the actual value of $A_v$ can vary all over the place as the amplifier ages and supply voltages change without affecting the relation between $V_0$ and $V_1$.* Because of the precision with which $G$ can be maintained, the circuit in Fig. 41.2 is usually thought of as an analog multiplier rather than an amplifier of gain $G$.

There is a much simpler way of looking at the circuit of Fig. 41.2 that lets us understand why Equation 41.3 should be true without going through a lengthy derivation. The crucial point is that the gain $A_v$ of the operational amplifier itself is high. This means that only a very tiny input voltage will produce an appreciable output voltage. Thus suppose we have an operational amplifier whose $V_0$ may swing over a range of $-10$ to $+10$ V. This is typical of solid-state amplifiers designed to operate from a pair of 15-V power supplies. If $A_v = 100,000$ (a typical value), *only a ten-thousandth of a volt at the input drives the output over its full range.* But this will not be allowed to happen, for as soon as $V_0$ gets away from zero, the feedback through $R_f$ will apply a correcting signal at the input. Thus, although the $-$ input is not grounded, it is maintained at almost exactly ground potential by the action of the circuit and is called a *virtual ground*. If the voltage at

this point can really be taken to be zero, we can then write

$$i = \frac{V_1}{R_1} = -\frac{V_0}{R_f}$$

from which Equation 41.3 immediately follows.

Figure 41.2 illustrates the basic circuit in which negative feedback is applied to an operational amplifier to produce a system capable of performing analog operations. There are many possible modifications for adapting the circuit to particular conditions or setting it up to perform operations other than simple multiplication by a fixed factor. A typical adder circuit is shown in Fig. 41.3. It is the same as Fig. 41.2 except that more than one input has been provided. If the amplifier's input really takes no current, the sum of the currents at their junction at the $-$ input must be zero. That is,

$$i_1 + i_2 + i_3 + i_f = 0 \qquad (41.4)$$

Because the $+$ input is grounded, the amplifier's input voltage is that between the $-$ input and ground, and if this is negligible (the $-$ input is a virtual ground), we have $i_1 = V_1/R_1$, $i_2 = V_2/R_2$, $i_3 = V_3/R_3$, and $i_f = V_0/R_f$. Substitution in Equation 41.4 yields

$$\frac{V_1}{R_1} + \frac{V_2}{R_2} + \frac{V_3}{R_3} = -\frac{V_0}{R_f}$$

or

$$V_0 = -\left(\frac{R_f}{R_1}V_1 + \frac{R_f}{R_2}V_2 + \frac{R_f}{R_3}V_3\right) \qquad (41.5)$$

Equation 41.5 shows that the negative of the output voltage $V_0$ is equal to the sum of the voltages $V_1$, $V_2$, and $V_3$, each multiplied by a factor that we are free to choose by choosing the resistances $R_1$, $R_2$, $R_3$, and $R_f$. In particular, if we want $-V_0$ to be just the sum of $V_1 + V_2 + V_3$, we have only to make all the resistances equal. Note that Equation 41.4 states a fundamental rule of circuit theory, namely that the alge-

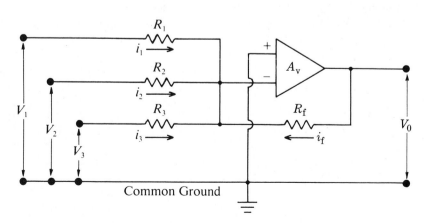

**Figure 41.3**  *The analog adder*

braic sum of all the currents entering a junction must be zero. Because of this, the − input in the adder circuit is sometimes called the *summing point*.

An operational amplifier may be used to perform integration by substituting a capacitor $C_f$ for the feedback resistor $R_f$ appearing in our earlier circuits. Figure 41.4(a) shows this arrangement. If the − input is a virtual ground, the voltage across the capacitor is the output voltage $V_0$, and the capacitor's charge $q$ is given by

$$q = C_f V_0$$

But again, if no current flows into the amplifier input, current $i$ must all flow into the capacitor. Therefore $q$ will be simply $i$ times the time, or if $i$ varies with time

$$q = - \int i\,dt$$

where $dt$ is a time interval so small that $i$ may be considered constant over it, and the minus sign comes from the fact that $i$ flows into the side of $C_f$ opposite to that connected to $V_0$. Putting these results together yields

$$V_0 = - \frac{1}{C_f} \int i\,dt \qquad (41.6)$$

so that the output voltage is proportional to the time integral of the input current. If we now look at the circuit of Fig. 41.4(b), we see that, provided the − input remains a virtual ground, $i$ has the constant value $V_1/R_1$. Consequently Equation 41.6 now becomes

$$V_0 = - \frac{V_1}{R_1 C_f} t \qquad (41.7)$$

where $t$ is the time in seconds measured from an initial instant at which the charge on $C_f$ and hence the output $V_0$ was zero. Note in Fig. 41.4(b) that the battery supplying $V_1$ has been connected to make this voltage negative so as to remove the negative sign in Equation 41.7 and make the output voltage positive. This voltage will thus *increase exactly linearly with time*, the slope of a graph of $V_0$ against $t$ being the

constant $V_1/R_1C_f$, which we can choose at will. Because this graph, which is a straight line sloping up to the right, looks like a ramp, the circuit of Fig. 41.4(b) is often called a *ramp generator*. According to Equation 41.7, the ramp could go on forever, but of course a limit is imposed by the maximum output voltage of which the operational amplifier is capable (usually about 10 V). The circuit may be reset and $V_0$ returned to zero by momentarily short-circuiting the capacitor.

One important application of operational amplifiers that does not employ feedback is service as a *Schmitt trigger*. The original Schmitt trigger is a bistable circuit characterized by an output voltage that can assume one of two values. Circuits of this general type are called "flip-flops" because they "flip" or switch their output voltage from one possible value to the other practically instantaneously without this voltage remaining at any intermediate value. The Schmitt trigger has two inputs, one of which is maintained at some reference voltage $V_{ref}$ while a signal voltage of some sort is applied to the other. When the signal voltage is above the reference voltage, the output voltage has one of its two values, and when the signal voltage is below $V_{ref}$, $V_0$ has its other value. Switching takes place as the signal voltage passes through the reference value. An actual Schmitt trigger employs two transistors in a special circuit incorporating a large amount of positive feedback, but an operational amplifier with no feedback can be arranged to provide essentially the same performance. The important point once again is that the operational amplifier's unfedback voltage gain $A_v$ is large. Taking 100,000 as a representative value, we again note that an input voltage $V_{in}$ of only 0.0001 V is enough to drive the output voltage $V_0$ to its maximum value of 10 V. In practice there are always enough noise voltages or other stray voltages around so that even with the amplifier inputs connected together and grounded, $V_0$ will not stay around zero but will "hang up" at either its maximum positive or its maximum negative value. The operational amplifier with no feedback thus exhibits two, and only two, stable output voltage values, just as the Schmitt trigger does.

(a) The Integrator

(b) The Ramp Generator

**Figure 41.4** *Analog integrator circuits*

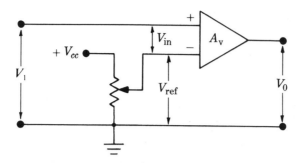

**Figure 41.5**  *The Schmitt trigger application*

Now let us assume that the − input is maintained at a positive reference voltage $V_{ref}$, which can, for example, be selected by a rheostat as shown in Fig. 41.5. A signal voltage $V_1$ is applied between the + input and ground. We note at once that if $V_1 = 0$, $V_{in} = -V_{ref}$ and the amplifier is driven far into saturation with $V_0$ "hung up" at its maximum negative value. This situation continues as $V_1$ is increased until suddenly, as $V_1$ gets very close (of the order 0.0001 V) to $V_{ref}$, the amplifier will suddenly find itself driven the other way and $V_0$ will switch to its maximum positive value. This changeover takes place extremely rapidly, the switching speed being a characteristic of the amplifier and its load. The speed with which $V_0$ can switch from one value to another value in response to an instantaneous change in $V_{in}$ is called the amplifier's *slewing rate* and is quoted in volts/second in data sheets supplied with commercial units.

The operational amplifier chosen for this experiment is the type 741, a very popular general-purpose commercial unit. It is made with integrated circuit techniques that allow 20 transistors and their associated circuitry to be packaged in a tiny plastic encapsulation having a standard size and connection tab arrangement called an 8-pin DIP. A top view of this package, drawn four times the actual size and showing the pin connections for the type 741, is given in Fig. 41.6. Note that this is a *top view* and that to observe the pin numbering correctly the orienting mark must be to your left. The connections for − input, + input, the negative supply voltage $-V_{cc}$, the positive supply voltage $+V_{cc}$, and the output voltage $V_0$ are clearly shown. In addition, there is no connection to pin 8, and we will not use the offset null pins 1 and 5. The term *offset* refers to the fact that the slight imbalances in practical differential amplifier circuits result in a zero output voltage corresponding not to a zero input voltage but to a $V_{in}$ slightly "offset" from zero. The type 741 provides for connection to an external circuit that allows the offset to be "nulled," that is, adjusted to exactly zero. This is needed only for certain very critical applications and will not concern us here.

## TYPE 741 CHARACTERISTICS

| | |
|---|---|
| Supply voltage $V_{cc}$ | ± 12 to ± 15 V |
| Range of output voltage $V_o$ | ± ($V_{cc}$−2) V |
| Voltage gain $A_v$ | Approximately 200,000 |
| Input resistance | Approximately $2 \times 10^6 \ \Omega$ |
| CMRR | 90 dB |
| Common mode input voltage range | ± 13 V |
| Maximum short-circuit output current | ± 25 mA |

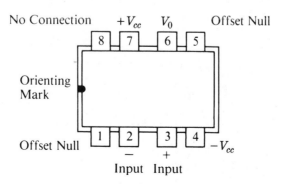

**Figure 41.6**  *Type 741 connection diagram (top view)*

## APPARATUS

1. Two type 741 operational amplifiers (8-pin DIP package)
2. Two 15-V regulated power supplies
3. Two good 1.5-V dry cells
4. dc voltmeter (0–15 V)
5. Single-pole, double-throw relay (12-V dc coil)
6. Type 2N2291 transistor
7. Two single-pole, single-throw switches
8. Decade resistance box (10,000 Ω)
9. Decade resistance box (1000 Ω)
10. Four carbon 1-W 5% tolerance resistors (470,000, 4700, 1500, and 1000 Ω)
11. Volume-control-type rheostat (5000 Ω wire-wound)
12. Oil or paper dielectric capacitor (2.0 μF)
13. Electrolytic capacitor (about 500 μF, 35-V rating)
14. Assorted hookup wires and connectors
15. Stopwatch or stop clock

## PROCEDURE

**1.** Wire the analog multiplier circuit shown in Fig. 41.7. Use the 10,000-Ω decade box for the feedback resistor $R_f$ and the 1000-Ω box for the input resistor $R_1$. Notice that the dry cell is connected with its positive side to ground so that the amplifier output will be positive. Figure 41.7 shows the metal decade-box cases connected to the common ground line, and you should do this if your boxes have terminals for this purpose. Common ground is established as the junction of the two power supplies, and the + input is connected to this line to fix the common mode voltage at zero. Set $R_f$ at zero and $R_1$ to 1000 Ω. Do not plug in or turn on the power supplies until your wiring has been checked by the instructor.

**2.** When your wiring has been checked and approved, turn on the power supplies. If you are using supplies whose output voltage is variable, start with the controls all the way down. Temporarily disconnect the 15-V voltmeter from the amplifier output and use it to measure the two power-supply voltages. Set each to 15 V, making sure the two voltages have the same magnitude ($|+V_{cc}| = |-V_{cc}|$). Note that the supply for $-V_{cc}$ has its positive side grounded,

which means that the voltmeter connections must be temporarily interchanged to allow measurement of this voltage. Then reconnect the voltmeter to the amplifier output and check that it reads zero. If it does not, see whether $R_f$ has been set to zero and $R_1$ to 1000 Ω. If the voltmeter still does not read zero, shut off power to the circuit and recheck your wiring. If trouble persists, call your instructor.

**3.** When you are sure your circuit is working properly, set $R_f$ to 1000 Ω and read the voltmeter. Record this reading in the data table. Repeat this procedure with $R_f = 2000, 3000, 4000, 5000$, and 6000 Ω.

**4.** Shut off the power to your circuit and then make your setup into a simple analog adder by installing the modification diagrammed in Fig. 41.8. This consists of removing the 1000-Ω decade box and connecting $R_1$, $R_2$, $K_1$, $K_2$, and the two dry cells to the − input as shown. Note that $R_1$ and $R_2$ are 1-W carbon resistors. Set the decade box at $R_f$ to 3000 Ω and make sure switches $K_1$ and $K_2$ are both open. Have your wiring checked by the instructor before applying power to the circuit.

**Figure 41.7** *Experimental circuit for the analog multiplier*

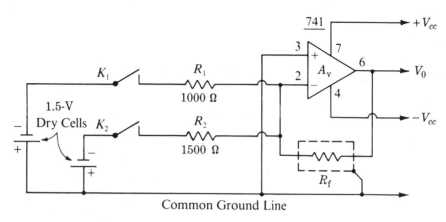

**Figure 41.8** *Circuit changes for the analog adder*

**5.** Turn on the power supplies. If necessary, check the supply voltages $+ V_{cc}$ and $- V_{cc}$ with the voltmeter as described in Procedure 2. When the voltmeter is reconnected to the amplifier output, it should read zero.

**6.** Close switch $K_1$ and read and record $V_0$. Open $K_1$ and close $K_2$, and again read and record $V_0$. Repeat with $K_1$ and $K_2$ both closed.

**7.** Shut off power to your circuit and rewire it as an analog integrator as shown in Fig. 41.9. $R_1$ is a 470,000-$\Omega$, 1-W carbon resistor and $C_f$ is a 2-$\mu$F capacitor with oil or paper dielectric. Use the same dry cell you used in the analog multiplier and note that again its positive terminal is grounded so that you will get a positive output voltage. Close $K_1$ to short-circuit $C_f$ and have your wiring checked by the instructor before applying power to the circuit.

**8.** Turn on the power supplies. If necessary, check $+ V_{cc}$ and $- V_{cc}$ with the voltmeter as before. Reconnect the voltmeter to measure $V_0$. It should read zero with $K_1$ closed.

**9.** Open $K_1$ and simultaneously start the stopwatch (or stop clock). You should observe a steady rise in $V_0$ as read on the voltmeter. Stop the stop-

watch when the voltmeter reads exactly 10 V. Reset $V_0$ to zero by closing $K_1$. Record the stopwatch reading.

**10.** Repeat Procedure 9 twice to obtain a total of three measurements of the time required for $V_0$ to rise from 0 to 10 V.

**11.** An operational amplifier can be used as a Schmitt trigger to reset the integrator-ramp generator to zero by short-circuiting $C_f$ automatically whenever $V_0$ rises above a preset value. Shut off power to your present circuit and add the second type 741 to make the complete circuit shown in Fig. 41.10. In this circuit the type 2N2219 transistor is normally "turned off" by having a negative voltage applied to its base from the output of the No. 2 operational amplifier. At the start of the operation, this amplifier's output is hung up at its maximum negative voltage because its $+$ input is at zero while its $-$ input is at some positive voltage $V_{ref}$ selected by the volume-control-type rheostat $R_2$. Operational amplifier No. 1 is connected as an integrator-ramp generator as in Procedures 7–10, $R_1$ and the dry cell being the source of constant current that makes the integrator output $V_0$ be a constantly rising voltage or "ramp." When $V_0$

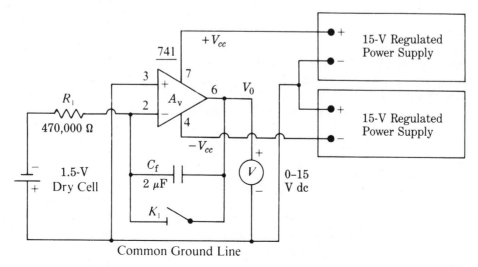

**Figure 41.9** *Experimental circuit for the analog integrator*

**Figure 41.10** *The sawtooth generator*

reaches $V_{ref}$, amplifier No. 2 switches, sending a large base current to the 2N2219, which is thereby "turned on" so that current passes through its collector-emitter circuit from the $+ V_{cc}$ power supply to the coil of relay $K_1$. The relay then closes, short-circuiting $C_f$ and resetting the integrator so that $V_0$ is returned to zero. Amplifier No. 2 then flips back to a negative output, turning off the transistor and deenergizing the relay. With the relay contacts open, $C_f$ can recharge and the cycle repeats. $C_0$ is a large electrolytic capacitor that keeps current flowing in the relay coil for a brief interval after the transistor is turned off. This allows the relay contacts to stay closed long enough to completely discharge $C_f$. $R_3$ limits the base current to the transistor and thus limits the collector-emitter current, protecting the transistor against an excessive current surge when it is turned on and starts charging $C_0$. A plot of the resulting $V_0$ as a function of time appears in Fig. 35.2, which illustrates why this wave is called a "sawtooth" and the circuit that produces it a "sawtooth generator."

When wiring your circuit, be careful to avoid mistakes in making connections to the operational amplifiers and the transistor. Figure 41.6 is the connection diagram for the type 741 and Fig. 40.5 for

the 2N2219. Leave your power supplies unplugged and with their switches off until your wiring has been approved by the instructor.

**12.** When your wiring has been checked, plug in and turn on the power supplies. Temporarily disconnect the voltmeter from the output of amplifier No. 1 and use it to check and set the supply voltages $+ V_{cc}$ and $- V_{cc}$ as described in Procedure 2. Then measure $V_{ref}$ and set it to $+ 4$ V with $R_2$. Return the voltmeter to the No. 1 amplifier output as shown in Fig. 41.10. You should see $V_0$ rise linearly until it gets to 4 V, at which point it will suddenly return to zero and start up again. If the circuit appears to be operating properly, time five complete cycles of the sawtooth wave. Do this by starting the stopwatch just as $V_0$ returns to zero and counting five such returns before stopping the watch. Record your result.

**13.** Shift the voltmeter from the output of amplifier No. 1 to the $-$ input of No. 2 to measure $V_{ref}$. Set this voltage to $+ 6$ V. Then return the voltmeter to the amplifier No. 1 output and again time five complete cycles of the observed wave. Record your result.

**14.** Repeat Procedure 13 with $V_{ref}$ set to $+ 8$ V and again with $V_{ref} = + 10$ V.

## DATA _____

**Multiplier**

| $R_f$, Ω | $V_0$, V | $R_f$, Ω | $V_0$, V |
|---|---|---|---|
| 1000 | | 4000 | |
| 2000 | | 5000 | |
| 3000 | | 6000 | |

Slope of graph _____ Dry-cell voltage from graph _____

**Adder**

| $K_1$ | $K_2$ | $V_0$, V | | Percent error |
|---|---|---|---|---|
| | | Measured | Predicted | |
| Open | Open | | | |
| Closed | Open | | | |
| Open | Closed | | | |
| Closed | Closed | | | |

**Integrator-Ramp Generator**

Time for 0–10 V rise:

Trial 1 _____ Trial 2 _____ Trial 3 _____ Average _____

Calculated value _____ Percent error _____

**Sawtooth Generator**

| $V_{ref}$, V | Time for 5 cycles, s | Period, s | | Percent error |
|---|---|---|---|---|
| | | Measured | Calculated | |
| 4 | | | | |
| 6 | | | | |
| 8 | | | | |
| 10 | | | | |

## CALCULATIONS _____

**1.** Plot the analog multiplier output voltage against $R_f$ using the data obtained in Procedures 2 and 3. Draw the best straight line through the plotted points.

**2.** Find the slope of the graph plotted in Calculation 1. From this slope, and assuming that $R_1$ is exactly 1000 $\Omega$, compute the dry-cell voltage.

**3.** Calculate what the analog adder output voltage $V_0$ should be for each condition of switches $K_1$ and $K_2$. Use Equation 41.5 with $R_1 = 1000$ $\Omega$, $R_2 = 1500$ $\Omega$, $R_f = 3000$ $\Omega$, and $V_1 = V_2 =$ the dry-cell voltage obtained in Calculation 2. There is, of course, no $V_3$.

**4.** Compare your predicted values of the analog adder output with the measured values by finding the percent error.

**5.** Calculate the average of your three measurements of the time for a 10-V rise in the integrator-ramp generator output.

**6.** Calculate the theoretical value of the time for a 10-V rise from Equation 41.7 using $R_1 = 470,000\ \Omega$, $C_f = 2.0\ \mu F$, and $V_1 =$ the dry-cell voltage obtained in Calculation 2.

**7.** Compare the theoretical value of Calculation 6 with the measured value of Calculation 5 by finding the percent error. Is this error consistent with the possible errors in the numerical values of $R_1$ and $C_f$ used in Calculation 6?

**8.** Divide each of the time measurements made in Procedures 12–14 by 5 to obtain the sawtooth period corresponding to each of your four values of $V_{ref}$.

**9.** Calculate the sawtooth period for each $V_{ref}$ from Equation 41.7, in which we note that $t$ is the period when $V_0 = V_{ref}$.

**10.** Compare your four measured values of the sawtooth period with the corresponding predicted values obtained in Calculation 9 by finding the percent error.

## QUESTIONS _____

**1.** Explain how the analog multiplier circuit of Fig. 41.2 can function as a voltage regulator for a power supply. *Hint:* Large operational amplifiers can supply considerable output current, and remember that $V_0$ is exactly $R_f/R_1$ times $V_1$.

**2.** Discuss the errors found in Calculation 4 for your results with the analog adder. Note in particular whether these errors are consistent with the resistance tolerances on $R_1$ and $R_2$ and your assumptions about the dry-cell voltages.

**3.** Suggest how the analog integrator circuit could be modified to make it perform differentiation instead of integration.

**4.** The sawtooth period should be given by Equation 41.7 with $t$ set equal to the period $T$, and $V_0$ to $V_{ref}$, so that

$$T = \frac{R_1 C_f V_{ref}}{-V_1}$$

Show that your results from Calculation 8 are in accord with this linear relation between $T$ and $V_{ref}$. *Hint:* You may want to plot $T$ against $V_{ref}$ on an extra sheet of linear graph paper.

**5.** Explain how a Schmitt trigger can be used to generate a square (or rectangular) wave from a sine-wave input. In particular, point out what determines the repetition rate and pulse length of the rectangular wave.

**6.** In the sawtooth circuit of Fig. 41.10, suppose the inductance of the relay coil can be neglected so that the coil can be represented by its resistance alone (about 200 $\Omega$). (a) What is the time constant of the relay coil and capacitor $C_0$? (b) If the relay contacts open when the coil voltage drops to 60% of its value at the instant the transistor turns off, how long after turn-off do the contacts remain closed? (c) What is the effect of making $C_0$ too large?

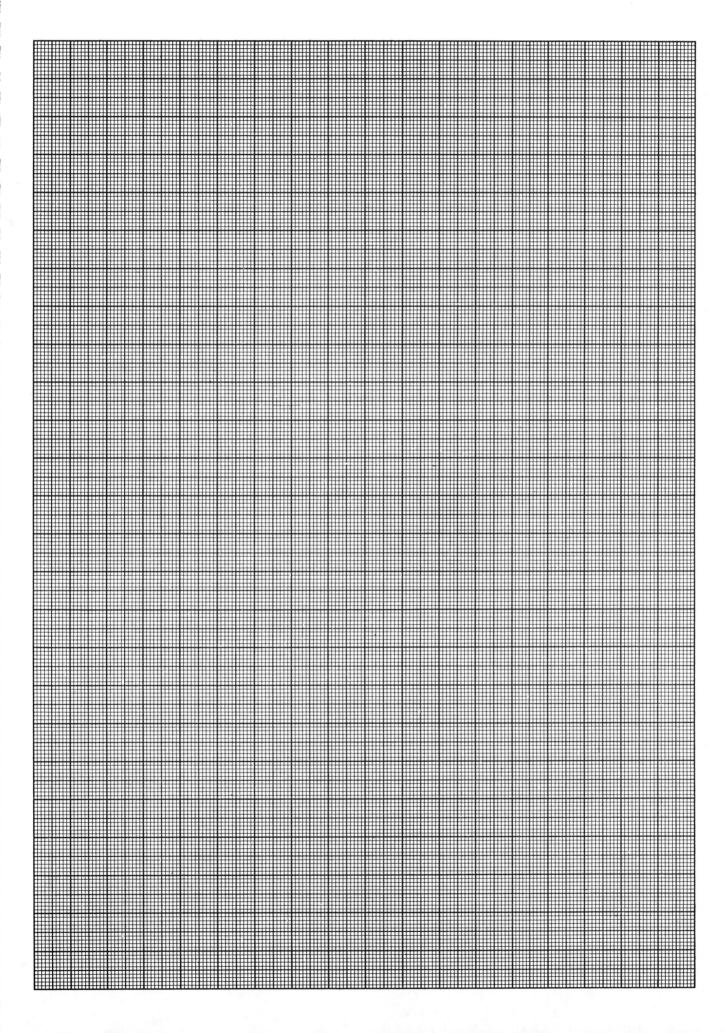

# Geometrical Optics 42

Although light is an electromagnetic wave, its wavelength is so short that when dealing with it in equipment whose dimensions are large by comparison (such as lenses, mirrors, etc.), we may ignore its wave nature. Light can then be discussed in terms of beams or rays that travel in straight lines (the law of rectilinear propagation) unless bent by reflection or refraction. When a light ray strikes the boundary between two media of different densities, all or part of it is reflected according to the law of reflection. If the medium into which the incident ray is directed is transparent, most of the light enters this second medium, but the ray is bent where it passes through the boundary if it does not strike normally (that is, at right angles to the boundary surface). This bending of light rays is called *refraction* and follows a third law. The purpose of this experiment is to study these three laws by tracing the pattern of light rays. We will find that a great deal of this work is geometrical in nature; hence the study of light rays in experimental setups in which their wave nature may be ignored is called *geometrical optics* and the laws of reflection, refraction, and rectilinear propagation are known as the laws of geometrical optics.

## THEORY

A light ray travels in a straight line if it is in a transparent medium of uniform density. This simple fact bears the somewhat grandiose name of *the law of rectilinear propagation*. If, however, a light ray strikes a boundary surface separating two transparent media of different density, some of the light will be reflected and some will proceed into the new medium but in an altered direction. The situation is illustrated in Fig. 42.1, which shows a ray in Medium 1 incident on a boundary separating Mediums 1 and 2 and producing a reflected ray coming back in Medium 1 and a refracted ray proceeding into Medium 2. Regular reflection takes place at a smooth surface such as the boundary shown in this figure. Such a surface might be the surface of a mirror (the boundary between glass and air) or of a lake (a boundary between water

and air). The law of reflection states that a single ray of light will proceed, after reflection, in a direction such that the angle of incidence equals the angle of reflection. These angles are the angles made by the respective rays with the *normal*, that is, the perpendicular drawn to the boundary surface at the point of incidence. In addition, the incident ray, the normal, and the reflected ray must all lie in the same plane. Figure 42.1 shows a ray incident on a boundary between two media, the normal, the angle of incidence $i$, the reflected ray, and the angle of reflection $s$. The law of reflection requires $i = s$.

The inter-medium boundary does not, of course, have to be plane. Consider, for example, a spherical mirror, that is, a mirror whose surface is a portion of a sphere. It can be shown fairly easily that if a concave spherical mirror is held facing a beam of parallel light rays, the reflected rays will (to a close approximation) converge to a point. This point is called the mirror's *principal focus, F* (see Fig. 42.2). Another important point is the *center of curvature, C,* which is the center of the spherical surface of which the mirror is a part. The *principal axis* of a mirror is a straight line drawn from the center of curvature through the focus to the mirror's surface. The distance from the center of curvature to the mirror along the principal axis is obviously a radius $R$ of the reflecting surface. The distance from the focus to the mirror along the principal axis is the mirror's *focal length, f,* and can be shown to be $\frac{1}{2}R$.

A spherical mirror can also be convex instead of concave, in which case it will diverge incoming paraxial rays instead of converging them, as shown in Fig. 42.3. In this case, the reflected ray *seems* to

**Figure 42.1** *Reflection and refraction at a plane surface*

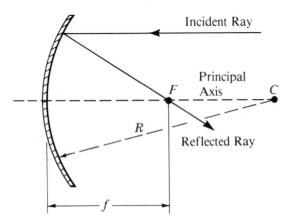

**Figure 42.2** *The concave spherical mirror*

come from a point *behind* the mirror. This point is therefore taken as the principal focus *F*, but since no rays actually go through this point, it is called a *virtual* focus. Moreover, because in this case the center of curvature *C* lies behind the mirror, the mirror's radius of curvature *R* is considered negative, and for similar reasons the focal length *f* (which remains equal to $\frac{1}{2}R$) is considered negative also. It should be noted that the occurrence of a negative focal length indicates a diverging optical element, mirror or lens.

Refraction is the bending of light rays when they pass obliquely from one medium to another (see Fig. 42.1). The direction of a particular ray after refraction is given by the law of refraction, often called Snell's law after its discoverer, Willebrord Snell (1591–1626). This law states that the ratio of the sine of the angle of incidence to the sine of the angle of refraction is equal to the ratio of the velocity of light in the first medium (the medium of incidence) to that in the second. As in the case of reflection, the incident ray, the refracted ray, and the normal to the boundary all lie in the same plane, and the angle of

refraction is the angle between the refracted ray and the normal, as shown in Fig. 42.1. Snell's law may be stated mathematically by the expression

$$\frac{\sin i}{\sin r} = \frac{v_1}{v_2} = \frac{n_2}{n_1} \qquad (42.1)$$

where   $i$ = angle of incidence
$r$ = angle of refraction
$v_1$ = velocity of light in Medium 1
$v_2$ = velocity of light in Medium 2
$n_1$ = index of refraction of Medium 1
$n_2$ = index of refraction of Medium 2

The index of refraction of a particular medium is the ratio of the velocity of light in vacuum to its velocity in the medium. Thus $n_1 = c/v_1$ and $n_2 = c/v_2$, where $c$ represents the velocity of light in vacuum. Hence $v_1/v_2 = n_2/n_1$, as indicated in Equation 42.1. Note that the velocity of light in air is very nearly the same as its velocity in vacuum, so that the index of refraction of air is very nearly 1. It may be taken as equal to 1 for most purposes.

A lens is a piece of glass or other transparent material bounded in its simplest form by spherical surfaces and used to converge or diverge rays of light passing through it. It is similar to a spherical mirror in this respect but is used to converge or diverge *transmitted* rays rather than rays that are reflected back on themselves, and it does this by refraction rather than reflection. Figure 42.4 illustrates this situation for a converging lens bounded by convex spherical surfaces. These surfaces have centers of curvature $C_1$ and $C_2$, and the principal axis of the lens is the line joining these points. The principal focus $F$ is the point on the left of the lens to which rays coming in from the right parallel to the principal axis are converged by the lens's action. $F$ lies on the principal axis at a distance $f$, the focal length, from the lens.

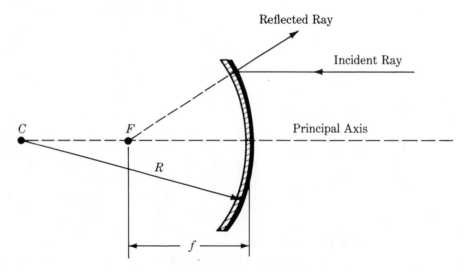

**Figure 42.3** *The convex spherical mirror*

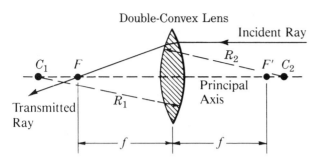

**Figure 42.4** *The double-convex converging lens*

As long as the same medium (for example, air) lies on both sides of the lens, it makes no difference which way the light passes through it, and parallel rays incident from the left will be converged to focal point $F'$ located on the principal axis at the same focal length $f$ to the right of the lens that $F$ was to the left. The focal length should properly be measured from the center of the lens, but we commonly deal with so-called thin lenses, whose thickness is so small that measuring $f$ from the point where the principal axis intersects the lens surface introduces negligible error.

The focal length of a thin lens must obviously be determined not just by the radii of its two surfaces but also by the index of refraction of the material of which the lens is made. The expression for $f$ in this case must therefore be much more complicated than the simple one already mentioned for the spherical mirror, both because the more complicated law of refraction is involved and because, as seen in Fig. 42.4, refraction takes place at two surfaces. Detailed analysis shows that

$$\frac{1}{f} = (n-1)\left(\frac{1}{R_1} - \frac{1}{R_2}\right) \qquad (42.2)$$

where $f$ is the focal length in meters, $n$ is the index of refraction of the lens material, and $R_1$ and $R_2$ are the radii of the two surfaces as defined in Fig. 42.4. Note that if $f$ is in meters, $R_1$ and $R_2$ must be in meters also.

This formula is known as *the lens maker's equation* because it is mostly of interest to someone who is handed a piece of glass or plastic having a certain index of refraction and is told to make a lens of a specified focal length. It will not concern us much in the present experiment, in which we will measure the focal length of a given lens, but the role of the radii of curvature $R_1$ and $R_2$ is worth some discussion.

The lens of Fig. 42.4 is called *double convex* because both surfaces are convex, and it should be noted that both contribute to converging the incident ray. In this situation the radius $R_1$ of the right-hand surface, whose center $C_1$ is in the space to the left of the lens, is considered positive, whereas the radius $R_2$ of the left-hand surface, with center $C_2$ on the right, is considered negative. The term $1/R_2$ in Equation 42.2 is therefore negative, with the result that the quantities $1/R_1$ and $1/R_2$ will add in this equation, thus expressing the fact that both surfaces contribute to the convergence of rays by increasing the quantity $1/f$, which means shortening the focal length. Notice that a *shorter* focal length implies *stronger* convergence, and the reciprocal $1/f$ is therefore called the *power* of the lens, as it is a direct measure of the lens's ability to converge rays. If $f$ is given in meters, the power $1/f$ is measured in *diopters*, whose dimensions are $m^{-1}$.

A lens can also diverge incoming rays by being made with concave surfaces. This situation is illustrated in Fig. 42.5, which shows a paraxial ray coming in from the right and being diverged by a thin, double-concave diverging lens. The dotted line indicates that the refracted ray *appears* to be coming from a point on the axis to the right of the lens, so that in the present case this point is considered the principal focus $F$, and its distance from the lens is the focal length $f$. A glance at Fig. 42.4 shows that the positions of $F$ and $F'$ are reversed as compared to their positions with the converging lens, and the focal length of a diverging lens is therefore taken to be negative just as it was with the diverging mirror. Note also that no ray actually comes from $F$ so that it

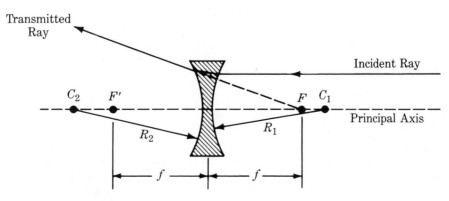

**Figure 42.5** *The double-concave diverging lens*

is a virtual focus. Thus the diverging lens, like the diverging mirror, has virtual focal points and negative focal lengths. And just as the focal length of the diverging mirror can be considered negative because it is one half of the negative radius of curvature, so the negative focal length of the diverging lens can be obtained from Equation 42.2. Thus, if we apply our convention about signs to $R_1$ and $R_2$ in Fig. 42.5, we find that now $R_1$ is negative and $R_2$ positive. Thus both the $1/R_1$ and $1/R_2$ terms in this equation will be negative, giving a negative power and hence a negative focal length for the lens. Note that $R_1$ and $R_2$ need not be equal, and in fact you will encounter plano-concave and plano-convex lenses, for which the radius pertaining to the flat surface is infinite so that its reciprocal in the lens maker's equation is zero. There are also concavo-convex lenses with one convex surface and one concave, an arrangement that seems contradictory, as one surface then wants to diverge incoming rays and the other to converge them. Such lenses have certain specialized applications which we will not consider.

Snell's law also shows that if a ray of light passes from one medium into another whose index of refraction is less than that of the first (for example, a ray from an underwater source passing from water into air), the angle of refraction $r$ is greater than the angle of incidence $i$ and the ray is bent away from the normal. As the angle of incidence is increased, eventually the refracted ray will graze the boundary surface. In this case $r = 90°$ and Equation 42.1 becomes $\sin i_c = n_2/n_1$, where $i_c$, the angle of incidence for which the refracted ray grazes the boundary, is called the *critical angle*. If the angle of incidence is made greater than the critical value, no light proceeds into the new medium and total reflection occurs at the bounding surface.

A prism is a solid piece of transparent material having flat surfaces. When a ray of light impinges on

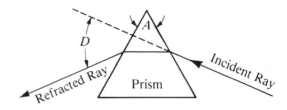

**Figure 42.6** *Deviation by a triangular prism*

a prism it is refracted on entry into the prism material and again when it exits through one of the other surfaces. Prisms of various shapes are used to deflect light rays in predetermined desired ways, but a common form is the triangular prism whose principal use is in optical spectrometers. Used this way it deviates the incident beam through an angle $D$, called the *angle of deviation*, as shown in Fig. 42.6. $D$ depends on the index of refraction of the prism material, and because the velocity of light in transparent media varies with the light's wavelength, the index of refraction and hence $D$ will be different for light of different colors. A prism spectrometer can thus be used to separate a beam of light into its component wavelengths. For light of a fixed wavelength, $D$ depends on the direction of the incident ray (the initial angle of incidence), but for one particular direction of incidence the angle of deviation becomes a minimum. Minimum deviation occurs when the path of the ray inside the prism is symmetrical. This means that the initial angle of incidence is equal to the final angle of refraction, and the path of the ray inside the prism is parallel to the face opposite the prism's apex angle $A$ (see Fig. 42.6). Under these conditions the index of refraction of the prism material is given by

$$n = \frac{\sin \frac{1}{2}(A + D)}{\sin \frac{1}{2}A} \qquad (42.3)$$

This constitutes one method of measuring index of refraction.

## APPARATUS

1. Low-power helium-neon laser
2. Drawing board
3. Plane mirror
4. Concave mirror
5. Convex mirror
6. Plano-convex lens
7. Semicircular lens
8. 60° triangular prism
9. Meter stick
10. Protractor
11. Drawing compass
12. White paper sheets for drawing board
13. White cards, masking tape, pins

## PROCEDURE

**1.** Tape a sheet of white paper to the drawing board. Place the plane mirror on the sheet near the center with the reflecting surface facing your right. Plug in and turn on the laser and allow a few minutes for it to warm up. *Note:* There are two important features of laser light. First, a laser emits a narrow beam that spreads out very little and is thus ideal for use as a "ray" in this experiment. Second, laser light is highly monochromatic, meaning that it contains, to a close approximation, just one wavelength. For a helium-neon laser this wavelength is 6328 angstrom units. An angstrom unit (abbreviated Å) is $10^{-10}$ m or 0.1 nm; thus the wavelength of the helium-neon laser may also be quoted as $6.328 \times 10^{-7}$ m, $6.328 \times 10^{-5}$ cm, or 632.8 nm. Although knowledge of the wavelength of the light used in this experiment is not necessary because we are doing geometrical optics, you should realize that the refractive index you will encounter for the lens and prism material will have the value corresponding to 6328 Å.

**2.** Place the laser on the laboratory bench just beyond the right edge of the drawing board and direct it so that the beam passes over the board just above the paper surface and strikes the mirror at an angle of incidence of about 30°. CAUTION: Although a low-power laser is used in this experiment so that the beam will not hurt your skin, *be very careful not to look into the beam.* Keep it down on the bench and take care not to shine it in anyone's face, including your own.

**3.** Using the mirror as a straightedge, draw a heavy line with a sharp pencil to represent the reflecting surface on the paper. Make a mark on the paper directly under the point where the laser beam strikes the mirror. Also make marks near the right-hand edge of the paper directly under the incoming (incident) beam and under the outgoing (reflected) beam. *Note:* In clear air the laser beam will be invisible because no part of the light travels to your eye. It can be made visible by blowing chalk dust or cigarette smoke into the beam path, in which case the small particles will scatter some of the light to your eye so that you can see it. There will be enough scattering at the mirror surface so that you will be able to see a red spot where the beam strikes and so be able to mark the paper immediately below it. To place a mark under the incident and reflected rays, hold your pencil vertical with the point down and move it sideways through the beam. You will see the beam spot on the pencil and will easily be able to tell when the pencil is centered in the beam path. You can then lower the point onto the paper to make the desired marks.

**4.** Remove the mirror from the paper and point the laser toward the wall or other backstop behind the bench to stop the beam. Do this whenever you are not actually using the laser. With the drawing compass and meter stick, construct the normal to the line representing the mirror at the point of incidence. Draw this normal as a broken line. Draw solid lines representing the incident and reflected rays. Use the meter stick as a straightedge and connect the point you marked under the incoming beam in Procedure 3 and the point of incidence to represent the incident ray. Connect the point under the reflected ray and the point of incidence to represent the reflected ray. Place arrowheads on these lines to indicate direction of

propagation. With the protractor, measure and record the angles of incidence and reflection.

**5.** Remove the sheet of paper from the drawing board and set it aside for inclusion in your report. Put down a new sheet and secure it with tape. Place the concave mirror on this sheet near its left edge and facing to the right. Note that for the two-dimensional work we are doing here the mirror is cylindrical rather than spherical. Trace the reflecting surface on the paper with a sharp pencil so that you get a heavy curved line representing the mirror. Now remove the mirror and locate its center of curvature on the paper as follows: draw any two chords to your curved line and construct their perpendicular bisectors. The intersection of these two lines is the mirror's center of curvature. Pick a point near the middle of the curved line and, using the meter stick as a straightedge, draw a straight line from this point through the center of curvature. This line will serve as the mirror's axis. Now construct two lines, one on each side of the axis, parallel to it, and separated from it by 2 in. These lines should extend from the right-hand edge of the paper to the curved line representing the mirror and will represent incoming rays. Put arrowheads on them to show that they are directed toward the mirror.

**6.** Replace the concave mirror on the paper so that its edge matches the curved line traced from it in Procedure 5, thus ensuring that the mirror is back in its original position. Direct the laser beam at the mirror so that it lies right above one of the incoming ray lines drawn in Procedure 5. Do this by making the spot where the beam strikes the mirror come right above the point where the ray line intersects the curved mirror tracing, while at the same time a pencil held vertically in the beam path near the right edge of the paper points to the line below it. Observe the reflected ray and mark a point directly under it near the right-hand edge of the paper as described in Procedure 3.

**7.** Repeat Procedure 6 with the laser beam directed along the other ray line.

**8.** Remove the mirror from the paper and direct the laser beam at the wall to stop it. Draw lines representing the reflected rays from the respective points of incidence on the mirror through the points marked in Procedures 6 and 7. These lines should intersect on the axis, the point of intersection being the mirror's principal focus. If they do not, mark the best average position on the axis and call this the focus. With the meter stick, measure and record the focal length and radius of curvature of the concave mirror. Put arrowheads on your reflected ray lines to indicate that they are directed away from the mirror.

**9.** Repeat Procedures 5–8 for the convex mirror. Note, however, that this mirror diverges the reflected rays so that, instead of *coming to* a focus, they appear to *come from* a focus behind the mirror. To handle this situation, modify your procedure as follows:

    a. Place the mirror near the center of the paper rather than near the left edge to leave room to draw projections of the rays and axis behind it.

    b. Extend the axis line across the paper behind the mirror.

    c. To locate the principal focus, draw the projections of the reflected rays as broken lines behind the mirror. These projections are called *virtual rays*, since there is really no light there. They should intersect at a point on the axis. This is the principal focus. As described in the Theory section, it is a virtual focus, and the focal length and radius of curvature are considered to be negative.

**10.** Remove the sheet of paper on the board for inclusion in your report and replace it with a clean sheet. Place the plano-convex lens on the paper near its center with the convex side facing to the right. With a sharp pencil trace the outline of the lens. Then remove the lens and construct the perpendicular bisector to the straight line representing the lens's flat side. Extend this perpendicular bisector all the way across the paper. It will represent the lens axis. Also construct two lines, one on each side of the axis, parallel to it and spaced away from it by 2 in. These lines should be drawn from the right edge of the paper to the convex surface of the lens and will represent incoming rays. Put arrowheads on them to show that they are directed toward the lens.

**11.** Replace the lens on the paper, centering it carefully in the tracing to ensure its return to its original position. Direct the laser beam at the lens, adjusting the laser so that the beam lies above one of the incoming ray lines drawn in Procedure 10. Do this as described in Procedure 6. Observe the refracted ray issuing from the left (plane) side of the lens. Mark the point where this ray crosses the axis. Also mark the paper directly below the point where the beam exits the lens.

**12.** Repeat Procedure 11 with the laser beam directed along the other ray line.

**13.** Remove the lens from the paper and point the laser beam at the wall to stop it. Draw lines representing the refracted rays through the points marked in Procedures 11 and 12. Note that the points where these rays intersect the axis should coincide, this being the lens's principal focus. If they do not, mark the best average position on the axis and call this the focus. Establish a point on the axis in the center of the lens tracing and measure the focal length as the distance from this point to the focus. Record your result.

**14.** Replace the sheet of paper on the board with a new sheet. Lay the semicircular lens on this sheet near its center with the flat side facing to the left. Aim the laser beam in from the right so that it strikes "normally" in the center of the curved surface of the lens, passes through the lens material without being refracted, and exits normally through the middle of the flat surface, again without refraction because the ray is normal to the boundary between the two media involved.

**15.** Stick a pin into the drawing board at the midpoint of the flat surface of the lens, to serve as a pivot about which the lens may be conveniently rotated. Note that because this is a semicircular lens, the pin is at the center of the curved surface. Carefully rotate the lens about the pin as an axis, noting that in this way the laser beam is kept normal to the curved surface at the point of incidence so that no refraction occurs as the beam enters the lens material. At the point of *exit*, however, the angle of incidence is being increased from zero, and the beam is bent *away* from the normal because it is passing from a *denser* medium (the lens material) to a *less dense* medium (air). Observe the refracted beam by holding a small white card in its path and noting the spot where it strikes. Continue rotating the lens until the refracted beam just grazes the flat face. You will now notice two interesting effects: (1) Just as you get to an angle of refraction of 90°, the refracted beam will disappear (the spot will vanish from the card) and (2) a strong reflected beam emerging normally from the curved side will appear and can be detected by observing the spot on your card when it is held in the appropriate position. This is total reflection occurring as you set the angle of incidence of the ray *inside the lens* on the flat surface to its critical value. With the lens in this critical position, trace its outline on the paper with a sharp pencil. Mark points under the incoming beam where it comes over the right edge of the paper and where it strikes the curved surface of the lens. Also mark points under the totally reflected beam where it exits the curved surface and where it passes over the paper's right edge.

**16.** Remove the lens from the drawing board and direct the laser beam toward the wall to stop it. Construct the perpendicular to the line representing the flat face of the lens at the point of incidence. This point should be the center of this line and should be marked for you by the pin, which may now be removed. The perpendicular should be drawn as a broken line and represents the normal to the lens's flat surface. Also draw the incident and totally reflected rays through the appropriate points marked in Procedure 15 and place arrowheads on them to show their directions. Note that these rays are radii of the curved lens surface inside the lens. Measure and record the angles of incidence and reflection at the flat surface.

**17.** Remove the sheet of paper on the board for inclusion in your report and replace it with a fresh sheet. Place the triangular prism on the paper near the center with one side toward you. Direct the laser beam at the prism as illustrated in Fig. 42.6 and observe the refracted beam by holding the white card in its path and noting the spot where it strikes. Carefully rotate the prism and watch how the spot moves on the card. You should be able to find an arrangement in which rotation of the prism in one direction causes the spot to move to an extreme position and then start to move back. Set the prism so that the spot is in its extreme position. This is minimum deviation.

**18.** Mark points under the incident beam where it comes over the right-hand edge of the paper on the drawing board and where it strikes the prism. Similarly, mark points under the refracted beam where it emerges from the prism and where it passes over the

left edge of the paper. With a sharp pencil, and being careful not to let the prism move, trace the prism's outline on the paper.

**19.** Remove the prism and shut off the laser. Draw the incident and refracted rays through the points established in Procedure 18 and place arrowheads on them to show their directions. Draw a line from the point of incidence of the incoming ray on the prism to the point at which the refracted ray exits the prism to represent the ray inside the prism material. See if this ray is parallel to the prism base (the face opposite the apex angle). Extend the incident ray line through the prism as a broken line so that you can measure the angle of deviation $D$ as shown in Fig. 42.6. With the protractor, measure and record this angle and the prism's apex angle $A$.

## DATA

**Plane mirror**

Angle of incidence _____   Percent error _____
Angle of reflection _____

**Curved mirrors**

| Mirror | Focal length | Radius of curvature | $\frac{1}{2}$ Radius of curvature | Percent error |
|---|---|---|---|---|
| Concave | | | | |
| Convex | | | | |

**Lens**

Focal length _____

**Critical angle**

Angle of incidence _____   Index of refraction of lens material _____
Angle of reflection _____

**Triangular prism**

Apex Angle $A$ _____   Index of refraction of prism
Angle of deviation $D$ _____   material _____

## CALCULATIONS

**1.** Compare the angles of incidence and reflection measured in Procedure 4 by finding the difference between them, treating this difference as an error, and calculating the percent error.

**2.** From your data in Procedure 8, calculate one half the radius of curvature of the concave mirror. This should be the mirror's focal length. Compare it with your measured focal length by finding the percent error.

**3.** Repeat Calculation 2 for the convex mirror using your data from Procedure 9.

**4.** Assuming the index of refraction of air to be 1, calculate the index of refraction of the semicircular lens material from the value of the critical angle measured in Procedure 16. *Note:* The critical angle is the angle of incidence measured in this procedure. It should be equal to the angle of reflection. Is it?

**5.** Use Equation 42.3 to obtain the index of refraction of the triangular prism material using your data from Procedure 19.

## QUESTIONS _____

**1.** What is meant by the chromatic aberration of a lens?

**2.** (a) What is meant by the spherical aberration of a mirror? (b) Why are the concave mirrors used in automobile headlights parabolic instead of spherical?

**3.** What is wrong with a lens that has astigmatism?

**4.** A ray of light shines from under water through the surface into the air above. Find the critical angle of incidence of this ray on the water-air interface. Take $n = \frac{4}{3}$ for water and 1 for air.

**5.** Under what conditions is the angle of refraction greater than the angle of incidence?

**6.** Show that the angle of deviation $D$ produced by a triangular prism is minimum when the prism is positioned so that the ray inside it is parallel to the prism's base.

**7.** (a) Compute the velocity of light in the triangular prism from your value of the index of refraction for this prism and the known value of the velocity of light in vacuum. (b) Repeat for the semicircular lens if its material has a different index of refraction.

**8.** A ray is incident on the plane surface of a transparent medium having index of refraction $n$. Derive an expression for the angle of incidence that will cause the reflected ray and the refracted ray (the ray transmitted into the medium) to be perpendicular to each other. The incident and reflected rays are in air, whose index of refraction can be taken to be 1. *Note:* The angle of incidence found in this question is called *Brewster's angle*.

**9.** Show that for a spherical concave mirror that is small enough so that the angle of incidence of any ray coming in parallel to the mirror's axis and intercepted by the mirror is small, the rays are converged to a point (the principal focus) that is one half the mirror's radius of curvature away from the mirror.

**10.** A ray of light impinges at an angle of incidence $i$ on the surface of a transparent slab of thickness $d$. The material of the slab has index of refraction $n$. (a) Show that the ray that emerges on the other side of the slab is parallel to the incident ray. (b) Derive an expression giving the lateral displacement of the emergent ray from the incident one in terms of $i$, $d$, and $n$.

**11.** A woman looks at herself in a plane mirror. (a) Why does her image appear to be behind the mirror? (b) Is this a virtual or real image? (c) Why is it reversed left-to-right but not top-to-bottom?

**12.** A straight stick is held so that it extends through the surface and down into a pool of clear, still water. Explain why the stick appears bent at the surface to an observer viewing it from above. *Hint:* Remember that the human eye knows nothing about reflection or refraction but judges an object's location on the assumption that rays coming to it from the object travel in straight lines.

**13.** A six-foot man's eyes are 4 inches below the top of his head. (a) How tall must a flat mirror mounted on the wall be so that the man can just see all of himself when he looks in it? (b) How high above the floor should the bottom edge of this mirror be placed?

**14.** Find out if your plano-convex lens is made of the same material as either the semicircular lens or the prism. If it is, determine the radius of the convex surface and use this radius and the appropriate index of refraction (from either Calculation 4 or 5) in the lens maker's equation to obtain a theoretical value of the plano-convex lens's focal length. Compare this result with the experimental value found in Procedure 13 and explain any significant difference.

# The Focal Length of Lenses; The Telescope and Microscope 43

The formation of images by lenses is one of the most important studies in geometrical optics. The purpose of this experiment is to observe the real images formed by various lenses and to verify the lens equation; in particular, to measure the focal length of some positive and negative lenses and the equivalent focal length of a combination of thin lenses. Combinations of lenses are used extensively in many kinds of optical instruments, and it is instructive to see how such a combination works in actual practice. As examples, this experiment also presents methods of constructing an astronomical telescope, an opera glass, and a microscope, and of determining the magnifying power of each.

## THEORY

When a beam of rays parallel to the principal axis of a lens impinges upon a converging lens, it is brought together at a point called the principal focus of the lens. The distance from the principal focus to the center of the lens is the focal length of the lens; the focal length is positive for a converging lens and negative for a diverging one. The relation between the object distance $p$, the image distance $q$, and the focal length $f$ of a thin lens may be determined by the application of some plane geometry to the diagram of Fig. 43.1, which shows a converging lens of focal length $f$ forming a real image at a distance $q$ to the right of the lens from an object placed at distance $p$ to the lens's left. Note that in this arrangement the rays are traveling left to right so that it is the right-hand focal point that is labeled as the principal focus $F$, although a lens in air is completely reversible, exhibiting the same focal length in either direction.

Three rays are shown proceeding from the object's tip: Ray no. 1 starts out parallel to the axis and must therefore pass through focal point $F$; ray no. 2 passes through the center of the lens and is therefore undeflected; and ray no. 3 passes through focal point $F'$ and must therefore exit the lens parallel to the axis. These three rays (and hence all others leaving the object's tip in such a direction as to be intercepted by the lens) intersect at a point which is the image of that tip. Rays from other points on the object will come together at corresponding points on the image, so that a complete real image of the object is formed as shown.

Now the triangles $ABF'$ and $CGF'$ are both right triangles and the angles at $F'$ in each one are vertical. Hence these triangles are similar, so that their corresponding parts are in proportion, and we can write

$$\frac{AB}{CG} = \frac{AF'}{CF'} = \frac{p - f}{f} \qquad (43.1)$$

Moreover, triangles $DEF$ and $CHF$ are also similar so that

$$\frac{DE}{CH} = \frac{DF}{CF} = \frac{q - f}{f} \qquad (43.2)$$

But because $AB$ and $CH$ are opposite ends of the rectangle $ABHC$, they are equal, and $DE = CG$ for the same reason. Hence $DE/CH = CG/AB$, and sub-

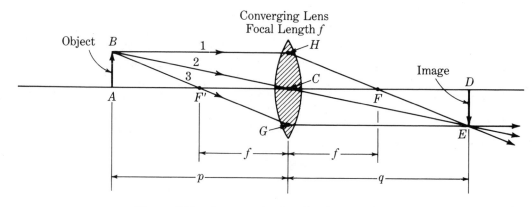

Converging Lens
Focal Length $f$

**Figure 43.1** *A converging lens forming a real image*

391

stitution in Equations 43.1 and 43.2 yields

$$\frac{f}{p-f} = \frac{q-f}{f}$$

or

$$(p-f)(q-f) = f^2 \qquad (43.3)$$

Equation 43.3, in which the object and image positions appear as their distances $p-f$ and $q-f$ from the two focal points, is called *Newton's form of the lens equation*. A little algebra readily transforms it to the more usual form

$$\frac{1}{p} + \frac{1}{q} = \frac{1}{f} \qquad (43.4)$$

The linear magnification produced by a lens is defined as the ratio of the length of the image to the length of the object. In Fig. 43.1 this is seen to be $-DE/AB$, where the minus sign has been inserted to indicate that the image is inverted with respect to the object. Because the triangles $DEC$ and $ABC$ in Fig. 43.1 are another similar pair, $DE/AB = DC/AC = q/p$, and so the linear magnification can be stated as the ratio of the image distance to the object distance. Thus,

$$\text{Magnification} = -\frac{q}{p} \qquad (43.5)$$

Notice that the magnification will be negative with a real, inverted image, but in the case of a virtual image the image distance $q$ will be negative. Hence the magnification will be positive as it should be with the erect image obtained under these circumstances.

An interesting example of this is the magnifying glass, which is simply a converging lens of the sort shown in Fig. 43.1 but is placed so that the object falls inside the focal length $f$. Thus $p$ is less that $f$ and the lens equation (43.4) shows that $q$ will then be negative and hence the image erect and virtual. Note that this is simply a consequence of the fact that the lens, although converging, does not have sufficient power (short enough focal length, i.e. $f > p$) to overcome the strong divergence of the rays coming

from the nearby object. Note also that we can then write

$$\frac{1}{q} = \frac{1}{f} - \frac{1}{p} = \frac{p-f}{fp} = -\frac{f-p}{fp}$$

and hence from Equation 43.5

$$\text{Magnification} = -\left(-\frac{fp}{f-p}\right)\bigg/ p = \frac{f}{f-p}$$

The denominator in this result is clearly less than the numerator and can be made quite small if $p$ is just a little less than $f$, so that a large magnification is obtainable.

The principal focal length of a converging lens may be determined by forming an image of a very distant object on a screen and measuring the distance from the screen to the lens. This distance will be the focal length, since the rays of light from a very distant object are very nearly parallel. In other words, $p$ in Equation 43.4 is infinite so that $1/p$ is zero and $q = f$. A better method is to place an object at a known distance $p$ from the lens, measure the resulting image distance $q$, and calculate the focal length from Equation 43.4.

Another method, which demonstrates the meaning of this equation very clearly, makes use of the following arrangement: An object and a screen on which an image of the object is to be focused are placed a known distance $D$ apart, and a converging lens of focal length $f$ is placed between them. Since $q$ is the distance from the lens to the image (that is, the screen), the sum $p + q$ must be equal to $D$. Hence $q = D - p$, and substitution in Equation 43.4 to eliminate $q$ produces a quadratic equation in $p$ whose solutions are

$$p_a = \frac{D}{2}\left(1 + \sqrt{1 - 4\frac{f}{D}}\right) \qquad (43.6)$$

and

$$p_b = \frac{D}{2}\left(1 - \sqrt{1 - 4\frac{f}{D}}\right) \qquad (43.7)$$

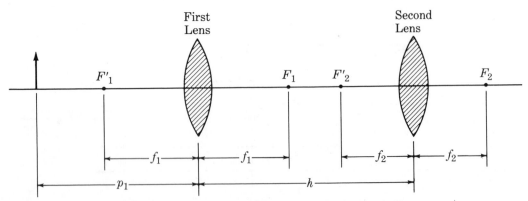

**Figure 43.2** *An optical system of two converging lenses*

There are thus two positions of the lens between the object and the screen for which an image on the screen will be in focus *provided D is at least four times the lens's focal length.* If *D* is just equal to 4*f*, the square root in Equations 43.6 and 43.7 disappears and $p_a = p_b = D/2$. That is, there is only one lens position, namely the midpoint of the object-screen distance, for which a focused image is obtained. This condition is easily observed and the corresponding value of *D* measured. The focal length of the lens is then found by dividing this result by 4.

The lens equation (43.4) may also be used to handle the situation where an optical system is formed in which light rays pass through two or more lenses in sequence. Figure 43.2 illustrates the sort of arrangement that might be encountered. The procedure is to consider the first lens in front of the object first and apply the lens equation to it as if it were acting alone. An image distance $q_1$ is found, and the image appearing at $q_1$ is then treated as the object of the second lens. In Fig. 43.2, $q_1$ will be positive, the real image formed by the first lens (which in this illustration is converging) appearing on the right of that lens. The object distance $p_2$ for the second lens is then $h - q_1$, where *h* is the space between the lenses, and the lens equation may then again be applied with $p_2 = h - q_1$ to find an image distance $q_2$, which gives the position of the final image *measured from the location of the second lens.* Notice that in the present arrangement, positive object distances are measured to the left of the pertinent lens in each case, positive image distances to the right, and converging lenses have positive focal lengths. If these sign conventions are adhered to, the above-described procedure will handle any optical system regardless of the lens spacing *h* or of whether the lenses are converging or diverging. Note in partiuclar that if $q_1$ is larger than *h*, the second lens's object distance $p_2$ will be negative.

A useful example of a two-lens system is the arrangement of two thin lenses that touch each other. In this case, $h = 0$, $p_2 = -q_1$, and two successive applications of the lens equation will show that the combination has an equivalent focal length *f* given by

$$\frac{1}{f} = \frac{1}{f_1} + \frac{1}{f_2} \qquad (43.8)$$

where $f_1$ is the focal length of the first lens and $f_2$ is the focal length of the second lens. This arrangement is useful because a concave lens by itself cannot form a real image, since it is a diverging lens; hence a different method must be used for measuring its focal length. This may be done by placing the negative lens in contact with a positive lens of shorter and known focal length, measuring the equivalent focal length of the combination experimentally, and then using Equation 43.8 to solve for the focal length of the negative lens.

Other well-known examples of multi-lens optical systems are the telescope and the microscope. There are three main types of telescopes: the astronomical telescope, the terrestrial telescope, and the opera glass. In the astronomical telescope, a so-called objective lens, which is a positive lens of long focal length, forms a real inverted image of a distant object in its focal plane. Note that *p* is really infinite for the astronomical telescope's objective, so that $q = f$ for this lens. The second lens or *eyepiece* is a positive lens of short focal length used as a magnifying glass to produce an enlarged virtual image of the real image formed by the objective. For this purpose, the eyepiece is placed so that the focal point of the objective lies between the eyepiece and its focal point. In other words, the arrangement is similar to that of Fig. 43.2 but with *h* less than $f_1 + f_2$ so that $F_2'$ lies to the left of $F_1$ to make $p_2 = h - q_1 < f_2$. This will result in a negative $q_2$ and hence a virtual image as already described in the discussion of the magnifying glass. The eye looks directly at the magnified virtual image produced by the eyepiece.

A disadvantage of the astronomical telescope, which is of little consequence when it is used to look at the stars, is that the image is inverted. This will be understood by noting that the eyepiece forms a virtual image that is not reinverted. In the terrestrial telescope, a third lens is introduced between the objective and the eyepiece. It is a converging lens and is placed beyond the image formed by the objective at such a distance as to form a second real image that is the inversion of the first. The two inversions thus produce a real, erect image that the eyepiece then views as in the astronomical telescope.

Because of its three lenses and the spacing required between them, the terrestrial telescope tends to be very long. This disadvantage is overcome without sacrificing the advantage of an erect final image in the opera glass, a telescope in which the eyepiece is a divergent lens. It is placed so that the rays from the objective strike it before they converge to form an image. Thus if Fig. 43.2 is to represent an opera glass, the second lens is diverging and is spaced from the first lens by a distance *h* less than $q_1$. In fact, *h* is made equal to the difference between the absolute values of the two focal lengths $f_1$ and $f_2$. For the divergent eyepiece used in this case, $f_2$ is shorter than $f_1$ and negative, hence $h = f_1 - |f_2|$ so that the spacing will not only be less than $q_1$ but less than $f_1$. The second lens therefore lies within the focal length of the first. Thus, the image that would have been formed by the objective serves as the virtual object ($p_2$ negative) for the diverging eyepiece, from which the rays diverge as if they came from an enlarged virtual image. This type of telescope has a small field of view, but it is clearly shorter than either of the other two types and, as already noted, has the advantage of giving an erect image of a distant object.

Like the astronomical telescope, the microscope uses two convergent lenses, one as an objective and one as an eyepiece, but in this case the objective is the lens with the shorter focal length. The object is placed at a distance only slightly greater than the focal length of this lens, so that a highly enlarged real image is formed. The longer focal length lens is used as the eyepiece and is placed as in the telescope so that the real image produced by the objective falls just inside its focal length. A further magnified virtual image is thus obtained. The microscope gives a much greater magnification than does the eyepiece used alone as a magnifier because in the microscope the eyepiece is used to view the enlarged real image that the objective forms from the object, rather than viewing the object directly.

The magnifying power of an optical instrument is defined as the ratio of the angle subtended at the eye by the image of an object viewed through the instrument to the angle subtended at the eye by the object viewed directly. This definition is somewhat different from that of linear magnification noted earlier (Equation 43.5), but it is an important characteristic of a telescope or a microscope. Thus much of the experimental work on these instruments will be devoted to measuring their magnifying power.

## APPARATUS

1. Optical bench
2. Illuminated object
3. Lens holders
4. Screen
5. Two converging lenses, A and B (of about 20-cm and 10-cm focal length respectively)
6. One diverging lens, C (of about −15-cm focal length)
7. Vertically mounted metric scale
8. Telescope magnification scale. This scale is made up of a strip of white paper 2 ft long with a series of thick black lines drawn horizontally across it at regular intervals of about 2 in. The lines are numbered to facilitate counting.
9. Vernier caliper
10. Desk lamp

## PROCEDURE

**1.** Measure the focal length of lens A directly by obtaining the image of a very distant object on the screen and measuring the image distance. The object may be a tree or a house about a block away. Record the focal length so determined.

**2.** Repeat Procedure 1 for lens B.

**3.** Determine the focal length of lens A by the use of the lens equation. Place the illuminated object at one end of the optical bench and the screen at a distance of about five times the focal length of the lens. With the object and screen fixed, find the position of the lens for which a sharp, enlarged image is produced on the screen. Make sure that the object, lens, and screen all lie along the same straight line (the principal axis of the lens) and that they are all perpendicular to the axis. Measure and record to a precision of 1 mm the object distance $p$ from the illuminated object to the lens and the image distance $q$ from the lens to the screen. Then use either the vernier caliper or the metric scale to measure the sizes of the object and image. Record these measurements to 0.5 mm or, if you can, to 0.1 mm.

**4.** Using the arrangement of Procedure 3, move the lens back and forth to find two positions for which the image on the screen is in focus. Move the screen a few centimeters closer to the object and again observe that a focused image is obtained for two different lens positions. Notice that these positions are closer together than they were before the screen was moved. Continue to move the screen closer to the object and to observe that there are two lens positions that give a focused image until these two positions coincide at the midpoint between the object and the screen. Measure and record the value of the object-screen distance $D$ corresponding to this condition.

**5.** Repeat Procedure 3 using lens B.

**6.** Repeat Procedure 4 using lens B.

**7.** Repeat Procedure 3 using the combination of lenses A and B in contact.

**8.** Repeat Procedure 3 using the combination of lenses B and C in contact.

**9.** Construct a simple astronomical telescope using the long-focus converging lens A as the objective and the short-focus converging lens B as the eyepiece. Special care must be taken to have the various optical parts at the same height and well lined up. In addition, the lenses must be perpendicular to the optic axis of the system. Mount the eyepiece near one end of the optical bench and the screen in front of it at a distance of its focal length away. Place the illuminated object at the far end of the laboratory table and point the optical bench at it. Mount the objective lens on the optical bench on the other side of the screen from the eyepiece and move it until it focuses a real inverted image of the object on the screen.

Then remove the screen and adjust the eyepiece so that on looking through it you can see a sharp inverted image of the illuminated object.

**10.** Measure the magnifying power of your telescope by taping the telescope magnification scale to a distant wall and pointing your telescope at it. Illuminate the scale with the desk lamp if necessary and readjust the eyepiece to bring the scale into sharp focus. Look through the telescope with one eye and directly at the scale with the other. The magnified image of the scale will be superimposed on the unmagnified scale. Count the number of divisions of the scale viewed directly that cover exactly one division of the magnified scale. This number gives the magnifying power of the telescope.

**11.** Construct an opera glass using the same lens for the objective as before and the diverging lens as the eyepiece. This time place the screen at one end of the optical bench, point the optical bench at the illuminated object at the far end of the laboratory table, and mount the objective lens so that a real inverted image of the object is formed on the screen. Mount the eyepiece on the same side of the screen as the objective lens and spaced from this lens by a distance equal to the difference of their focal lengths. Remove the screen and adjust the eyepiece so that on looking through it you can see the largest clear and erect image that can be formed.

**12.** Determine the magnifying power of the opera glass as in Procedure 10.

**13.** Construct a microscope using lens B as the objective and lens A as the eyepiece. Place the illuminated object at one end of the optical bench and mount the short-focus lens in front of it at a distance

of a little more than this lens's focal length. Mount the screen on the other side of this lens and slide it along the optical bench until a real inverted image is sharply focused on it. Then mount the eyepiece on the other side of the screen from the objective and at a distance from the screen equal to the eyepiece's focal length. Finally remove the screen and adjust the eyepiece so that when you look through it you see a sharp inverted image of the illuminated object.

**14.** Replace the illuminated object with the vertical metric scale; position the latter so that it is the same distance from your microscope objective as the illuminated object was. Illuminate the scale with the desk lamp. Look through your microscope and make small adjustments to the eyepiece position to bring the scale divisions into sharp focus. Place the vernier caliper against the eyepiece and adjust the jaws until they span exactly five millimeter divisions as seen in the microscope. Remove both lenses and mount the caliper on the eyepiece support without disturbing the setting of its jaws. Look at the scale directly through the space between the jaws, keeping your eye in the same position it was in when looking through the microscope, and note the length of scale bracketed by these jaws. You will have trouble distinguishing the millimeter divisions at this distance, but you should be able to see the centimeter divisions quite clearly. Read the length of the scale seen between the caliper jaws, estimating to 0.1 cm. This is the length subtended at the eye by the same angle that subtended 5 mm (0.5 cm) in the image. This length in centimeters divided by 0.5 cm is therefore the magnifying power of the microscope.

## DATA _____

| Lenses | Focal length measured directly | Object distance $p$ | Image distance $q$ | Distance $D$ for single lens position | Size of object | Size of image |
|--------|-------------------------------|---------------------|--------------------|---------------------------------------|----------------|---------------|
| A | | | | | | |
| B | | | | | | |
| A and B | ✕ | | | ✕ | | |
| B and C | ✕ | | | ✕ | | |

| Lenses | Focal length from lens equation | Focal length from Equation 43.8 | Focal length from *D* for single lens position | Magnification | |
|---|---|---|---|---|---|
| | | | | Size of image / Size of object | $-\dfrac{q}{p}$ |
| A | | | | | |
| B | | | | | |
| A and B | | | | | |
| B and C | | | | | |
| C | | | | | |

Magnifying power of the astronomical telescope _____

Magnifying power of the opera glass _____

Magnifying power of the microscope _____

## CALCULATIONS

**1.** Calculate the focal length of lens A from the data of Procedure 3 and the focal length of lens B from the data of Procedure 5. Compute the magnification produced in each case from the ratio of the size of the image to the size of the object and from the ratio of the image distance to the object distance. Enter these results in the boxes provided for them in the second data table.

**2.** Calculate the focal lengths of lenses A and B from the values of the object-screen distance *D* found in Procedures 4 and 6, respectively.

**3.** Calculate the focal length of the combination of lenses A and B from the data of Procedure 7. Compare the result with the value obtained from theory, using Equation 43.8.

**4.** Calculate the focal length of the combination of lenses B and C from the data of Procedure 8. By using this value of the focal length of the combination, compute the focal length of the negative lens C, using Equation 43.8.

## QUESTIONS _____

**1.** Explain in your own words what is meant by a real image and a virtual image.

**2.** Derive Equation 43.4 from Equation 43.3.

**3.** (a) What is meant by the spherical aberration of a lens? (b) How can this aberration be reduced in a simple lens?

**4.** Derive Equation 43.8. Assume that the two lenses in contact with each other are so thin that the combination is thin also.

**5.** State what is meant by the *power* of a lens and write Equation 43.8 in the form it would take if powers were used instead of focal lengths. Does your result remind you of something you encountered in your study of electricity?

**6.** In a film projection apparatus it is desired to produce pictures 12 ft wide on a screen 50 ft from the lens; the size of the picture on the film is 1 in. wide. What must be the focal length of the projection lens used?

**7.** Substitute $q = D - p$ in Equation 43.4 and show how the solutions 43.6 and 43.7 are obtained. What happens mathematically if $D$ is less than $4f$ in these equations and what does this tell you about what you would observe physically in this case?

**8.** Draw a ray diagram similar to Fig. 43.1 but with the object distance $p$ less than the focal length $f$ so that the lens functions as a magnifying glass. (b) Show from your diagram that the image is erect and virtual and that the magnification is $f/(f - p)$.

**9.** If a telescope is accurately focused on a distant object, in what direction must the eyepiece be moved to focus on a near object? Explain.

**10.** What is the basic difference between the astronomical telescope and the microscope?

**11.** Draw a ray diagram, approximately to scale, showing the illuminated object, the real image formed by the objective, and the virtual image formed by the eyepiece of your astronomical telescope. Represent the object by a small arrow.

**12.** The magnifying power of an astronomical telescope 24 in. long is equal to 15. Determine the focal length of the objective lens and that of the eyepiece. Assume that the telescope is focused on a distant object.

**13.** Redraw Fig. 43.2 for the case of an opera glass.

**14.** Explain how the method of Procedure 14 measures the magnifying power of the microscope in accordance with the definition of magnifying power.

**15.** For both the astronomical telescope and the opera glass, the distance between the lenses is approximately equal to the *algebraic* sum of their focal lengths. Explain why this is so, remembering that the focal length of a diverging lens is negative, so that in the opera glass the lens separation is the *difference* between the *values* of the focal lengths.

# Analysis of Light by a Prism Spectrometer

# 44

Light is made up of many different colors, each of which has a definite wavelength. When a beam of light is passed through a glass prism, it is broken up into a spectrum. Each chemical element produces its own characteristic spectrum, so that by analyzing the light given off by a substance it is possible to identify the elements it contains. The object of this experiment is to introduce you to the prism spectrometer and to use this important optical instrument to study the bright-line and continuous spectra obtained from certain light sources.

## THEORY

In general, light is made up of many component wavelengths, each of which corresponds to a definite color. When a beam of light is passed through a glass prism, refraction takes place when the beam enters the prism and again when it leaves, with the result that the beam is bent through an angle $D$, called the angle of deviation. This angle is a function of the index of refraction $n$ of the glass as discussed in the Theory section of Experiment 42 (Fig. 42.6 and Equation 42.3). Since $n$ is different for different wavelengths (because they propagate with different velocities in transparent materials such as glass), each component wavelength in the incident beam is bent through a different angle. The incident beam is thus separated into beams of its component colors (wavelengths) that may then be displayed as a *spectrum*. We say that the incoming light has been *dispersed* into a spectrum, and the property exhibited by the prism of separating this light into its constituents is called *dispersion*.

Spectrum analysis is the decomposition of a beam of light into its constituent wavelengths and the examination of the image so formed. A spectrometer (or spectroscope, as instruments not provided with means for recording the analyzed light are often called) is an optical instrument for producing and analyzing spectra. A diagram of a prism spectrometer is shown in Fig. 44.1, and a typical laboratory instrument appears in Fig. 44.2. It consists of three essential parts: a collimator, a prism, and a telescope. The collimator is a tube with a converging lens at one end and a fine slit of adjustable width at the other. The length of the tube is equal to the focal length of the lens so that the collimator renders the rays of light from the slit parallel. The prism deviates these rays and disperses them into a spectrum. The telescope contains an objective lens, which brings the rays of light to a focus in its focal plane, and an eyepiece, through which the image of the spectrum is viewed. If the light is all of one color, or monochromatic, a single image of the slit appears. If a source contains several colors, several images of different color will appear side by side, each being an image of the slit formed by one component of the light. The telescope can be rotated about the prism so that light emerging at different angles can be viewed, and the angular position of each slit image can be read off from the divided circle under the telescope mount.

**Figure 44.1** *Prism spectrometer*

**Figure 44.2** *A typical laboratory prism spectrometer*

There are three main types of spectra: continuous, bright-line, and absorption. A continuous spectrum is produced by light from an incandescent solid or liquid. It contains all wavelengths and hence appears as a continuous gradation of colors, merging with one another with very gradual changes from red to violet. The bright-line spectrum is produced by exciting an element in the gaseous state by means of an electric discharge or by heating. Such a spectrum consists of bright lines, each an image of the spectrometer slit formed by light of a definite color, separated by dark spaces. The number and relative positions of the lines depends on the element used and the method of excitation employed. An absorption spectrum is produced when light from a source furnishing a continuous spectrum is passed through some cooler absorbing medium. The resulting spectrum is crossed by dark spaces or lines that form the absorption spectrum of the material.

Wavelengths of visible light are usually expressed in angstrom units (abbreviated Å) or, in more recent literature, in nanometers. An angstrom unit is defined as $10^{-10}$ m and thus equals $10^{-8}$ cm. A nanometer is $10^{-9}$ m. Hence a nanometer is 10 Å.

**TABLE 44-1 HELIUM LINES**

| Wavelength, Å | Color |
|---|---|
| 7065 | red |
| 6678 | red |
| 5876 | yellow |
| 5048 | green |
| 5015 | green |
| 4922 | blue-green |
| 4713 | blue |
| 4471 | deep blue |
| 4388 | blue-violet |
| 4026 | violet |
| 3964 | violet |
| 3889 | violet |

**TABLE 44-2 MERCURY LINES**

| Wavelength, Å | Color |
|---|---|
| 6907 | red |
| 5790 | yellow |
| 5770 | yellow |
| 5461 | green |
| 4916 | blue-green |
| 4358 | blue |
| 4078 | violet |
| 4047 | violet |

The principal lines in the helium spectrum are given in Table 44-1. The principal lines in the mercury spectrum are given in Table 44-2. In these spectra other fainter lines are also present, but the ones listed are the brightest and easiest to find.

## APPARATUS

1. Spectroscope
2. Mercury discharge tube
3. Helium discharge tube
4. Hydrogen discharge tube
5. Sodium light source
6. Desk lamp
7. Discharge-tube power supply
8. Stand and clamp for holding the discharge tubes (unless part of the power supply)

## PROCEDURE

**1.** The spectrometer should be already adjusted when you come to the laboratory, with the collimator and telescope properly focused and aligned and the prism correctly mounted and leveled. CAUTION: Do not handle the spectrometer until the instructor has demonstrated its operation to you. This instrument is easily thrown out of adjustment, and needless delay or even permanent damage can be caused by careless handling. Be extremely careful when using it.

**2.** Mount the helium discharge tube in its holder and place it immediately in front of the slit at the end of the spectrometer's collimator. The spectrometer should be positioned so that the slit faces directly into the center of the discharge tube. Connect the leads from the power supply to the discharge-tube terminals and turn the supply on. CAUTION: The discharge tube operates at a high voltage, and its power supply can deliver enough current to give a painful shock. Do *not* touch either the tube or the wires leading to it while the power supply is in operation.

**3.** Adjust the angular position of the telescope so that the helium spectrum is visible when you look into the eyepiece. A small adjustment of the eyepiece position may be necessary to bring the spectrum into sharp focus. Note that the spectrum consists of a number of vertical lines of different colors. Each of these lines is an image of the collimator slit formed by the pertinent wavelength, but because most optical spectrometers use such slits, the word "line" has come to mean a particular wavelength in a spectrum. Now adjust the spectrometer for minimum deviation. To do this, move the telescope so that the bright yellow line at 5876 Å is roughly in the center of the field of view. Slowly turn the prism table in such a direction as to diminish the angle of deviation while watching this line. It should move first one way and then the other as the prism is rotated. If it moves out of the field of view, the telescope position must be readjusted to follow it. Lock the prism table in the position corresponding to the extreme position of the yellow line. This is the orientation giving minimum deviation for light of this wavelength. The prism also gives maximum dispersion when set this way, which is why it is so adjusted for use in a spectrometer. *Note:* When you rotate the prism table, every line in the spectrum will move through a minimum deviation position, but the table setting for minimum deviation will be slightly different for different lines, that is, different wavelengths. The yellow line is chosen because it is at about the middle of the visible spectrum. Therefore the prism setting for minimum deviation of this line represents a good average for the entire spectrum. Do not change the prism table position after the minimum deviation setting just described has been found.

**4.** Adjust the position of the telescope so that the first red line of the helium spectrum falls exactly on the intersection of the eyepiece cross hairs. Note that directly under the arm supporting the telescope is a clamping device that rotates with the arm and permits the telescope to be clamped to the base by means of a radial screw. A tangent screw on the arm allows a fine adjustment of the telescope setting; however, the tangent screw can be used only after the telescope has been clamped with the radial screw. Rotate the telescope until the first red line is near the vertical cross hair and then clamp it by means of the radial screw. Use the tangent screw to set the intersection of the cross hairs exactly on the red line. Record the angular setting of the telescope to the nearest minute of arc, reading the divided circle with the help of the vernier. Also record the first red line's color and known wavelength (see Table 44-1).

**5.** Repeat Procedure 4 for the other lines of the helium spectrum, taking readings of the telescope setting for each of the lines that can be recognized.

**6.** Turn off the discharge tube power supply and replace the helium discharge tube with the mercury tube. Turn the supply on and repeat Procedures 4 and 5 for the mercury spectrum.

**7.** Repeat Procedure 6 using the hydrogen tube. Four lines should be observed and telescope settings obtained for them. Record the longest wavelength (red) line as $H_\alpha$, the next as $H_\beta$, and so on.

**8.** Turn off the power supply, remove the discharge tube, and set up a sodium light source in front of the spectroscope slit. Turn on the source and wait for it to warm up so that a brilliant yellow light is produced. Look through the telescope and adjust it so that the yellow sodium doublet falls exactly on the intersection of the cross hairs. Record the telescope setting.

**9.** Examine the spectrum of the light from a tungsten lamp. Record the setting of the telescope for the upper and lower limits of the visible spectrum.

## DATA

| Spectrum | Color | Known wavelength | Setting of telescope | Wavelength from calibration curve |
|---|---|---|---|---|
| Helium | | | | |
| | | | | |
| | | | | |
| | | | | |
| | | | | |
| | | | | |
| | | | | |
| | | | | |
| | | | | |
| | | | | |
| | | | | |
| | | | | |
| | | | | |
| Mercury | | | | |
| | | | | |
| | | | | |
| | | | | |
| | | | | |
| | | | | |
| | | | | |
| | | | | |
| | | | | |
| Sodium | | | | |
| Visible | | | | |
| | | | | |

### Hydrogen Spectrum

| Line | Color | Setting of telescope | Wavelength from calibration curve | Reciprocal of wavelength | $1/n^2$ |
|---|---|---|---|---|---|
| $H_\alpha$; $n = 3$ | | | | | |
| $H_\beta$; $n = 4$ | | | | | |
| $H_\gamma$; $n = 5$ | | | | | |
| $H_\delta$; $n = 6$ | | | | | |

Measured value of Rydberg constant from slope _____

Balmer series limit _____

Measured value of Rydberg constant from intercept _____

Accepted value of Rydberg constant _____

## CALCULATIONS _____

**1.** Draw a graph with wavelengths in angstrom units as abscissas and the settings of the telescope as ordinates for the lines of the helium spectrum from the data of Procedures 4 and 5 and the wavelengths given in Table 44-1. The smooth curve so obtained is the calibration curve of the instrument. The wavelengths of the lines observed in other spectra may be determined from it.

**2.** From your calibration curve, read off the wavelengths of the mercury lines corresponding to the telescope readings obtained in Procedure 6. Tabulate the results and see how they compare with the known values given in Table 44-2.

**3.** Repeat Calculation 2 for the hydrogen readings obtained in Procedure 7. No table of known wavelengths is given for the hydrogen lines in this experiment. However, the hydrogen spectrum observed with a glass discharge tube and prism is part of the so-called *Balmer series* in hydrogen. The wavelengths of lines in this series are given by the Balmer formula

$$\frac{1}{\lambda} = R_H\left(\frac{1}{2^2} - \frac{1}{n^2}\right) = \frac{R_H}{4} - \frac{R_H}{n^2} \tag{44.1}$$

where $\lambda$ is the wavelength in angstrom units, $R_H$ is the famous Rydberg constant, and $n$ is an integer having the value 3 for the longest wavelength ($H_\alpha$) line in the series, 4 for the next longest ($H_\beta$), and so on. Calculate $1/\lambda$ for the four lines you observed and plot the results as ordinates against the corresponding values of $1/n^2$ as abscissas. To obtain more accurate results, construct the graph using as large a scale as possible; that is, utilize the entire sheet of graph paper. Extend the line to $1/n^2 = 0$.

**4.** According to Equation 44.1, your graph in Calculation 3 should be a straight line of slope $-R_H$. Obtain a measured value of the Rydberg constant in this way.

**5.** Equation 44.1 also indicates that $1/\lambda$ tends to a maximum value $R_H/4$ and hence $\lambda$ to a minimum value $4/R_H$ as $n$ gets very large. This minimum value of $\lambda$ is called the *series limit*. Find it from the $1/n^2 = 0$ intercept on the graph in Calculation 3. Obtain a second measured value of the Rydberg constant from this intercept and compare it with that found in Calculation 4 and with the accepted value listed in Table I at the end of the book. Enter the accepted value in the space provided in the data sheet so that it is displayed along with your two measured values. Watch your units.

**6.** Measure the wavelength of the sodium doublet by using the setting of the telescope in Procedure 8 and reading off the wavelength from the calibration curve. Compare your results with the known value (use 5893 Å, the mean of 5890 Å and 5896 Å, which are the wavelengths of the two lines forming the so-called doublet).

**7.** From the data of Procedure 9 and the calibration curve determine the range of the visible spectrum in angstrom units.

## QUESTIONS

**1.** Compare the appearance of the bright-line spectrum with that of the continuous spectrum and explain the difference.

**2.** Of what use is the calibration curve and why is it needed?

**3.** Compare your results for the range of the visible spectrum with the known value, which you should look up and record in the spaces provided in the data table.

**4.** Does the index of refraction of glass increase or decrease with increasing wavelength? Explain your answer.

**5.** The Balmer formula, Equation 44.1, predicts spectrum lines in hydrogen corresponding to all integer values of *n* equal to or greater than 3. Why, then, did you see only four lines in Procedure 7?

**6.** (a) What are the frequencies in hertz of the limits of the visible spectrum whose wavelengths you found in Calculation 7? (b) How many times the lower limit frequency is the upper limit frequency? Is it as much as twice (an octave)?

**7.** (a) Find out what the Doppler effect is and explain it in your own words. (b) Show how this effect can be applied to light waves in determining the speed of stars toward or away from the earth.

**8.** Calculation 6 indicates that the yellow light from your sodium lamp consists of two closely spaced lines (a "doublet") with respective wavelengths of 5890Å and 5896 Å rather than a single line even though you could only see one line in your spectrometer. Explain why this is so.

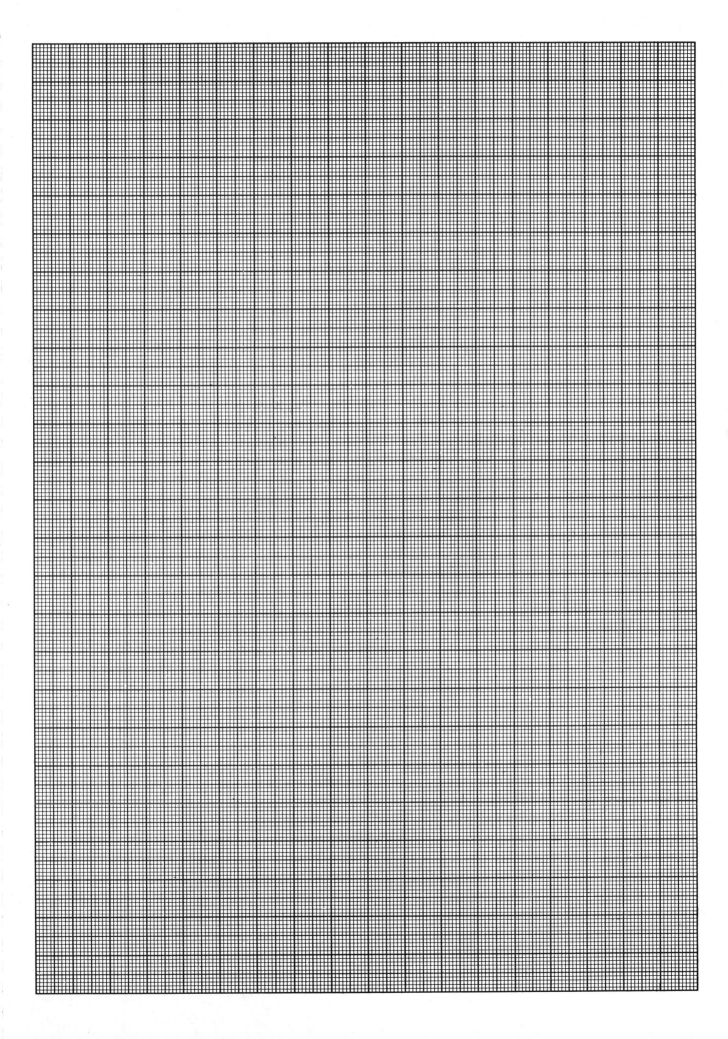

# Interference of Light 45

One of the important methods of measuring the wavelength of light is to allow monochromatic light to be reflected from the two surfaces of a very thin film of varying thickness. The light reflected from one surface will *interfere* with that reflected from the other to produce interference fringes. The light's wavelength can then be calculated from the dimensions of the film and the number of fringes. The object of this experiment is to study the interference of light, to measure the wavelength of the yellow light from a sodium lamp, and to determine the radius of curvature of a convex lens from an observation of the interference pattern called Newton's rings.

## THEORY

Interference is the effect produced when two or more wave trains are superimposed so as to either reinforce or cancel each other; the effect's magnitude depends on the relative phases of the two trains. With light, interference occurs when two monochromatic beams having identical frequencies are superimposed, darkness (destructive interference) being produced when the two beams are united so that the positive vibration peaks of one match the negative ones of the other (a 180° phase difference). Conversely, a bright spot (constructive interference) is seen when the two waves are matched peak to peak (zero phase difference). In most interference experiments, the two beams are produced by splitting a single beam to ensure that the frequencies will be the same. The phase difference is obtained by making the two beams traverse paths of different length before being recombined. A path difference of half a wavelength produces a relative phase shift of 180°, hence a path difference of a whole wavelength or any integer number of wavelengths produces no phase shift at all. Therefore, a path difference of any integer number of wavelengths (an even number of half wavelengths) gives rise to constructive interference, whereas a difference of an integer number of wavelengths plus a half wavelength (an odd number of half wavelengths) gives rise to destructive interference.

One method of producing interference is to allow monochromatic light to fall upon two pieces of plate glass placed at a very small angle to each other (see Fig. 45.1). Light will be reflected from each of the glass surfaces. The light reflected from the lower surface of the upper plate and the light reflected from the upper surface of the lower plate will interfere, the second reflected beam traveling a longer distance, namely, across the air film and back. Since the air film varies regularly in thickness, the interference pattern produced will consist of a series of parallel bright and dark lines that are called *interference fringes*. However, one point should be noted in connection with this method: reflection of the light at the upper surface of the lower glass plate (where the light strikes a denser medium) causes a 180° phase shift, which is equivalent to an extra half wavelength of path for this beam. Consequently, with this setup the conditions for constructive and destructive interference are reversed; an actual path difference of an even number of half wavelengths results in destructive interference while an odd number gives the constructive effect.

The wavelength of the yellow light from a sodium lamp may be determined by using interference fringes. The experimental arrangement for doing this with two glass plates that make a small angle with each other is shown in Fig. 45.2. Note that the plates touch along one edge and are separated by a thin steel strip of thickness $T$ at the other. If $L$ is the distance from the touching edge to the edge of the

**Figure 45.1** *Interference from two plane surfaces*

**Figure 45.2** *Apparatus for viewing interference fringes*

strip and $y$ is the thickness of the air wedge along a line parallel to and at a distance $x$ from the touching edge, then by similar triangles

$$\frac{y}{x} = \frac{T}{L} \quad \text{or} \quad y = \frac{xT}{L} \quad (45.1)$$

Now at any place where the path difference is an odd number of half wavelengths, the two reflected rays will be in phase because of the phase reversal at the lower plate, and a bright line or fringe will appear at this value of $x$. Conversely, where the path difference is an even number of half wavelengths, a dark line will be seen. Hence for a dark line, $2y = m\lambda$ where $m$ is an integer and $2y$ is the round-trip distance across the air wedge. Substitution in Equation 45.1 then yields

$$x_m = \frac{mL\lambda}{2T} \quad (45.2)$$

where $x_m$ is the distance of the $m^{\text{th}}$ dark line from the touching edge. The distance between adjacent dark lines is clearly

$$x_{m+1} - x_m = \frac{(m+1)L\lambda}{2T} - \frac{mL\lambda}{2T} = \frac{L\lambda}{2T} \quad (45.3)$$

from which we get

$$\lambda = \frac{2T}{L} \text{(spacing between adjacent fringes)} \quad (45.4)$$

Note that attention must be paid to the units in this relation. If $T$, $L$, and the fringe spacing are all in centimeters, $\lambda$ will be in centimeters also and must be multiplied by $10^7$ for nanometers or $10^8$ for angstrom units.

Another very interesting example of interference produced by reflection from the two sides of an air wedge bounded by glass surfaces is called *Newton's rings*. The arrangement, which appears in Fig. 45.3, is seen to be similar to that of Fig. 45.2 except that the upper glass plate and steel strip have been replaced by a convex lens whose spherical lower surface touches the lower glass plate at its central point and forms an air wedge whose thickness increases as we go radially outward from that point. Again destructive interference will occur when $2y = m\lambda$ and will be seen as a dark ring of radius $r_m$ about the central point, where $r_m$ is the radial distance from this point to where the lower lens surface has curved up to the distance $m\lambda/2$ above the upper surface of the glass plate. To find $r_m$ we need a relation between $r$ and $y$ as defined in Fig. 45.3 analogous to Equation 45.1 for the arrangement of Fig. 45.2. A study of Fig. 45.3 shows that

$$(R - y)^2 + r^2 = R^2 \quad (45.5)$$

which, when multiplied out, reduces to

$$r^2 = 2Ry - y^2 \quad (45.6)$$

**Figure 45.3** *Apparatus for viewing Newton's rings*

where $r$ is the radial distance outward from the point where the lens contacts the glass plate and $R$ is the lens's radius of curvature. Because $y \ll R$, the term $y^2$ in Equation 45.6 is negligible compared to $2Ry$, and we can write $r^2 \approx 2Ry$ from which there follows

$$r_m^2 = m\lambda R \qquad (45.7)$$

Here, as already noted, $r_m$ is the radius of the $m^{th}$

dark ring. This result shows that a plot of the squares of the radii of Newton's rings against the corresponding values of the index $m$ should be a straight line of slope $\lambda R$. The wavelength of light can then be found if the radius of curvature of the lens is known, or, conversely, if $\lambda$ is known, the value of $R$ for the lens could be determined (as, for example, if a source of light of a known wavelength were used).

## APPARATUS _____

1. Sodium light source
2. Three optically flat glass plates, about 7 cm square
3. Measuring microscope
4. Thin strip of steel about 0.0075 cm thick

5. Converging lens, 20- to 25-cm focal length
6. Centimeter scale
7. Micrometer caliper
8. Stand and clamps
9. Sheet of black optical paper

## PROCEDURE _____

**1.** Measure the thickness of the thin strip of steel with the micrometer caliper and record the reading.

**2.** Be sure that the glass plates are clean. Be careful not to get finger marks on the surfaces that are to be placed together. Lay the sheet of optical paper on the microscope base and bend it up to form a backstop for the incident light as shown in Fig. 45.2. Hold the paper down on the base by laying one of the

glass plates on it. Then carefully lay the other plate down flat upon the first with the best surfaces in contact, so that the upper plate projects slightly at the thin end of the air film, again as shown in Fig. 45.2. Separate the plates at the other end by the thin steel strip, placing the edge of the strip parallel to the ends of the glass plates.

**3.** Measure the length $L$ of the air film (the distance between the edge of the steel strip and the line along which the two plates rest on each other) and record this reading.

4. Place the support stand beside the measuring microscope, mount the third glass plate in the clamp, and attach this clamp to the stand with a right-angle clamp so that the plate is held under the microscope objective but above the other two plates at an angle of 45° to the horizontal. See Fig. 45.2, which shows the complete experimental arrangement. Set up the sodium light source so that it shines horizontally into the 45°-plate, which serves as a partial mirror to reflect the light down onto the two plates forming the air film but allows the light reflected from the film to pass through to the microscope. Note that much of the light from the source passes through the 45°-plate and is absorbed by the optical paper backstop, and some of the light reflected from the air film is reflected back to the source instead of being transmitted. These losses result in a decrease in the intensity of the observed pattern, but as the light source is very strong, there is plenty of intensity remaining for observation.

5. Turn the source on and wait for it to warm up so that a brilliant yellow light is produced. Then look through the microscope and adjust its focus (by sliding the microscope barrel up and down in its mount) until you see a sharp image of the interference fringes. They consist of a series of dark parallel lines across the microscope's field of view and are perpendicular to the length of the air film (parallel to the edge of the steel strip). Adjust the eyepiece focus until you also see a sharp image of the microscope cross hairs, and rotate the microscope barrel until one of the cross hairs (the double one if your microscope is so equipped) is parallel to the dark fringes.

6. Your measuring microscope has a calibrated micrometer screw that allows you to traverse the cross hairs across the fringe pattern through distances that can be measured to high precision. Begin by setting the parallel cross hair right on top of one of the dark fringes toward the left side of the image. If your microscope has a double cross hair, set it so that the dark fringe falls in the center of the space between the pair. Double cross hairs are sometimes provided because this setting can then be made with higher precision. Observe and record the reading of your microscope's micrometer dial. Then traverse the cross hairs until you come to the tenth dark fringe to the right of the one you started from. Set the parallel cross hair exactly on this fringe and again read the micrometer dial setting. Note that you will have turned the micrometer dial through many complete revolutions in traversing ten fringes. The best way to take account of this is to count the number of times you turned the dial through zero, multiply this num-

ber by the number of dial divisions in one revolution (usually 10 large or 100 small divisions), and add on the final dial setting to obtain the final reading to be recorded.

7. Repeat Procedure 6 twice to obtain a total of three sets of measurements of the distance spanned by ten dark fringes.

8. Carefully remove the upper one of the two horizontal glass plates and the steel strip. Wipe off the lens with a damp cloth or lens paper to be sure it is free of lint, dust, and finger marks, and place it on the glass plate that is still under the microscope. If the lens is plano-convex instead of double convex, be sure the convex side is down, that is, in contact with the plate surface, and position the lens directly below the microscope. Look through the microscope and adjust its focus until you see a sharp image of the series of dark circles that constitute Newton's rings. Be careful about this observation: If you did not get the lens in the right position under the microscope, you may miss the ring pattern entirely, and in any case you may have to move the lens so that the center of the pattern is in the center of your field of view. Moreover, when you have the pattern in focus, there should be a dark spot at the center. If there is not, it means that either the lens or the glass plate is still dirty, even though the dirt may be just a speck of dust. Again clean them carefully until you get a sharp pattern of Newton's rings with a dark spot in the middle.

9. Measure the diameter of the first Newton's ring (the first dark circle around the central dark spot) by moving the cross hairs in the measuring microscope until the vertical one is tangent to the circle on the left. This cross hair is the double one if your microscope is so equipped, and in this case you should position the pair so that the dark, rather thick circular line just comes into the space between them. In the case of a single cross hair, position it so that it is tangent to the center line of the circle rather than to its outer edge. Read and record the micrometer dial setting. Then traverse the cross hairs to the right until the vertical one is tangent to the circle on the right in the same manner as on the left. Count the number of times you turned the micrometer dial through zero, note the final dial setting, and record the final reading as the number of turns multiplied by the divisions per turn plus the final setting, as you did in Procedure 6.

10. Repeat Procedure 9 twice to obtain a total of three sets of measurements of the diameter of the first dark ring.

11. Repeat Procedures 9 and 10 for the second, third, fourth, and fifth rings.

12. Carefully remove the lens from your apparatus and measure its focal length by the method of Procedure 1 in Experiment 43.

## DATA _____

Thickness $T$ of the thin steel strip    _____        Length $L$ of the air film        _____

**Micrometer dial readings for the width spanned by ten fringes:**

| Trial | Initial setting | Final setting | Difference | | Deviation |
|-------|-----------------|---------------|------------|--------|-----------|
|       |                 |               | Image | Actual | |
| 1 | | | | | |
| 2 | | | | | |
| 3 | | | | | |
| Average | | | | | |

Measured wavelength of sodium
   light        _____        Percent error        _____

**Diameters of Newton's rings:**

| Ring no. $m$ | Trial | Initial setting | Final setting | Diameter | | Actual radius | Actual radius$^2$ |
|--------------|-------|-----------------|---------------|----------|--------|---------------|-------------------|
|              |       |                 |               | Image | Actual | | |
| 1 | 1 | | | | | | |
|   | 2 | | | | | | |
|   | 3 | | | | | | |
|   | Average | | | | | | |
| 2 | 1 | | | | | | |
|   | 2 | | | | | | |
|   | 3 | | | | | | |
|   | Average | | | | | | |
| 3 | 1 | | | | | | |
|   | 2 | | | | | | |
|   | 3 | | | | | | |
|   | Average | | | | | | |
| 4 | 1 | | | | | | |
|   | 2 | | | | | | |
|   | 3 | | | | | | |
|   | Average | | | | | | |
| 5 | 1 | | | | | | |
|   | 2 | | | | | | |
|   | 3 | | | | | | |
|   | Average | | | | | | |

Value of $R$ from slope of graph        _____        Measured focal length of lens        _____
Calculated focal length of lens        _____        Percent error        _____

## CALCULATIONS

**1.** Subtract the initial settings of the micrometer dial from the final settings recorded in Procedures 6 and 7. Note that the final setting is the dial reading plus the total number of divisions turned through from the zero *before* the initial dial setting, so that subtraction of this initial setting gives you the total number of dial divisions turned through in traversing the ten fringes. Be careful of your units. For example, for the Gaertner M101 or M101A microscope equipped with the M202 eyepiece, each *small* division on the micrometer dial represents a cross-hair motion of 0.0025 mm. If you are using a different microscope, your instructor will give you its calibration. Enter your differences between initial and final dial settings along with the proper units in the *image* column in the pertinent table in the data sheet.

**2.** The measurements made in Procedures 6 and 7 and hence the width spanned by ten fringes found in Calculation 1 are all distances along the *image* of the interference pattern formed by the microscope's objective lens. They are therefore all magnified by the magnification produced by that lens and must be divided by that magnification to get the actual distances along the interference pattern itself. Divide each of your results from Calculation 1 by this factor to get the actual distances, and record them in the appropriate column. For the Gaertner M101 and M101A microscopes the magnification factor is 3.2; if you are using a different microscope, this factor will be supplied by the instructor.

**3.** Compute the average of your three actual distances spanned by ten fringes. Then compute the deviation of each distance from the average and find the average deviation (a.d.).

**4.** Find the spacing between adjacent fringes by dividing your result from Calculation 3 by 10, and use this spacing and your data from Procedures 1 and 3 in Equation 45.4 to determine the wavelength of the light from your sodium lamp.

**5.** Use the a.d. found in Calculation 3 to find the error to be attached to your result in Calculation 4. Now compare this result with the known value, 5893 Å, for the average wavelength of the sodium doublet by noting whether this known value falls within the limits of error you just determined. Also calculate the percent error by which your measured value differs from the known or accepted value.

**6.** Subtract the initial micrometer dial settings from the final setting obtained in Procedures 9–11 to obtain the diameters of the five Newton's rings you measured just as you did for the set of ten fringes in Calculation 1. Enter your results as the diameters of the images.

**7.** Divide your results from Calculation 6 by the objective magnification as you did in Calculation 2 to obtain the actual ring diameters. Then divide each actual ring diameter by 2 to get the actual radius $r_m$ and square each of these radii to get $r_m^2$. Enter all these results in the appropriate data table boxes.

**8.** Calculate the average radius and an average of the radius squared for each ring from $m = 1$ to $m = 5$.

**9.** Plot your values of $r_m^2$ as ordinates against the corresponding values of the index $m$ as abscissas. According to Equation 45.7, this graph should be a straight line of slope $\lambda R$, where $\lambda$ is the wavelength of the light used and $R$ is the radius of curvature of the lens. Note that the origin should be a point on your graph.

**10.** Find the slope of the graph you made in Calculation 9 and from it determine $R$, using 5893 Å for $\lambda$. Then use the lens maker's equation (Equation 42.2) to find the focal length of your lens and compare this result with the focal length measured in Procedure 12 by calculating the percent error. *Note:* If your lens is plano-convex, the radius of the flat side is infinite and its reciprocal zero. If your lens is double convex, assume that both sides have the same radius of curvature. Also assume that your lens is made of crown glass for which the index of refraction is given in Table XII at the end of the book unless your instructor gives you a better value. Note that you should not be surprised if you find an appreciable difference between your calculated and measured values of the lens's focal length, especially if you are uncertain about the index of refraction of the lens material.

## QUESTIONS _____

**1.** Explain how the method used in this experiment could be employed to test whether a surface is optically plane or not.

**2.** Explain why the fringes sometimes run diagonally across the plates.

**3.** Suppose that white light were used to produce the interference fringes in this experiment. What would be the appearance of the fringes produced?

**4.** Two plane glass plates touch at one edge and are separated at the opposite edge by a strip of tinfoil. The air wedge is examined with sodium light reflected normally from its two surfaces, and a total of 35 dark interference fringes are observed. Compute the thickness of the tinfoil.

**5.** How is this experiment evidence for a wave theory of light? Explain.

**6.** Explain why interference patterns caused by light reflected from the top surface of the upper plate or from the bottom surface of the lower plate in Fig. 45.1 are not observed.

**7.** In the setup to observe Newton's rings, why should you see a black spot in the center of the pattern?

**8.** Did you get good agreement between the value of your lens's focal length obtained from observing Newton's rings and the directly measured value? If not, discuss why you might expect a disagreement.

**9.** Prove that the area between two adjacent Newton's rings is a constant regardless of which pair of adjacent rings is chosen.

**10.** A certain double convex, crown-glass lens has a focal length of 30 cm. Find the radius of the tenth dark Newton's ring viewed with sodium light ($\lambda = 5893$ Å).

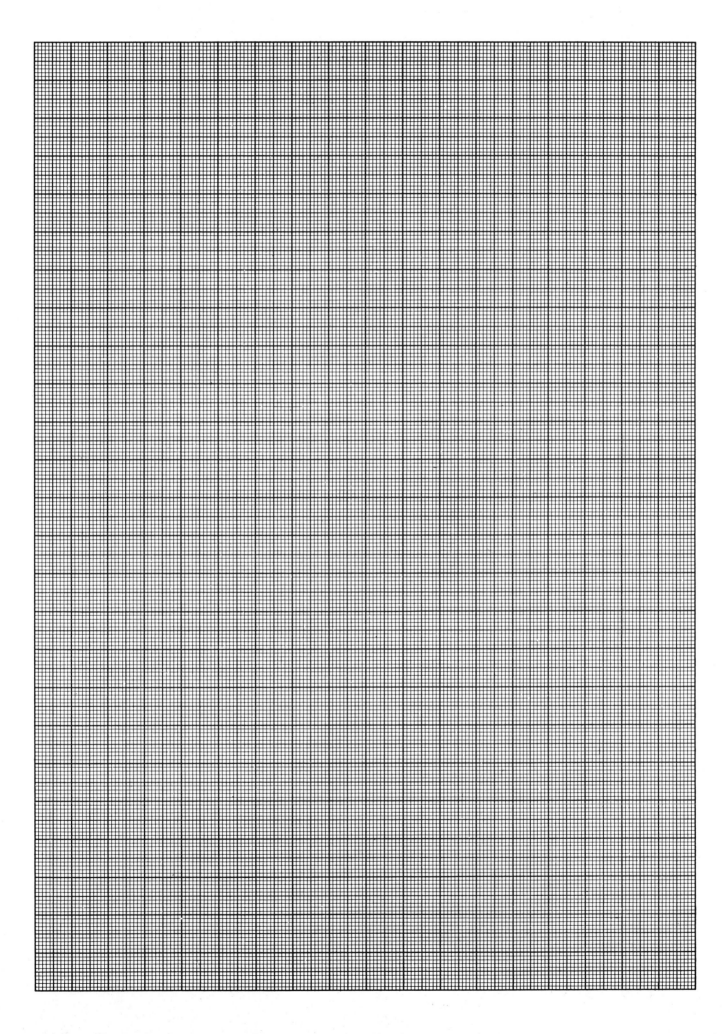

# The Diffraction Grating 46

The diffraction grating is a most important and interesting optical device. It is among the simplest devices for producing spectra so that the wavelengths contained in a given light beam may be measured and has the advantage of making calibration with a known spectrum unnecessary. The object of this experiment is to study the operation of a diffraction grating and, in particular, to see how it may be used to measure the wavelength of light from a helium-neon laser. The wavelength produced by such a laser is known to great accuracy, so that an excellent check on a grating's ability to measure wavelength can be made.

## THEORY

The diffraction grating provides the simplest and most accurate method for measuring wavelengths of light. It consists of a very large number of fine, equally spaced parallel slits. There are two types of diffraction gratings: the reflecting type and the transmitting type. The lines of the reflection grating are ruled on a polished metal surface and the incident light is reflected from the unruled portions. The lines of the transmission grating are ruled on glass; the unruled portions of the glass act as slits. Gratings usually have about 10,000 to 20,000 lines per inch. The grating used in this experiment is a transmission grating replica.

The principles of diffraction and interference are applied to the measurement of wavelength in the diffraction grating. Let the broken line in Fig. 46.1 represent the edge view of a magnified portion of a diffraction grating, and let a beam of parallel rays (a plane wave) of monochromatic light impinge on the grating from the left. By Huygens's principle, the light spreads out in every direction from the apertures of the grating, each of which acts as a separate new source of light. Because the grating slits are very narrow, they act as line sources and the emitted secondary wavelets are, as Fig. 46.1 shows, cylindrical waves. The envelope of these wavelets determines the position of the complete advancing wave. The figure shows the instantaneous positions of several successive wavelets after they have advanced beyond the grating. Lines drawn tangent to these wavelets connect points in phase; hence they represent the new wave fronts. One of these wave fronts is tangent to wavelets that have all advanced the same distance from the slits; this wave front is parallel to the original one. A converging lens placed in the path of these rays would form the central image. Another wave front is tangent to wavelets whose distances from adjacent slits differ by one wavelength. This wave front advances in direction 1 and forms the first-order spectrum. The next wave front is tangent to wavelets whose distances from adjacent slits differ by two wavelengths. This wave front advances in direction 2 and forms the second-order spectrum. Spectra of higher orders will be formed at correspondingly greater angles.

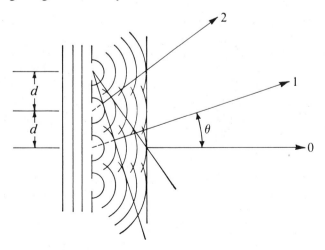

**Figure 46.1** *The diffraction grating*

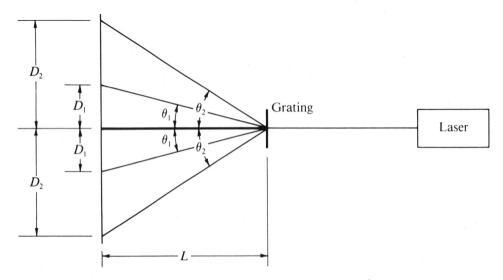

**Figure 46.2** *Experimental arrangement for the diffraction grating*

Looking at Fig. 46.1, we note that ray 1 has been drawn perpendicular to its wave front at the wave front's point of tangency with a Huygens wavelet that is coming from the third slit down from the top. A projection of this ray backward will pass through the center of the slit, because a line drawn perpendicular to a tangent at the point of tangency is a radius of the circle in question. In the right triangle of ray 1, the associated wave front, and the line representing the grating, we see that the angle between the wave front and the grating is the same as $\theta$, the angle between rays 1 and 0. Consideration of this triangle reveals that $\sin \theta = 2\lambda/2d = \lambda/d$, where $d$ is the distance between slits. Similarly, for ray 2, which for convenience has been drawn to the center of the second slit down, we note that $\sin \theta = 2\lambda/d$, where now $\theta$ is the angle between rays 2 and 0. In general, we will find that $\sin \theta = n\lambda/d$, or

$$n\lambda = d \sin \theta \qquad (46.1)$$

where $n$ is an integer (0, 1, 2, 3, etc., not to be confused with the index of refraction) and is called the *order* of the spectrum. In this equation, $\lambda$ is of course the wavelength of the light and will be in the same units as $d$, the spacing between the grating lines. The wavelength of visible light is usually expressed in angstrom units; one angstrom unit is $10^{-8}$ cm. However, the nanometer ($10^{-9}$ m, hence $10^{-7}$ cm or 10 Å) is becoming increasingly popular in expressing optical wavelengths. The grating constant $d$ is obtained by noting the number of lines per inch or per centimeter marked on the grating and taking the reciprocal to find the spacing between lines. This spacing will then be in inches or centimeters, so that a conversion must be made to obtain $\lambda$ in appropriate units.

In this experiment, a helium-neon laser will be used to provide a narrow beam of highly monochromatic light ($\lambda = 6328$ Å). A plan view of the experimental arrangement is shown in Fig. 46.2. The laser beam strikes the grating normally and the direct ray proceeds in a straight line to the screen. The first-order ($n = 1$) diffracted rays are deflected through angles $\theta_1$ on either side of the central ray and strike the screen at points spaced $D_1$ on either side of the central spot. Similarly, the second-order ($n = 2$) diffracted rays each make angle $\theta_2$ with the central ray and strike the screen at points separated from the central spot by $D_2$. These angles are determined by measuring $D_1$, $D_2$, and $L$, the distance between the grating and the screen, and noting that $\tan \theta_1 = D_1/L$ and $\tan \theta_2 = D_2/L$.

Because the laser light contains only one wavelength, only one pair of spots will appear on the screen in each order. However, if the experiment is run with light that is not monochromatic, there will be as many pairs of spots in each order as there are different wavelengths in the light from the source, the diffracting angle $\theta$ for each wavelength being given by Equation 46.1. We will observe these results when white light from a tungsten lamp is substituted for the laser light.

## APPARATUS

1. Low-power helium-neon laser
2. Diffraction grating
3. Optical bench
4. Long-focus converging lens (about 25 cm)
5. Ground-glass screen
6. Lens holders and optical bench clamps
7. Meter stick
8. Masking tape
9. Tungsten lamp
10. Stand and clamps for mounting the light source

# PROCEDURE _____

**1.** Set up the optical bench and mount the ground-glass screen at one end. Mount the grating about 30 cm in front of it using one of the lens holders and optical bench clamps. The laser is then mounted so that the beam is directed normally at the grating, as shown in Fig. 46.2.

**2.** Turn the laser on and observe the central spot on the screen. CAUTION: The low-power laser used in this experiment cannot hurt you even if the beam is allowed to fall on your skin for a prolonged period. However, *you must not look into it or allow anyone else to do so.* Keep the laser beam down below eye level and do not let it shine in anyone's face including your own.

**3.** With the usual size ground-glass screen and a grating of about 15,000 lines per inch placed 30 cm away, the first-order spots will appear on the screen but the second-order ones will miss it on either side. To display both orders and facilitate measurement of the separations $D$, tape the meter stick to the screen so that the central spot falls approximately at the 50-cm mark and the diffracted spots are displayed right on the stick. Adjust the orientation of the screen so that it and the meter stick are perpendicular to the central ray. Do this by rotating the screen in its mount until the two second-order spots are equidistant from the central spot.

**4.** Read and record the positions of the two first-order spots on the meter stick.

**5.** Similarly, read and record the positions of the two second-order spots. These spots will be fainter than the first-order ones; you may need to darken the room in order to see them clearly.

**6.** Measure and record the distance $L$ from the grating to the surface of the meter stick. If the length graduations on the optical bench are used for this purpose, allowance must be made for the thickness of the meter stick and one half the thickness of the ground-glass screen.

**7.** Shut off the laser, replace it with the tungsten lamp, and mount the converging lens on the optical bench between the lamp and the grating. Align the lamp and the lens so that the light passing through the lens falls on the middle of the grating. The lens should be moved back and forth along the optical bench until the central spot observed on the screen is a focused image of the lamp filament. The lamp must be mounted at a distance greater than the lens's focal length behind the grating to make this adjustment possible.

**8.** Examine the continuous spectrum displayed on the meter stick in first order, noting the relative positions of the different colors. The second-order spectrum will probably be too faint to be seen. Record the positions of the inner ends of the first-order spectra observed on either side of the central white spot.

**9.** Record the positions of the outer ends of the displayed spectra.

# DATA _____

Number of lines per inch on the
    grating      _____

Grating constant $d$ in centimeters    _____

Distance $L$ from grating to meter
    stick in centimeters      _____

| | Meter stick reading | | Calculated values | | | | | | |
|---|---|---|---|---|---|---|---|---|---|
| **Source** | **Right** | **Left** | **Difference** | **D** | **tan θ** | **θ** | **sin θ** | **λ** | **% Error** |
| Laser (first-order) | | | | | | | | | |
| Laser (second-order) | | | | | | | | | |
| Lamp (violet end) | | | | | | | | | |
| Lamp (red end) | | | | | | | | | |

Range in wavelength of visible
    spectrum      _____

## CALCULATIONS _____

**1.** Compute the grating constant (that is, the distance between the lines on the grating) in centimeters from the number of lines per inch stated on the grating.

**2.** Find $D_1$, the displacement of the first-order spot from the central spot, by computing the difference between the two spot positions recorded in Procedure 4 and dividing by 2.

**3.** Calculate $\theta_1$, the deviation of the first-order diffracted laser beam, from your values of $D_1$ and $L$.

**4.** Calculate the laser light wavelength by using Equation 46.1 with $n = 1$, $\theta = \theta_1$ (from Calculation 3), and the grating constant $d$ found in Calculation 1. Note that with $d$ in centimeters, $\lambda$ will come out in centimeters and must be multiplied by $10^8$ to give angstrom units.

**5.** Find the percent deviation of your computed value of the laser light wavelength from the known value (6328 Å).

**6.** Repeat Calculations 2–5 for the second-order diffracted laser light, using the data of Procedure 5.

**7.** Repeat Calculations 2–4 for the violet end of the continuous spectrum obtained with the tungsten lamp, using the data of Procedures 7 and 8.

**8.** Repeat Calculations 2–4 for the red end of the continuous spectrum, using the data from Procedure 9.

**9.** Calculate and record the range in wavelengths of the visible spectrum as measured in this experiment.

## QUESTIONS _____

**1.** Why does a grating produce several spectra, while a prism produces only one?

**2.** (a) Discuss the difference between the continuous spectrum obtained with a grating and one obtained with a prism. (b) Which color is deviated most in each case? Why?

**3.** A monochromatic beam of light impinges normally on a grating and forms an image of the source slit 10 cm from the central image on a screen. The screen in 50 cm from the grating and is set parallel to it. The grating has 10,000 lines per inch. What is the wavelength of the light?

**4.** Why are the higher-order spectra more accurate than the first-order spectrum in determining wavelengths of light?

**5.** A grating having 15,000 lines per inch is used to produce a spectrum of the light from a mercury arc shining through a slit. What is the angle between the first-order green line and the second-order green line? See Experiment 44 for the wavelength of the mercury green line.

**6.** Consider two rays coming from adjacent slits in a grating of grating constant $d$ and intersecting at the point where they strike a distant screen. Use the rule for constructive interference (see Experiment 45) to derive Equation 46.1. *Note:* Because the distance from the grating to the screen is very much larger than the separation between the slits, the two rays are almost parallel.

**7.** (a) Why in the present experiment did you not observe a third-order diffracted beam of laser light? (b) What is the highest order in which you would expect to be able to view the mercury green line (see Experiment 44 for $\lambda$) when using a grating having 15,000 lines per inch?

**8.** Is it possible for the red end of the visible spectrum viewed in first order to overlap the violet end viewed in second order? Assume that the visible spectrum runs from 7000 to 4000 Å.

**9.** What would be the effect of the plane of the grating not being perpendicular to the beam of light coming from the laser? If you can, show how Equation 46.1 should be modified to take account of an angle $\phi$ between this beam and the normal to the grating plane.

**10.** What is the effect on the observed spectrum of using a larger grating, that is, one having the same grating constant but a greater total number of lines? Assume an incident beam whose cross section covers the grating's entire active surface. *Hint*: How does the pattern produced in a Young's experiment, which, in effect, uses a grating with only two slits, differ from the pattern you observed with your grating of many thousands of slits?

# Polarized Light $47$

Experiments on interference and diffraction show that light is composed of waves, but they do not offer any evidence as to whether these waves are transverse or longitudinal. However, experiments on polarization show that light waves must be transverse. Polarization, which refers to the direction assumed by the wave vector in the plane perpendicular to the direction of propagation, is only possible with transverse waves. The purpose of this experiment is to study some of the fundamental phenomena of polarized light.

## THEORY

Light waves consist of transverse electromagnetic vibrations that travel with finite velocity. The electric vibrations and the magnetic vibrations are perpendicular to the direction of propagation of the light and also perpendicular to each other. In ordinary light, the electric vibrations occur in all directions in a plane perpendicular to the direction of propagation as shown in Fig. 47.1(a). This light is unpolarized. It is possible to produce plane-polarized light, that is, light in which the electric vibrations all occur in some one direction perpendicular to the light ray, that is, to the direction of propagation. Vertical plane polarization is illustrated in Fig. 47.1(b) and horizontal plane polarization in Fig. 47.1(c). Notice that by definition of a transverse wave, in all these cases the direction of the electric vibrations (the electric vector **E**) lies in a plane perpendicular to the ray. Because the electric vibrations are restricted to this plane, it is called the *plane of polarization*.

Light may be plane-polarized by reflection from a nonconducting reflector or by passing through a doubly refracting crystal. When light is reflected from the surface of a transparent dielectric medium such as glass or water, it is polarized to some extent if the angle of incidence is greater than $0°$ but less than $90°$. The *reflected* light is *completely* polarized if the ray reflected from the medium is at right angles to the refracted ray proceeding into the medium. Calculation of the angle of incidence $i_p$ for which this condition is satisfied may be carried out by reference to Fig. 47.2. Clearly, $r_1 + r_2 + 90° = 180°$ or $r_1 + r_2 = 90°$. But by the law of reflection, $r_1 = i_p$, and by the law of refraction (Snell's law, Equation 42.1) $\sin i_p = n \sin r_2$. Then, since $r_2 = 90° - r_1 = 90° - i_p$, $\sin i_p = n \sin(90° - i_p) = n \cos i_p$ and there results

$$n = \frac{\sin i_p}{\cos i_p} = \tan i_p \qquad (47.1)$$

This is *Brewster's law*, and the angle of incidence $i_p$ whose tangent is equal to the index of refraction of the medium and for which the reflected beam is com-

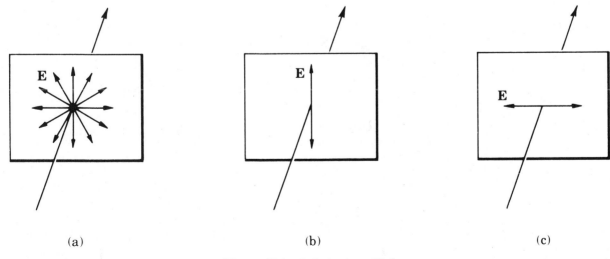

      (a)                            (b)                           (c)

**Figure 47.1** *Polarization of light*

pletely polarized is called *Brewster's angle*. For a piece of flat, unsilvered glass it is about 57°. Note that the direction of polarization of the reflected beam is parallel to the reflecting surface (perpendicular to the plane of the paper in Fig. 47.2). Note also that although this beam is completely polarized, the refracted beam in the dielectric medium is not. This situation may be understood by considering that the incident (unpolarized) light is made up of two components, one polarized perpendicular to the plane of the paper in Fig. 47.2 and the other lying in that plane. The former is largely reflected, although a part proceeds into the new medium, but at Brewster's angle *all* of the component polarized in the plane of the paper is transmitted into the medium. Thus the transmitted beam is "enriched" in this latter component, while the reflected beam contains none of it.

In a doubly refracting crystal such as calcite, light propagates at two different velocities depending on whether it is polarized along or perpendicular to the so-called optic axis. The electric vibrations of a light ray propagating along the optic axis are perpendicular to this axis even if the light is unpolarized, because light waves are transverse. Such a ray travels with a speed $v_\perp$ regardless of its polarization. However, if a ray is directed perpendicular to the optic axis in a doubly refracting crystal, we must consider two components. One is polarized perpendicular to the optic axis and travels with the same speed $v_\perp$ as does a wave traveling along the optic axis and thus is called the *ordinary ray*. The other component, polarized along the optic axis, travels with a different speed $v_\parallel$ and is called the *extraordinary ray*. The crystal then exhibits two indices of refraction, $n_\perp = c/v_\perp$ and $n_\parallel = c/v_\parallel$. If a ray of unpolarized light travels through a doubly refracting crystal in a direction making some angle (other than 0° or 90°) with the optic axis, there will still be an ordinary ray polarized perpendicular to this axis that obeys Snell's law in the usual manner. In this case, however, the

extraordinary ray's electric vibrations are neither parallel nor perpendicular to the optic axis and thus involve both indices of refraction. Therefore, this ray does not obey the simple form of Snell's law and is consequently bent away from the ordinary ray. Hence a doubly refracting (also called a birefringent) crystal has the property of dividing a beam of unpolarized light into two beams plane-polarized at right angles to each other.

Some crystals have the property of transmitting a light beam polarized along the optic axis and absorbing one polarized at right angles to this axis. Unpolarized light incident perpendicular to the optic axis can be plane-polarized in this manner. This property is called *dichroism*. Tourmaline is the most common example of a dichroic crystal.

Natural crystals are usually too small to be used in an experiment where a wide beam of polarized light is needed; for this purpose, Polaroid sheets are used. In a Polaroid, very minute crystals are distributed uniformly and densely in a cellulose film mounted between two glass plates. The individual crystals are needle-shaped and submicroscopic in size and are turned so that their axes are parallel by stretching the film before it is mounted on glass. Like tourmaline, Polaroids are dichroic and produce polarized light in the same manner.

A Polaroid has an optic axis in the plane of the sheet along which the electric vibrations of a normally incident light ray must lie in order for the ray to get through without absorption. If there is an angle $\theta$ between the optic axis and the direction of the electric vibrations, as shown in Fig. 47.3, then the electric vector **E** in this direction may be broken up into a component $E_\perp = E \sin \theta$, which is perpendicular to the optic axis and gets absorbed by the Polaroid, and a component $E_\parallel = E \cos \theta$, which is parallel to the

**Figure 47.2** *Brewster's angle*

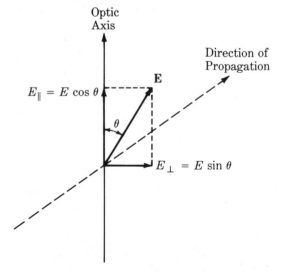

**Figure 47.3** *Components of **E** with respect to the optic axis*

optic axis and is therefore transmitted with little or no absorption. Since the intensity of a light beam is proportional to the square of the amplitude of the electric vibrations, the ratio of the intensity of the polarized light transmitted by the Polaroid to the incident intensity is

$$\left(\frac{E_\parallel}{E}\right)^2 = \cos^2\theta \qquad (47.2)$$

This result is called the law of Malus after its discoverer, E. L. Malus (1775–1812), and states that if plane-polarized light is normally incident on a Polaroid or other dichroic material, the transmitted intensity is proportional to the square of the cosine of the angle between the optic axis and the direction of polarization. Note that the intensity becomes zero when this angle is 90°.

## APPARATUS

1. Two Polaroid disks, each mounted in a holder in which it can rotate
2. Optical bench with clamps and holders
3. Light source with lens to produce a parallel beam of light
4. Desk lamp
5. Photronic cell connected to a microammeter
6. Calcite crystal
7. Pile of thin glass plates
8. Black glass mirror
9. Piece of cellophane
10. Thin piece of mica
11. Unannealed glass plate
12. Annealed glass plate
13. Piece of celluloid
14. Iris diaphragm

## PROCEDURE

**1.** Place the light source at one end of the optical bench and, if a separate lens is used, mount it one focal length from the source. Place the photronic cell at the other end of the bench in such a position as to receive the light. Place one of the Polaroid disks on the optical bench in front of the photronic cell so that the light has to pass through it to get to the cell. This Polaroid is the analyzer. Rotate it and notice the reading of the microammeter. If the microammeter goes off scale, decrease the size of the beam of light by placing a diaphragm in front of the lens, regulating the size of the beam so that the microammeter reads nearly full scale. The light coming directly from the lamp is unpolarized. The intensity of the light transmitted by the analyzer should be nearly independent of the angle that its axis makes with the vertical.

**2.** Place the second Polaroid disk on the optical bench in front of the lens and about 4 in. away from it, in such a position that all the light passes through it. This is the polarizer. Place it with its axis in the vertical direction.

**3.** Now rotate the analyzer and take readings of the microammeter for different angular settings. Begin with the axis vertical (and hence parallel to the polarizer's axis), and take readings of the microammeter for angles 0°, 10°, 20°, 30°, etc., up to 180°. You will observe that the intensity of the light transmitted by the analyzer depends upon the angle between the axis of the analyzer and that of the polarizer.

**4.** Remove the photronic cell. Allow the light to pass through both Polaroids, leaving a space of about 6 in. between them. Set the two Polaroids for extinction, that is, so that the minimum amount of light passes through them. Now place a piece of cellophane in the space between the Polaroids, look at the cellophane through the analyzer, and observe the field of view. Fold the cellophane into several thicknesses and observe the effect on the color produced. Describe what you see in the space below.

**5.** Place a thin piece of mica in the space between the Polaroids, look at the mica through the analyzer, and observe the field of view. Again describe what you see in the space below.

**6.** Repeat Procedure 5 using the unannealed glass plate.

**7.** Repeat Procedure 5 using the annealed glass plate.

**8.** Repeat Procedure 5 using a piece of celluloid. Twist the celluloid slightly out of shape and observe the regions where strains are produced.

**9.** Remove the polarizer, light source, and lens from the optical bench. Place the black glass mirror in a holder on the optical bench and allow the unpolarized light from the desk lamp to fall on it. Rotate the mirror holder so as to make the angle of incidence about 55°. Observe the light reflected from the mirror by looking at it through the analyzer Polaroid. By rotating this Polaroid, notice whether the reflected light is polarized. Report your results in the space below.

**10.** Repeat Procedure 9 using a pile of thin glass plates as the mirror. Determine whether the reflected light is polarized and also whether the transmitted light is polarized. State your findings in the space below.

**11.** Make a dot on a piece of paper. Place a calcite crystal over the dot and look through it. You will observe two images. Rotate the crystal and notice what happens to the two dots. Now look at the dots through the calcite and a Polaroid. Rotate the Polaroid and notice the result. The two dots represent two beams of light coming from the calcite crystal. Determine whether these two beams are polarized and find the relative direction of vibration of the two beams. Report your results in the space below.

**12.** If the sun is shining, look at the blue light from the sky through the analyzer. Notice particularly the light coming from a direction roughly perpendicular to the sun's direction. Determine whether this light is polarized and report your results in the space below.

## DATA _____

| Analyzer angle $\theta$ | $\cos^2 \theta$ | Microammeter reading | Analyzer angle $\theta$ | $\cos^2 \theta$ | Microammeter reading |
|---|---|---|---|---|---|
| 0° | | | 100° | | |
| 10° | | | 110° | | |
| 20° | | | 120° | | |
| 30° | | | 130° | | |
| 40° | | | 140° | | |
| 50° | | | 150° | | |
| 60° | | | 160° | | |
| 70° | | | 170° | | |
| 80° | | | 180° | | |
| 90° | | | | | |

## CALCULATIONS _____

**1.** The percentage polarization is given by the expression

$$P = 100 \frac{I_{max} - I_{min}}{I_{max} + I_{min}}$$

where $P$ is the percentage polarization, $I_{max}$ is the largest value of the microammeter reading observed, and $I_{min}$ is the smallest value observed. Calculate the percentage polarization from your readings in Procedure 3.

**2.** Calculate and record the value of $\cos^2 \theta$ for each of the angles between the polarizer and analyzer optic axes used in Procedure 3.

**3.** Plot a graph using the microammeter readings obtained in Procedure 3 as ordinates and the values of $\cos^2 \theta$ as abscissas. Note that $\cos^2 \theta$ will be the same for $\theta = 80°$ and $100°$, for $\theta = 70°$ and $110°$, etc., and you should therefore get the same microammeter readings for the two angles in each of these pairs. Differences in these readings represent errors in the experiment. Plot all points and draw the best straight line through them.

## QUESTIONS

**1.** Mention some of the uses of polarized light.

**2.** Explain the difference between transverse waves and longitudinal waves.

**3.** What is meant by a plane-polarized beam of light?

**4.** (a) Describe the construction of a nicol prism. (b) Explain how it produces a polarized beam of light.

**5.** A so-called quarter-wave plate is a slice cut from a calcite or other birefringent crystal in such a way that the optic axis lies in the plane of the slice. The name comes from the fact that the thickness of the slice is made equal to $\lambda/4$, where $\lambda$ is the wavelength of the light to be used. Look up and learn what is meant by *circular polarization* and how a quarter-wave plate can be used to produce it. Then explain in your own words how a quarter-wave plate works.

**6.** (a) Could one have elliptical polarization? (b) If so, discuss ways in which elliptically polarized light might be produced. *Hint:* Review the discussion of lissajous figures in Experiment 35 and make the analogy.

**7.** Two Polaroids are set for extinction, that is, with their optic axes at right angles, and a third Polaroid is then placed between them. (a) Explain how the third Polaroid can be adjusted so that light can pass through the set of three even though no light can get through the original pair. (b) What setting of the third (middle) Polaroid results in maximum transmitted intensity?

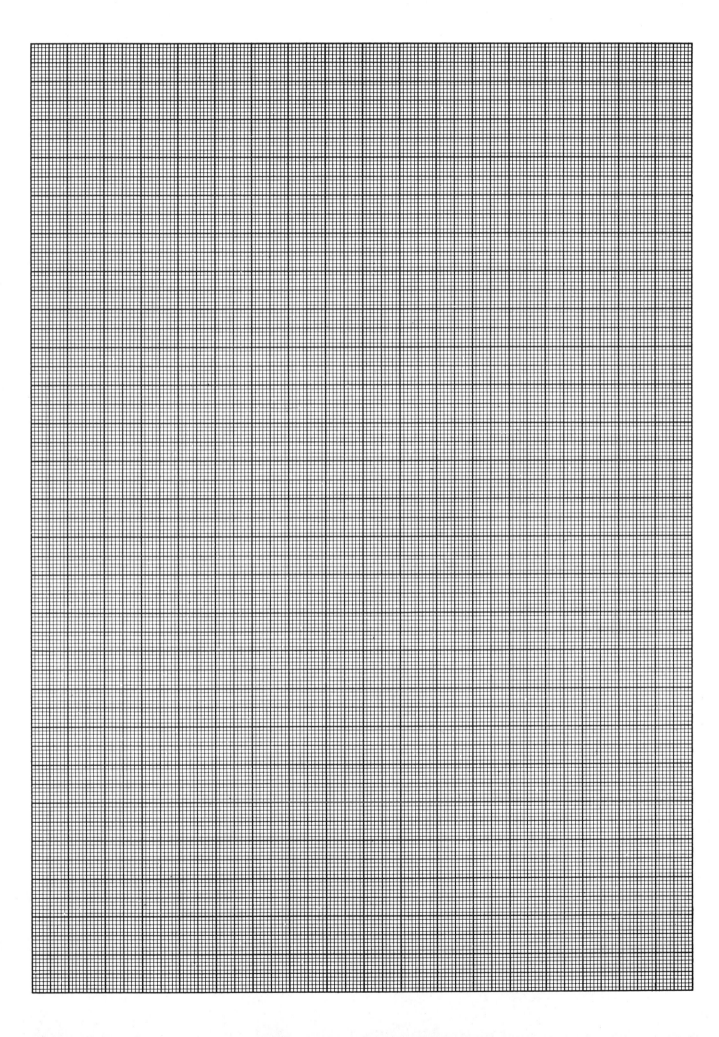

# Radioactivity and the Geiger Counter 48

The discovery of radioactivity has revolutionized our concept of the nature of the physical world and has given us the means by which we can study the innermost structure of an atom, that of the nucleus itself. Radioactive disintegration involves the actual transmutation of elements and the formation of new isotopes. This experiment contains a discussion of the nature of radioactivity and of the radiations from radioactive substances. One of the most important relations involved is the inverse square law of intensity. One purpose of this experiment is to show that the intensity of beta and gamma radiation varies inversely with the square of the distance from the source. Several methods are used to detect the radiation from radioactive sources; some of them will be discussed in this experiment. One of the most important—the Geiger-Müller counter—will be introduced and its operating characteristics studied.

## THEORY

In 1896 the French physicist Henri Becquerel (1852–1908) discovered that minerals containing uranium gave out a radiation that resembled X rays, since it could penetrate an envelope of black paper and affect a photographic plate inside. These rays could also produce ionization in air and were thus able to discharge electrified bodies. They could easily be detected with a sensitive electroscope and their intensity measured. This property of uranium was later found to be possessed by a number of the heavy elements, and the name radioactivity was applied to it. It was soon found that radioactivity is an atomic phenomenon—the activity of any given compound of uranium is proportional to the amount of uranium in the substance and does not depend on the physical or chemical condition of the substance.

Soon more powerful sources of radioactivity were discovered. Madame Marie Curie (1867–1934) and her husband, Pierre (1859–1906), began to suspect the existence of other radioactive elements in the uranium ores. After two years of hard labor, during which nearly a ton of uranium ore was worked over chemically, they discovered two new chemical elements, polonium and radium. Both of these elements were found to be thousands of times more radioactive than uranium.

The nature of the radiations from radioactive substances was investigated by Ernest Rutherford (1871–1937) and his collaborators. They found that the radiation consisted of three different kinds of rays. These were called $\alpha$ (alpha), $\beta$ (beta), and $\gamma$ (gamma) rays. They differ in their power of penetrating matter and of producing ionization and also in their behavior under the influence of electric or magnetic fields.

It is easy to show that these rays are different in nature. Suppose a piece of radioactive material is placed at the bottom of a small hole drilled in a lead block, as shown in Fig. 48.1. A photographic plate is placed at a convenient distance above the block, and the whole apparatus is highly evacuated. This arrangement produces a narrow beam of rays emerging from the top of the block. A strong magnetic field is then applied at right angles to the plane of the figure. When the photographic plate is developed, three distinct spots are found. One of these was produced by the alpha rays, which were deflected slightly to one side; from the known direction of the magnetic field, we know that the alpha rays consist of positively charged particles. The second spot was produced by the beta rays, which were deflected more strongly in the opposite direction; this shows that they consist of negatively charged particles. The third spot was produced by the gamma rays, which were not deflected at all; this shows that they do not carry any electrical charges.

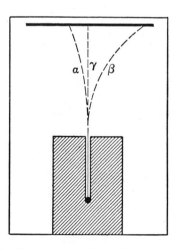

**Figure 48.1** *Separation of alpha, beta, and gamma rays by a magnetic field*

441

In 1907 Rutherford showed that *alpha particles* are the same as the nuclei of helium atoms. Each particle has a mass about four times that of a hydrogen atom and each carries a positive charge equivalent to two electron charges. Alpha particles are emitted with a speed of about one twentieth the velocity of light. The *beta particles* have the same mass and the same charge as electrons; they are in fact electrons emitted from the nuclei of radioactive atoms with speeds that nearly approach the velocity of light. The *gamma rays* consist of electromagnetic waves of extremely short wavelength, about one hundredth that of X rays, and are similar to X rays in their properties.

It was found that no physical or chemical process, such as raising the temperature or producing a new chemical compound with some other radioactive substance, could change the radioactivity of a given sample. As a result, it was concluded that radioactivity is a nuclear process and that the emission of a charged particle from the nucleus of an atom results in leaving behind a different atom, occupying a different place in the periodic table. Radioactivity, then, involves the transmutation of elements.

When an alpha particle is given off, the original atom loses four units of mass and two units of charge and thus becomes a new atom of lower atomic weight and lower atomic number. When a beta particle is given off, the new isotope produced will have the same mass number as the parent atom, since the mass of an electron is negligible in comparison with the mass of the nucleus, but the atomic number will be raised by one, since the change produced is equivalent to increasing the net positive charge on the nucleus. No new elements are produced in the emission of gamma rays, since both the atomic number and the atomic weight of the element remain the same. Gamma rays are produced during transitions from higher to lower nuclear energy states of the same isotope. The emission of alpha particles may be explained by assuming that some of the nuclear protons and neutrons may exist together as alpha particles within the nucleus. Explaining the emission of beta particles is not so simple. According to the present belief, electrons do not form part of the nuclear structure. However, when a beta particle is emitted, a neutron within the nucleus becomes a proton and emits an electron.

The occurrence of radioactive disintegration is statistical in nature. For any individual atom of a radioactive substance, there is a definite probability that it will disintegrate within a certain time interval. This statement means that when a very large number of atoms are involved, the emission obeys the laws of probability. The rate of disintegration is proportional to the number of atoms present. Accordingly, when the mathematical relation is set up and solved, the number of atoms of a particular radioactive species decreases exponentially according to the relation

$$N = N_0 e^{-\lambda t} \qquad (48.1)$$

where $N$ is the number of atoms present at any time $t$, $N_0$ is the number present at $t = 0$, $e$ is the base of natural logarithms, and $\lambda$, often called the decay constant, is the probability per unit time that a particular atom will disintegrate.

A convenient way of representing the rate of radioactive decay is by the *half-life* of a radioactive element, which is defined as the time required for half of a given quantity of the element to disintegrate into a new element. The rate at which a given quantity of a radioactive element decays is found by measuring the intensity of radiation given off by the sample over a period of time and plotting a graph like the one shown in Fig. 48.2. This figure shows the decay curve for the radioactive element polonium ($Po^{210}$). Thus, for polonium the intensity drops to half of its original value in 140 days; in another 140 days the intensity again drops to half the value it had at the beginning of this time interval, and so on. Since it would take an infinite time for the intensity to drop to zero, the rate of decay is usually specified by the half-life of the atom, and this is one of the important constants to be measured. The same law of decay applies to all the other radioactive elements, except that each one has its own half-life and its own decay constant. The half-life may be extremely short, such as less than one millionth of a second for polonium-212 ($Po^{212}$, sometimes called thorium C′), or it may be quite long, such as 1620 years for radium itself.

The intensity of the rays emitted by a radioactive substance decreases as the distance from the source is increased. Because the rays spread out in all directions, their intensity follows *the inverse square law of radiation*. This law states that the radiation received on a small surface held perpendicular to the rays

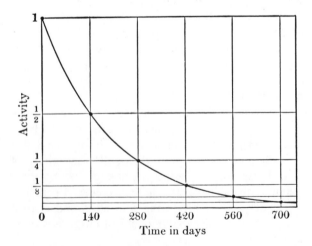

**Figure 48.2** *Decay curve for the radioactive element polonium (Po²¹⁰)*

varies directly with the source's intensity and inversely as the square of the surface's distance from the source. This law applies strictly for a point source or, in practice, when the distance is very great compared to the source size.

Experimentally, the radiations given off by radioactive substances may be detected and measured in several ways. These include the use of photographic plates, the ionization chamber, the Wilson cloud chamber, and the Geiger-Müller counter. With a photographic plate, the rays that penetrate the sensitive emulsion produce a latent image that can be developed. The density of the film is a measure of the intensity of radiation. The Wilson cloud chamber is another instrument used to study penetrating radiation. The cloud chamber works on the principle that when air saturated with water vapor is expanded suddenly, the cooling effect causes the formation of a cloud of tiny water drops. Drops of water form around dust particles or any ionized particles present in the chamber. When a ray of penetrating radiation passes through the chamber, it produces ions along its path. The air is then suddenly expanded and liquid droplets are formed on the ions. These water drops appear momentarily as a white line that may be seen or photographed.

Another method of measurement is to use an ionization chamber. When a gas is exposed to radioactive radiations, it becomes ionized, and the amount of ionization produced is a measure of the intensity of the rays. The ionization chamber consists of a closed metal box provided with a thin aluminum or plastic window and containing an insulated metal electrode maintained at a fairly high potential (either negative or positive) with respect to the box itself. If some penetrating radiation enters the chamber, it will ionize the gas therein, producing a large number of electron-ion pairs. Under the influence of the applied voltage, the positive ions go one way and electrons the other, so that an electric current flows through the chamber. The amount of ionization and hence the magnitude of the current is proportional to the intensity of the radiation, which can thus be measured by measuring the current.

The Geiger-Müller counter is one of the most important instruments of modern physics, giving us the most convenient method of measuring penetrating radiation. A Geiger counter consists of a metal tube fitted inside a thin-walled glass cylinder, with a fine tungsten wire stretched along the axis of the tube. The arrangement is illustrated in Fig. 48.3. The cylinder contains a gas such as air or argon at pressure of about 10 cm of mercury. A difference of potential of about 1000 V is applied between the wire and the metal cylinder, the positive voltage being applied to the wire and the negative to the cylinder. This difference of potential is slightly less than that necessary to produce a discharge. When any ionizing radiation, such as a high-speed particle from a radioactive source, enters the tube, ions will be produced by the freeing of electrons from the gas molecules. These ions and electrons are accelerated by the electric field and produce more ions by collision. The process repeats itself and the ionization current is rapidly built up. However, the current drops to zero just as rapidly. As the current flows through the external resistance $R$, it produces a potential drop across $R$ and this causes the net voltage across the tube to fall below the extinction value. The current then ceases to flow. But this sudden pulse of current, which produces a sudden pulse of voltage across $R$, can be amplified electronically and made to operate a mechanical counter or to produce a click on a loudspeaker. In this manner, the passage of a single particle through the counter can be registered.

The electrical pulses from the Geiger tube are usually fed to a counting circuit. In some instruments the individual pulses are actually counted. In others, a count-rate meter is used that integrates the pulses and gives a reading in terms of counts per minute (cpm). In the present experiment, however, some of the counting rates will be too slow to give a good reading on such a meter. Therefore, a counter which counts the incoming pulses over a set period of time, such as 1 minute, will be used.

Different types of Geiger counter tubes are used for the detection of different kinds of particles. When the walls of the tube are made comparatively thick, the tube is used for the more penetrating radiations, such as X rays or gamma rays. If the tube is to be used for less penetrating radiations, such as alpha or

**Figure 48.3**   *The Geiger counter (schematic diagram)*

**Figure 48.4** *A typical Geiger counter tube and scaler-timer*

beta particles, then very thin aluminum or glass bubble windows are used to permit the particles to enter the tube. One type of Geiger counter uses a thin-walled stainless steel tube for the detection of beta and gamma rays.

The proper operation of the tube as a counter is dependent upon the voltage applied to it. If the voltage is too low, the passage of radiation through the tube will not cause a current pulse. On the other hand, if the voltage is too high, there will be a discharge within the tube even without any radiation. Hence it is very important to determine the characteristic curve of the Geiger tube that is to be used in a set of measurements. This curve shows the effect of the voltage applied to the tube on its operation as a counter and clearly indicates the voltage needed for proper results. Each tube has its own characteristic curve and its own best operating voltage. The curve also provides an understanding of how the Geiger counter operates.

The *characteristic curve* of a Geiger tube is obtained by measuring its counting rate when placed near a radioactive source of moderate and constant intensity. The distance between the tube and the source is kept fixed. The counting rate is then determined for each of a series of increasing applied voltages. At low voltages, no counts are produced. As the voltage is increased, a particular voltage is reached at which the tube just begins to produce counts, at a slow rate. This is the *threshold voltage* or the starting voltage. As the applied voltage is increased, the counting rate also increases, until a region is reached where the counting rate is practically constant for an increase in voltage of 100 to 200 V. This region is known as the *plateau*, or the flat portion of the curve. As the applied voltage is increased still further, the counting rate increases rapidly, because of spurious counts. The *operating voltage* for the Geiger tube is usually taken at approximately the center of the plateau region.

When the intensity of radiation produced by a beta-gamma source is to be measured with a Geiger counter, the counter tube is placed at a definite distance from the source and the count rate observed. For a study of the variations in intensity with distance, the source is first placed a short distance from the tube, and this distance is then increased in regular steps. The corresponding intensity of radiation reaching the tube at each distance is recorded.

# APPARATUS

1. Geiger counter tube in holder (probe)
2. Electronic counter (scaler)
3. Calibrated mounting board
4. Source of beta-gamma radiation
5. Stopwatch or stop clock (not needed if a timing circuit is built into the scaler)

# PROCEDURE

**1.** Make sure that the scaler unit is turned off and that its high-voltage controls are turned fully counterclockwise to the zero position. The instrument's line cord may then be plugged into a 115-V, 60-Hz ac outlet.

**2.** Place the holder containing the Geiger tube on the mounting board so that the window faces along the board toward where the radiation source will be placed. The Geiger tube cable should be plugged into the appropriate jack on the rear of the scaler.

**3.** Turn the power switch on and observe that the numerical display lights up reading some random number. Press the "stop" and "reset" switches to reset the display to zero.

**4.** Place the source of beta-gamma radiation in its holder about 20 cm from the Geiger tube so that it faces the tube's window (see Fig. 48.4). If your scaler has a built-in timer, set it to the manual position so that the timing circuit is inoperative. Then press the "count" switch to put the scaler into operation.

**5.** Slowly increase the high-voltage controls until counting begins as indicated on the numerical display. Raise the voltage another 50 volts and observe that counting is proceeding at a good rate. Then stop the count by pressing the "stop" switch and reset the display to zero with the "reset" switch.

**6.** If your scaler has a built-in timer, set it to count for 1 minute. Otherwise, prepare the stop clock to time the counting for this period. Press the "count" switch and count the Geiger tube pulses for 1 minute. Repeat this procedure adjusting the distance between the source and the tube until about 2000 counts per minute are obtained. Keep the positions of the source and the tube fixed for Procedures 7–10.

**7.** Return the high-voltage controls to zero, reset the display to zero, and disable the built-in timer by setting its control to "manual." Press the "count" switch and slowly turn up the high voltage until counting just starts. The voltage at which these pulses first appear is called the *threshold voltage* for your Geiger tube. Record its value.

**8.** Stop the count and reset the display to zero. Set the timer or arrange your stop clock to count pulses for one minute. Press the "count" switch and thus determine the counts per minute obtained at the threshold voltage.

**9.** Increase the high voltage by 20 volts and record this new value. Reset the display to zero, press the "count" switch and again observe the num-

ber of counts obtained in one minute. Record this count rate.

**10.** Repeat Procedure 9 several times, increasing the voltage by 20 V each time and recording the counts per minute obtained with each voltage. Notice that after the first big increase in the count rate, the readings remain quite constant over a fairly wide range of voltage. This is the Geiger tube's plateau region. Then, with a further increase in voltage, there may be a considerable increase in the counting rate. Do *not* increase the voltage any further but lower it quickly to avoid damaging the tube and stop taking data.

**11.** From the data of Procedures 7–10, choose a voltage in the middle of the plateau region. Record this as the operating voltage for your Geiger tube. Set the scaler's high-voltage controls to this value.

**12.** Move the beta-gamma source in its holder so that it faces the Geiger tube window and is exactly 5 cm away. Reset the scaler to zero and then count pulses for 1 minute. Record the number of counts per minute so obtained.

**13.** Repeat Procedure 12 for tube-to-source distances of 6, 7, 8, 9, 10, 12, 14, 16, 19, 22, and 25 cm. If the number of counts registered in 1 minute gets small, increase the counting time to 2 or 5 minutes as required, but don't forget to divide the count by 2 or 5 to obtain the counts per minute. Record the count rate for each distance.

**14.** Measure the background radiation. To do this, remove the radioactive source from the vicinity of the Geiger tube. It should be placed at least 10 ft away so that the intensity of the radiation produced by it at the Geiger tube will be negligible. You will notice that the tube still continues to count at a slow rate. These pulses are produced by the background radiation due mostly to cosmic rays. Count the number of pulses obtained in a 5-minute interval and record this value.

## DATA

Threshold voltage _____   Operating voltage _____

Plateau region from _____ to _____

| Geiger tube voltage | Count rate, cpm | | Geiger tube voltage | Count rate, cpm |
|---|---|---|---|---|
| | | | | |
| | | | | |
| | | | | |
| | | | | |
| | | | | |
| | | | | |
| | | | | |
| | | | | |

Cosmic ray background:

Counts in 5 min _____   Counts per min _____

| Distance, cm | Intensity of radiation, cpm | | Distance, cm | Intensity of radiation, cpm |
|---|---|---|---|---|
| 5 | | | 12 | |
| 6 | | | 14 | |
| 7 | | | 16 | |
| 8 | | | 19 | |
| 9 | | | 22 | |
| 10 | | | 25 | |

Value of slope from the graph _____   Percent deviation from the theoretical value _____

## CALCULATIONS

1. Plot a graph showing the relation between the count rate and the voltage applied to the Geiger tube. Plot the voltage along the x-axis and the corresponding values of the count rate along the y-axis. Choose the scales on the coordinate axes in such a way that the curve will extend over almost all of a full sheet of graph paper. Indicate the position of each point on the graph by a small pencil dot surrounded by a small circle. Then draw a *smooth* curve through the plotted points.

2. Observe the plateau region of your curve; that is, notice the region where the count rate remains nearly the same for a wide range of voltage. This is the region where the curve is nearly horizontal. Record the width of this plateau region; that is, record the value of the voltage at the beginning and at the end of the flat portion of the curve.

3. Mark the operating voltage selected in Procedure 11 on your curve. Confirm that it is located at or near the middle of the plateau region.

4. Using the data of Procedures 12 and 13, plot the values of the distance of the source of radiation from the Geiger counter against the corresponding values of the radiation intensity in counts per

minute on a sheet of 2-cycle full logarithmic graph paper. Read the instructions given in the Introduction for plotting such a graph. Indicate the position of each point on the graph by a small pencil dot surrounded by a small circle. Then draw the best-fitting straight line through the plotted points. This line does not necessarily have to pass through all the points, but it should have, on the average, as many points on one side of it as on the other.

**5.** Theoretically, if the radiation intensity is inversely proportional to the square of the distance between the source and the Geiger tube, then the logarithm of the intensity should be a linear function of the logarithm of the distance with slope $-2$. Find the slope of the graph drawn in Calculation 4 and compare it with this theoretical value by computing the percent deviation.

**6.** From the data of Procedure 15, calculate the number of counts per minute produced by the background radiation by dividing the counts recorded in 5 minutes by 5.

# QUESTIONS _____

**1.** Discuss the differences between the ionization chamber and the Geiger-Müller tube and the advantages and disadvantages of each.

**2.** Suppose that the voltage applied to the Geiger tube fluctuated by $\pm 20$ V about the value of the operating voltage that you determined. What percent fluctuation in the count rate would this produce?

**3.** Strictly speaking, the number of counts per minute produced by the background radiation should have been subtracted from the count rates recorded in Procedures 12 and 13 in order to obtain the count rates due to the source alone. How big an error was introduced by neglecting to do this?

**4.** A source of gamma radiation is of such intensity as to produce 30,000 cpm on a Geiger counter placed 5 cm from the source. At what distance would the source produce 30 cpm?

**5.** Find the relation between the half-life of a radioactive isotope and its decay constant $\lambda$.

**6.** An example of radioactive decay is given by the radioactive gas radon, a product of radium. Radon loses its activity quite rapidly, its half-life being 3.82 days. A practical use is made of this fact in that thin-walled glass or metal tubes filled with radon are used for therapeutic work. These tubes are enclosed in hollow steel needles that may be inserted into a diseased tissue, which is thus exposed to the radiation given off by the radon. The short life of the radon is a safeguard against overexposure. Calculate the activity of the radon in these tubes after 19.1 days as compared with the initial activity.

**7.** Your instructor will tell you if your Geiger tube has an operating voltage recommended by its manufacturer. If so, compare this voltage with the one you chose in Procedure 11.

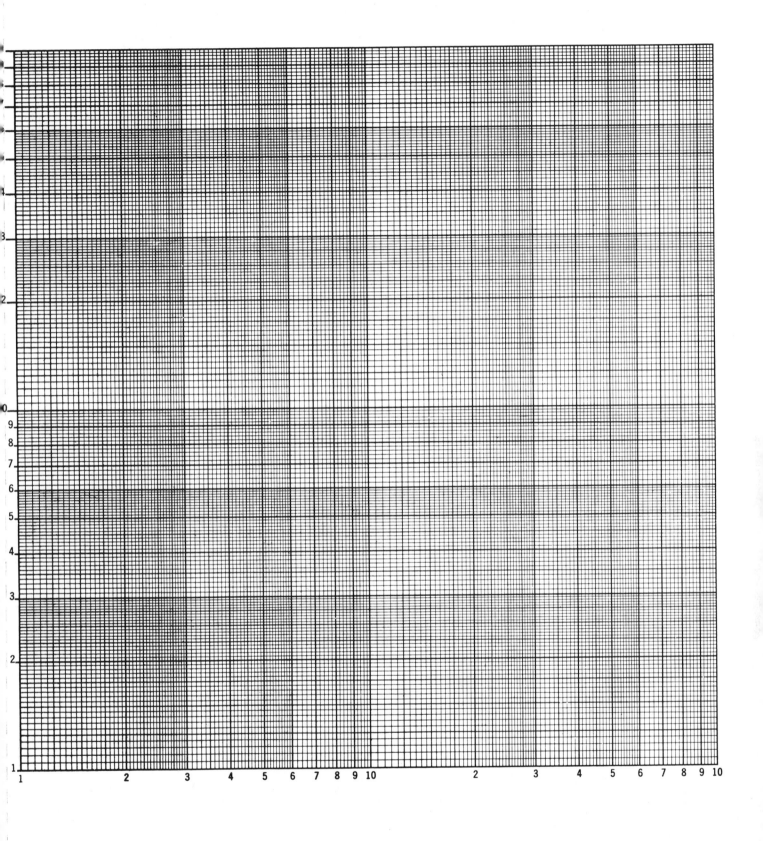

# The Absorption of Gamma Rays 49

Radioactive substances emit one or more radiations known as alpha rays, beta rays, and gamma rays. These rays differ widely in their penetrating power. Thus, a convenient way of distinguishing them is by measuring their intensity after passing through different thicknesses of material; that is, by measuring their absorption in matter. The absorption of beta rays is used in industrial applications for the auto- matic control of the thickness of metal or plastic films. Absorption measurements of gamma rays are used to determine the energy of the rays. The purpose of this experiment is to study the absorption of gamma rays by aluminum plates and to determine the linear absorption coefficient and the mass absorption coefficient of aluminum for the gamma rays used.

## THEORY

The radiation from radioactive materials, such as uranium, thorium, or radium, is found to contain three different kinds of rays: alpha rays, beta rays, and gamma rays. These rays can be distinguished from each other by their penetrating power, by their relative ionizing power, and by their deflection in a magnetic or electric field. Under the influence of a transverse electric or magnetic field, the alpha rays are bent slightly in one direction, the beta rays are bent more strongly in the opposite direction, and the gamma rays are not affected at all. The alpha rays are completely absorbed by a few centimeters of air, or by an aluminum foil 0.006 cm thick, or by a sheet of ordinary writing paper. The beta rays are about 100 times more penetrating than the alpha rays; they are absorbed by a sheet of aluminum about 0.3 cm thick. The gamma rays are the most penetrating of all; they can be detected after passing through a block of iron 30 cm thick. The ionizing power of the three kinds of rays is roughly inversely proportional to their penetrating power.

The *alpha rays* consist of a stream of positively charged particles, each having a mass about four times that of a hydrogen atom and each carrying a positive charge equivalent to two electron charges. In fact, an alpha particle is identical to the nucleus of a helium atom. Alpha particles are emitted with a velocity of about one twentieth the velocity of light. *Beta rays* consist of a stream of negatively charged particles traveling at very high speed. Actually, they are nothing but high-speed electrons emitted by radioactive substances with velocities varying from about 0.2 to 0.9995 the velocity of light. *Gamma rays* are electromagnetic radiations of the same nature as X rays, but of shorter wavelength and higher penetrating power. The wavelengths of gamma rays have been measured by crystal diffraction methods

and more indirectly from measurements of the velocities of ejected photoelectrons. Their wavelength varies from about 0.005 to 1.40 Å. All radioactive elements emit either alpha or beta rays, and either of these rays may be accompanied by gamma radiation.

The measurement of the penetrating power of gamma rays is carried on in the following manner: the intensity of radiation produced by a gamma-ray source is measured by means of a Geiger counter, without any absorber in the path. Then layers of various thicknesses of the absorber are placed between the source and the Geiger tube. The intensity of radiation for each thickness of the absorber is then recorded.

The intensity of a beam of gamma rays after passing through a certain thickness of the absorber is given by the relation

$$I = I_0 e^{-\mu d} \qquad (49.1)$$

where $I_0$ is the original intensity, $I$ is the intensity after passing through the absorber, $d$ is the thickness of the absorber in centimeters, $e$ is the base of natural logarithms, and $\mu$ is the *linear absorption coefficient*, which is a constant characteristic of the absorbing material and of the wavelength of the gamma rays. The law of absorption is illustrated in Fig. 49.1.

From the above equation, by taking the natural logarithms of both sides, we get the relation

$$\ln I = \ln I_0 - \mu d \qquad (49.2)$$

This is the equation of a straight line. If the values of the natural logarithm of the intensity $\ln I$ are plotted along the $y$-axis, and the values of the thickness of the absorber are plotted along the $x$-axis, then the intercept on the $y$-axis is the natural logarithm of the original intensity, and the slope of the straight line is the linear absorption coefficient $\mu$. The negative sign

*(a) Linear Scale*

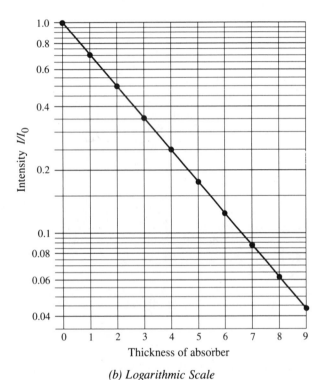

*(b) Logarithmic Scale*

**Figure 49.1** *Absorption of gamma rays*

for the slope merely shows that the line slopes downward from left to right. As described in the Introduction, Equation 49.2 may also be written

$$\ln \frac{I}{I_0} = -\mu d \qquad (49.3)$$

and the ratio $I/I_0$ plotted against the absorber thickness $d$ on semilogarithmic paper, which automatically does the job of taking the logarithm. A straight line is then obtained without the need for looking up the logarithm of each measured value of $I/I_0$.

The absorption of gamma rays by a certain element is simply proportional to its density in the absorbing body, irrespective of chemical combination or physical state of the element. It is convenient, then, to introduce another absorption coefficient, called the *mass absorption coefficient*. This is defined by the relation

$$\mu_m = \frac{\mu}{\rho} \qquad (49.4)$$

where $\mu_m$ is the mass absorption coefficient, $\mu$ is the linear absorption coefficient, and $\rho$ is the density of the absorber. The use of the mass absorption coefficient is convenient because its value is the same, for a given element, irrespective of physical state or chemical combination of the element. However, it does depend on the wavelength of the gamma rays.

A convenient way of representing the penetrating power of a radiation is to use the half-value thickness. This is defined as the thickness of the absorber necessary to reduce the intensity of radiation by a factor of two. The half-value thickness is easily read off directly from a plot of Equation 49.3. It will be the value of $d$ for which the intensity of radiation is one half of the original value, that is, for which $I/I_0 = 0.5$.

When a sensitive instrument is used to measure radiation, it is found that it will show a small reading

**Figure 49.2** *Typical apparatus assembly showing the Geiger tube on a stand above the absorbing plates*

even when no radioactive sources are in the vicinity. This small reading, known as the background, is due to cosmic rays coming from outer space and to minute traces of radioactive materials present in the laboratory. To obtain an accurate measurement of the intensity of radiation produced by a sample, a measurement is first made with the sample in position, then with the sample removed. The intensity of radiation due to the sample is obtained by subtracting the second reading, which is the background measurement, from the first.

## APPARATUS

1. Geiger counter tube in holder (probe)
2. Electronic counter (scaler)
3. Source of beta-gamma radiation
4. Stand for Geiger tube
5. A set of 12 aluminum plates, each about 6 mm thick and about 10 cm square
6. Vernier caliper
7. Stopwatch or stop clock (not needed if a timing circuit is built into the scaler)

## PROCEDURE

**1.** Measure the thickness of a stack of 10 of the aluminum plates with a vernier caliper and record the value.

**2.** Make sure that the electronic counter (often called a scaler) is switched off and that its high-voltage controls are turned down to zero. The instrument's line cord may then be plugged into a 115-V, 60-Hz ac outlet.

**3.** Place the holder containing the Geiger tube on the stand and plug the cable from the tube into the appropriate jack on the scaler as shown in Fig. 49.2.

**4.** Switch the scaler on and set the high-voltage controls for the proper operating voltage for your Geiger tube. If you have done Experiment 48 and are now using the same tube, the proper operating voltage is the one determined in that experiment. Otherwise, the instructor will recommend an operating voltage for your tube or tell you to determine it by doing Procedures 4–11 in Experiment 48.

**5.** Press the "stop" and "reset" switches on your scaler to zero the display. Then place the source of beta-gamma radiation at the bottom of the stand so that it faces the Geiger tube window. If your scaler has a built-in timing circuit, set it to count for 1 minute. Otherwise, get the stop clock ready to time the counting period for this interval. Press the "count" switch, count the pulses from the Geiger tube for 1 minute, and record this result as the count rate (counts per minute) obtained with no aluminum absorbing plates inserted.

**6.** Insert one aluminum plate in the stand between the Geiger tube and the beta-gamma source so as to shield the tube from the source. Zero the display on the scaler and again count the pulses coming from the Geiger tube for 1 minute. Record your result as the count rate with one plate as the absorber.

**7.** Repeat Procedure 6 with 2, 3, 4, etc., plates used as the absorber, until you have inserted all 12 plates. If the number of counts collected in a 1-minute interval gets small (less than 200) as the number of plates is increased, use a 5-minute counting period and divide the result by 5. Record the count rate (counts per minute) obtained with each number of plates.

**8.** Measure the background radiation by removing the radioactive source and the aluminum plates from the vicinity of the Geiger tube. The source should be placed at least 10 ft away so that the intensity of radiation it produces at the Geiger tube will be negligible. Notice, however, that the tube still continues to count at a slow rate. Press the "stop" and "reset" switches to clear the display and count the pulses from the tube for a 5-minute interval. Record the value so obtained.

## DATA

Geiger counter operating voltage _____

Thickness of 10 aluminum plates _____    Thickness of 1 plate _____

Cosmic ray background:

    Counts in 5 minutes _____    Counts per minute _____

**3.** Find the fractional intensity observed with each number of absorbing plates by dividing the intensity observed with the plates in position by the intensity observed with no plates. Use the corrected values of intensity. Record your results.

**4.** Plot an absorption curve for gamma rays on semilogarithmic graph paper. Read the instructions given in the Introduction for plotting a graph of this kind. Plot the absorber thickness (the number of plates) along the linear axis and the fractional intensity along the logarithmic axis. Indicate the position of each point on the graph by a small pencil dot surrounded by a small circle. Then draw the best-fitting straight line through the plotted points. *Note:* If the gamma rays used are made up of different wavelengths, the absorption curve will not be a straight line because each wavelength will have its own characteristic value of $\mu$. In this case, the best fitting straight line will give an average value of $\mu$.

**5.** From the slope of your absorption curve, determine the linear absorption coefficient of aluminum for the gamma rays used. Be careful with your units.

**6.** From Equation 49.4 and your value of the linear absorption coefficient, calculate the mass absorption coefficient of aluminum for the gamma rays used. The density of aluminum is given in Table IV at the end of the book.

**7.** Determine the half-value thickness of aluminum for the gamma rays used. Read the number of plates corresponding to $I/I_0 = 0.5$ from the absorption curve plotted in Calculation 4. Then multiply this number (not necessarily an integer) by the thickness of one plate to obtain the half-value thickness in centimeters. Express your result in centimeters of aluminum.

**8.** From Equation 49.3, calculate the half-value thickness of aluminum for the gamma rays used in this experiment. Use the value of the linear absorption coefficient obtained in Calculation 5. Compare this half-value thickness with the one obtained in Calculation 7 by finding the percent difference.

## QUESTIONS

**1.** From Equation 49.3, derive a relation between the half-value thickness and the linear absorption coefficient.

**2.** Bismuth-214 (Bi$^{214}$, sometimes called radium C) sends out a group of beta rays whose linear absorption coefficient in aluminum is 13.2 cm$^{-1}$. A beam of these beta rays is passed through an aluminum plate 3.03 mm thick. Find what fraction of the beam is absorbed by the plate. For the absorption of beta rays, use the same relation as was used for the absorption of gamma rays.

**3.** A beam of gamma rays of energy 1.5 MeV (million electron volts) is passed through an iron plate 10 cm thick. If the linear absorption coefficient of iron for these rays is 0.40 cm$^{-1}$, find what fraction of the beam is transmitted by the plate.

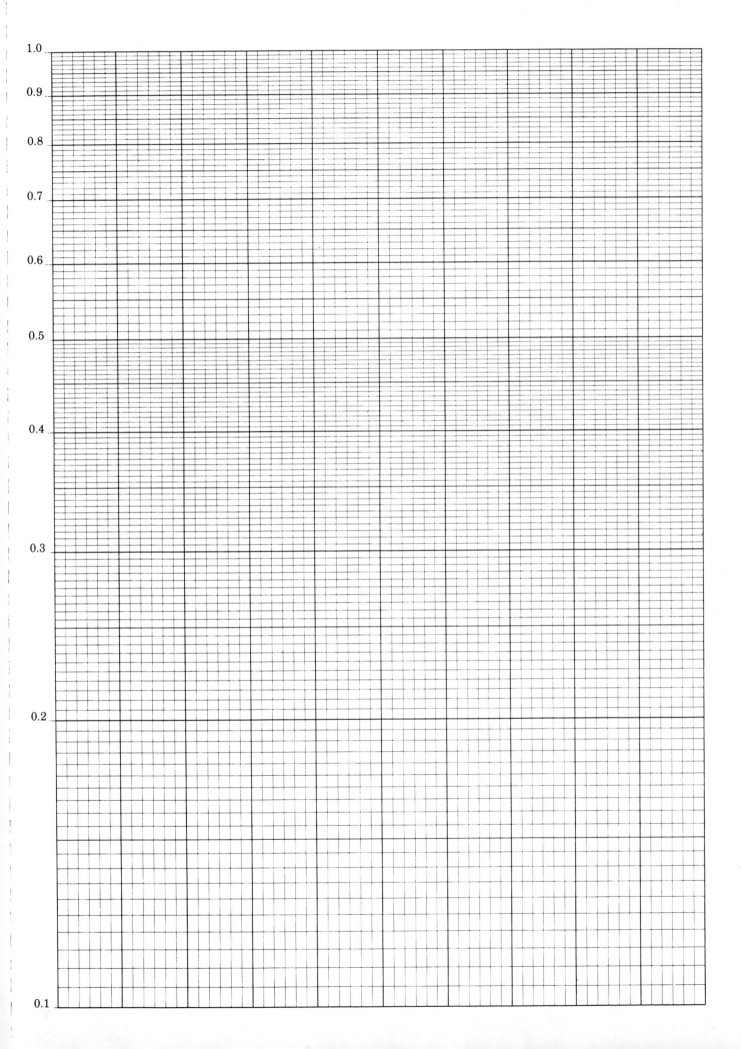

# Half-Life of a Radioactive Isotope 50

In this modern age, everyone is affected by the utilization of radioactive materials; hence some understanding of the nature of radioisotopes and their potential uses is very essential. Radioactive isotopes are widely used in a variety of practical applications in industry, medicine, agriculture, and in the physical and biological sciences. Some of the techniques that can be employed enable us to perform experiments that seem almost uncanny and permit the detection of amounts of materials far smaller than is possible by ordinary chemical means. One of the most important properties of radioisotopes is their rate of radioactive decay, which is measured in terms of their half-life. The object of this experiment is to determine the half-life of a radioactive isotope from a series of activity measurements taken over a suitable period of time, and to calculate this isotope's decay constant.

## THEORY

The nucleus of an atom is made up of a certain number of protons and neutrons. The number of protons in the nucleus is equal to the atomic number of the element, and all atoms of the same element have the same atomic number. But the atomic weight of an element is equal to the sum of the protons and neutrons in its nucleus. Atoms of the same atomic number but different atomic weights are called *isotopes*. Certain combinations of protons and neutrons produce *stable* isotopes, or isotopes that do not change into other elements. Other combinations produce unstable, or *radioactive*, isotopes. The radioactive isotopes reach a stable state by a process of decay that involves the emission of an alpha or a beta particle, usually accompanied by gamma radiation. This spontaneous emission of radiation by the nucleus is called radioactivity. Hence a *radioactive isotope* is an isotope that will spontaneously disintegrate into a new element by the emission of radiation. Some isotopes of the same element are stable, such as carbon-12 and -13, and some are radioactive, such as carbon-11 and -14. Radioactive atoms have the same number of protons as the stable atoms of the same element but a different number of neutrons in the nucleus. They behave chemically and biologically the same as the stable atoms of the same element. This is what makes the tracer technique possible with radioisotopes. Since the mass of a radioactive isotope is slightly different from that of a stable isotope of the same element, their separation involves very special techniques. For example, radioisotopes may be isolated by means of a mass spectrometer, or an ultra-high-speed centrifuge, or by gaseous diffusion techniques.

Some radioactive isotopes occur in nature, but most of those that are now being used have to be produced artificially. Those that occur in nature are isotopes of the heavy elements that are naturally radioactive, or they are products of the disintegration of these elements. Naturally occurring radioisotopes have half-lives so long that appreciable amounts remain in the earth's crust today. But since the discovery of artificial radioactivity, hundreds of new radioisotopes have been produced. In fact, radioisotopes of all the known elements have been made, including several new elements heavier than uranium and thus known as the *transuranic elements*. Radioisotopes can be produced artificially in one of two ways. One method is by bombarding stable elements with high energy atomic particles such as electrons, protons, neutrons, deuterons, or alpha particles. The bombardment results in adding one or more particles to the nucleus of the element, or in taking one or more particles away. In either case, the result is a change in the mass number or in the atomic number of the element. Thus a new isotope is formed, either of the same, or of a different element, and it is very likely to be a radioactive isotope. The other method is by breaking elements into pieces in a process known as *nuclear fission*. In the first process, if charged particles are used to bombard the elements, a cyclotron or some other type of high-voltage accelerator has to be employed. If neutrons are used, an atomic reactor is employed. Similarly, in producing radioisotopes by nuclear fission, an atomic reactor has to be used. The fission fragments are elements near the middle of the periodic table, and they are likely to have unstable combinations of protons and neutrons; hence they themselves are very likely to be radioactive.

As the number of manufactured radioactive isotopes has increased, very many uses have been found for them. They are used in *industry* to measure radiation intensity, to study diffusion and mixing operations, in thickness gauges for paper and plastics, and in toxicology studies to check the poisonous effects of many products. Radioisotopes are used in *agriculture* to study metabolism, nutrition, soil fertility, and the action of insecticides. They are used in *clinical research* to study vitamins, hormones, the action of drugs such as antibiotics and anesthetics, and the action of viruses, bacteria, and toxicants. They are used in *medicine* in diagnosis, such as in tumor detection, and in the treatment of thyroid and heart disorders, cancer, and leukemia. In *physics,* radioisotopes are employed in studying the nature of radiations and of radioactivity, and in *physical chemistry* in studying reaction rates, molecular structure, and diffusion rates in solids and in solutions. In *biological chemistry* they are useful in research on genetics, metabolic transformations, and the mechanism of enzyme reactions.

In *analytical chemistry* different radioisotopes can be detected from their half-lives and from the different radiations they emit. In quantitative analysis, minute traces of some elements present in compounds or mixtures can be detected by these methods. In some instances, quantities smaller than a billionth of a gram can be detected. In *inorganic chemistry* radioisotopes are used as *tracers* to detect impurities. In this way the degree of separation of contaminants from the desired substance can be measured. The tracer technique has been one of the most important developments in the studies made with radioisotopes. By means of this technique, the progress of tagged atoms (radioactive atoms) can be traced through a system. To follow the movements of individual atoms or molecules in a system, one must be able to identify them. If the system is nonhomogeneous, the changes in composition can be measured. If the system is homogeneous, the motion of individual atoms or molecules can be followed by using radioactive isotopes and detecting the emitted radiation with a Geiger counter. For example, by using radioactive sodium, it is possible to trace the motion of certain organic and inorganic chemicals passing through the human body, or through plants. The rate at which the bloodstream circulates through the body is measured by feeding a person a salt solution containing radioactive sodium. As the solution enters the bloodstream, it is distributed to all parts of the body. The presence of the radioactive sodium in the body can be demonstrated by detecting the radiation given off by the sodium with a Geiger counter. The circulation of the blood through different parts of the body can thus be followed and the rate of circulation measured. A similar technique is used in the study of blood cells, viruses, and bacteria, and in many other practical applications which have already been mentioned.

Even though the radioactivity of some materials, such as radium, appears to be quite constant, it has been found that all radioactivity becomes weaker as time goes on. This *radioactive decay*, or loss of activity, is quite apparent in some substances, such as polonium, for example. The rate at which a given quantity of a radioactive element decays is found by measuring the intensity of radiation emitted by the sample at regular intervals over a period of time and plotting a graph like the one shown in Fig. 48.2. This figure shows the decay curve for the radioactive element polonium. Notice that the intensity of radiation drops to half of its original value in 140 days; in another 140 days the intensity again drops to half the value it had at the beginning of this time interval; and so on. Since it would take an infinite time for the intensity to drop to zero, the rate of decay is specified in terms of the time required for the activity of a given sample to drop to half its original value. This is the half-life of the radioactive isotope and is a very convenient way of expressing its decay rate. The same law of decay applies to all the other radioactive isotopes, except that each one has its own characteristic half-life.

The half-life of a radioactive isotope is the time required for the disintegration of half of the atoms in a sample of the isotope. This is also the time required for a given quantity of radioactive material to lose half of its activity. Each isotope has a characteristic half-life that can be determined experimentally. Some isotopes have a very long half-life and some have an extremely short one. For example, the half-life of uranium-238 is 4.50 billion years, that of phosphorus-32 is 14.3 days, while that of polonium-212 is less than one millionth of a second.

The *rate* at which a radioactive material decays is an immutable physical property of each radioisotope and is characteristic of that isotope. The occurrence of radioactive disintegration is statistical in nature. For any individual atom of a given radioisotope, there is a definite probability that it will disintegrate within a certain time interval. This probability does not depend on the past history of the atom nor on the surrounding atoms. Under these conditions, the laws of statistics apply, and when we are dealing with a very large number of atoms, there is a definite probability that a specific proportion of disintegrations will occur in a given time interval.

We define the *decay constant* $\lambda$ as the probability that any individual atom of a particular radioisotope will decay per unit time. Thus $\lambda\,dt$ is the probability that an atom will decay in the short time interval $dt$. If $N$ is the number of such atoms present at time $t$, then on the average there will be $N\lambda\,dt$ disintegrations during the interval $dt$. This means a decrease $dN$ in the number $N$ in this interval, so that

$$dN = -N\lambda\,dt \qquad (50.1)$$

where the negative sign represents the fact that the change $dN$ is a *decrease.*

Equation 50.1 may be put in the form

$$\frac{dN}{N} = -\lambda\, dt$$

and then integrated as follows:

$$\int_{N_0}^{N} \frac{dN}{N} = -\int_{0}^{t} \lambda\, dt$$

$$\ln N - \ln N_0 = -\lambda t$$

$$\ln \frac{N}{N_0} = -\lambda t \qquad (50.2)$$

$$\frac{N}{N_0} = e^{-\lambda t} \qquad (50.3)$$

Note that if Equation 50.3 is multiplied through by $N_0$, Equation 48.1 (already mentioned in Experiment 48) is obtained. This is the expression for radioactive decay. In these equations, $N$ is the number of radioactive atoms present at any time $t$, $N_0$ is the number present at some initial instant taken as $t = 0$, $e$ is the base of natural logarithms, and $\lambda$ is the decay constant for the particular radioisotope. Notice that with $t$ in seconds, the dimensions of $\lambda$ must be seconds $^{-1}$.

The determination of the half-life of a radioactive isotope is very important. In performing experiments with short-lived isotopes, the time that it takes to complete the experiment may be long enough to decrease appreciably the total number of radioactive atoms present and, at the same time, the intensity of radiation produced. This loss of activity has to be taken into account in making accurate measurements. Similarly, in making a chemical analysis of radioisotopes, a determination of the half-life of the radioactive sample will help to identify the isotope present, since each isotope has its own characteristic half-life.

The relation between the half-life of a radioactive isotope and its decay constant may be found by setting the ratio $N/N_0$ in Equation 50.2 equal to $\frac{1}{2}$ and solving for $t$. Since $\ln 0.5 = -0.6931$, $t_{\frac{1}{2}} = 0.6931/\lambda$. For isotopes with very long half-lives,

such as thousands of years, $\lambda$ is easily measured directly by counting the number of disintegrations in some short time (such as a minute) and using Equation 50.1, in which $-dN$ is the number of disintegrations counted, $dt$ is the (short) counting interval, and $N$ is the number of atoms in the sample, obtained by weighing it and dividing by the atomic weight in appropriate units. Note that $dt$ must be in seconds in order to get $\lambda$ in seconds$^{-1}$. This procedure cannot be used for isotopes with short half-lives because $N$ changes by an appreciable amount during the counting period ($dt$ is not very short compared to the half-life). In this case, the activity of a given sample is measured at regular intervals over a period of hours, days, or weeks, that is, over several half-lives of the given isotope. The activity, in counts per minute, is then plotted as a function of the time, and a smooth curve is drawn through the plotted points. The half-life, being the time for the activity (that is, the count rate) to drop to one half its initial value, can then be read off from the curve directly. Alternatively, the logarithms of the activity measurements may be plotted as a function of the time, and the best-fitting straight line then drawn through the plotted points. Because (as Equation 50.1 shows) the activity as represented by the count rate $dN/dt$ is proportional to $N$, this procedure is tantamount to plotting Equation 50.2 with $\ln N/N_0$ and $t$ as variables. Hence the slope of the straight line is $\lambda$, from which the half-life can be calculated as already described. The need to calculate the logarithms of each count-rate reading is avoided by using semilogarithmic graph paper, as described in the Introduction, and this is what is usually done.

In this experiment, the half-life of radioactive iodine, $I^{131}$, will be determined by taking measurements of the activity of a sample at regular intervals over a period of several weeks. The net count rate will be plotted as a function of time on semilogarithmic paper. From the resulting straight line, both the decay constant and the half-life will then be determined.

**Figure 50.1**  *A Geiger-Müller tube in its holder,*
*mounted on a sample mount*

## APPARATUS

1. Geiger counter tube in holder (probe)
2. Electronic counter (scaler)
3. Source of beta-gamma radiation
4. Short-lived radioisotope such as radioactive iodine ($I^{131}$)
5. A holder such as a calibrated mounting board or a sample mount
6. Stopwatch or stop clock (not needed if a timing circuit is built into the scaler)

## PROCEDURE

**1.** Make sure that the scaler is switched off and its high-voltage controls are turned down to zero. The instrument's line cord may then be plugged into a 115-V, 60-Hz ac outlet.

**2.** Place the holder containing the Geiger tube on the mounting board or sample mount so that its window faces the position where you will be putting the radioactive sources. The cable from the tube should be plugged into the appropriate jack on the scaler.

**3.** Switch the scaler on and set the high-voltage controls to the proper operating voltage for your Geiger tube. If you have done Experiment 48 and are now using the same tube, the proper operating voltage is the one determined in that experiment. Otherwise the instructor will recommend an operating voltage for your tube or tell you to determine it by doing Procedures 4–11 in Experiment 48.

**4.** Place the source of beta-gamma radiation, the long-lived reference source, on the mounting board about 20 cm from the Geiger tube and facing the tube's window. Press the "stop" and "reset" switches on the scaler to clear the numerical display and then set the timer to count for 1 minute. Repeat this procedure, moving the source closer to the Geiger tube each time until you get a count rate of about 2000 cpm. Record this position exactly so that the source may be replaced at the same spot later. Also record the actual count rate obtained with the source in this position. It will be the count rate for the long-lived reference source. *Note:* If you are using a sample mount instead of a mounting board as shown in Fig. 50.1, place the reference source in a tray positioned to give the approximate count rate noted above. Record the tray position and the actual count rate as already described.

**5.** Remove the long-lived radioactive source from the vicinity of the Geiger tube. It should be placed at least 10 ft away so that the intensity of radiation it produces at the tube will be negligible. Now place the short-lived source of beta-gamma radiation, the radioactive iodine, on the mounting board about 20 cm from the Geiger tube. Press the "stop" and "reset" switches on the scaler to clear the display and set the timer to count for 1 minute. Repeat this procedure moving the source closer to the Geiger tube each time until you get a count rate of between 1000 and 2000 cpm. Record this position exactly so that the source may be replaced at the same spot later. Also record the actual count rate obtained with the source in this position. It will be the count rate for the radioactive iodine. *Note:* If the iodine source is in the form of a vial, it may be used in this form if a sufficiently large reading can be obtained on the Geiger counter and the source position can be determined sufficiently precisely to allow the vial to be replaced in exactly the same position later. Better results can be obtained with a planchet source. In this case a sample mount should be used and the planchet placed in a tray to be inserted in the mount at the same level each time. The planchet should have been already prepared by the instructor. It may be kept in a small beaker, which in turn is kept in a wide-mouthed glass container with a metal cap, and may be handled with forceps.

**6.** Measure the background radiation by removing all radioactive sources from the vicinity of the Geiger tube as noted at the start of Procedure 5. Notice that the tube still continues to count at a slow rate. Count the pulses coming from the tube for a period of 5 minutes and record your result. Also record the day and time of all the observations made in the last three procedures.

**7.** Repeat Procedures 1–6 after 2 or 3 days. Be sure to place the radioactive sources in exactly the same positions determined for them earlier in Procedures 4 and 5. Repeat Procedure 6 exactly as before. Record the day and time of all these observations.

**8.** Repeat Procedure 7 one week after the date of the original observations. Record the day and time of the present ones. Note that if the count rate with the radioactive iodine is getting too small you can increase the counting time to 2 minutes and then divide the result by 2 to get the counts per minute.

**9.** Repeat Procedure 7 after two weeks. Record the day and time of these observations.

**10.** Repeat Procedure 7 after three weeks. Record the day and time of these observations. If the iodine activity is getting low, it may be necessary to increase the counting time to 5 minutes and divide the resulting count by 5 to get a statistically reasonable value for the average count rate.

## DATA _____

Type of Geiger counter used _____     Position of reference source _____
Operating voltage _____     Position of radioactive iodine _____

| Time of observations | | | Intensity of radiation from reference source, cpm | | |
|---|---|---|---|---|---|
| Date | Time | Elapsed time, days | Count rate | Net rate | Correction factor |
|  |  |  |  |  |  |
|  |  |  |  |  |  |
|  |  |  |  |  |  |
|  |  |  |  |  |  |
|  |  |  |  |  |  |

| Time of observations | | | Intensity of radiation from radioactive iodine, cpm | | | |
|---|---|---|---|---|---|---|
| Date | Time | Elapsed time, days | Count rate | Net rate | Corrected values | Fraction $I/I_0$ of initial intensity |
|  |  |  |  |  |  |  |
|  |  |  |  |  |  |  |
|  |  |  |  |  |  |  |
|  |  |  |  |  |  |  |
|  |  |  |  |  |  |  |

| Time of observations | | Background radiation | |
|---|---|---|---|
| Date | Time | Counts in 5-minute interval | cpm |
|  |  |  |  |
|  |  |  |  |
|  |  |  |  |
|  |  |  |  |
|  |  |  |  |

Decay constant of radioactive
  iodine _____

Half-life of radioactive iodine:

  From decay curve _____     Percent error _____
  From decay constant _____     Percent error _____

## CALCULATIONS

**1.** Divide your readings of the cosmic ray background by 5 to get the intensity of the background radiation in counts per minute for each set of measurements. Use the nearest integral value in each case.

**2.** Calculate the net count rate for the long-lived reference source, at the time of each measurement, by subtracting the background intensity at that time from the corresponding count rate of the reference source.

**3.** Calculate the net count rate for the radioactive iodine, at the time of each measurement, by subtracting the background intensity at that time from the corresponding count rate of the radioactive iodine.

**4.** If the net count rate of the reference source is not the same at the time of all the measurements, then the values of the net count rate of the radioactive iodine must be corrected by multiplying each one by the proper correction factor. Calculate the ratio of the net count rate of the reference source at the time of the first set of measurements to the net count rate of the reference source at the time of each subsequent set. These will be the correction factors. Correct the net count rate of the radioactive iodine by multiplying each value of the net count rate by the correction factor for that time of measurement. These will be the corrected values of the net count rate for the radioactive iodine.

**5.** For each set of observations, calculate the time elapsed since the first set was taken. Calculate the time in days, express it in decimals, rounding off the figure to the nearest hundredth of a day, and record the values.

**6.** Find the fractional intensity observed each time the radioactive iodine was counted by dividing each corrected net count rate by the initial net corrected count rate observed on the first day. Clearly, the fractional intensity for this first measurement must be 1.000. Record all your results.

**7.** Plot a decay curve for radioactive iodine on semilogarithmic graph paper. Read the instructions given in the Introduction for plotting a graph of this kind. Plot the time in days along the linear axis and the corresponding values of the fractional intensity along the logarithmic axis. Indicate the position of each point on the graph by a small pencil dot surrounded by a small circle. Then draw the best-fitting straight line through the plotted points.

**8.** Read the half-life of radioactive iodine directly from the decay curve plotted in Calculation 7 by noting the point on the curve where the fractional intensity $I/I_0$ is equal to 0.5. This is the half-life of radioactive iodine as determined from the decay curve. Compare it with the known value of 8.04 days by finding the percent error.

**9.** From the slope of your decay curve, find the decay constant $\lambda$ for radioactive iodine.

**10.** Using the relation between $t_{\frac{1}{2}}$ and $\lambda$, calculate the half-life of radioactive iodine from your result in Calculation 9 and compare this second value of $t_{\frac{1}{2}}$ with the known value of 8.04 days by finding the percent error.

## QUESTIONS _____

**1.** In the case of a short-lived radioisotope, the time needed to complete an experiment may be long enough to decrease appreciably the total number of radioactive atoms present, so that the intensity of radiation will be decreased. How long could you work with a given sample of radioactive phosphorus ($P^{32}$) before it would have decayed sufficiently to make a 0.5% difference between the initial activity and the activity after a certain time $t$? The half-life of $P^{32}$ is 14.3 days.

**2.** A solution containing 10 microcuries of $P^{32}$ is obtained on the first of the month. What fraction of the radioactive phosphorus remains after 10 days? The half-life of this radioisotope is 14.3 days.

**3.** A source containing 10 $\mu$Ci of radioactive cobalt ($Co^{60}$) is stored away. What percentage of the cobalt will disintegrate in 1 yr? The half-life of radioactive cobalt is 5.27 yr.

**4.** A radioactive isotope has a half-life $t'_{\frac{1}{2}}$ and decays to a new isotope, which is also radioactive. This latter isotope has a half-life of $t''_{\frac{1}{2}}$, decaying in turn to a third (stable) isotope. At an initial instant $t = 0$ a certain sample contains only $N_0$ atoms of the original isotope. Find an expression for the number of atoms of the stable isotope present as a function of time.

**5.** A certain radioactive isotope can decay in either of two different ways characterized by decay constants (probabilities of disintegration per unit time) $\lambda_1$ and $\lambda_2$, respectively. Derive an expression for the half-life $t_{\frac{1}{2}}$ of this isotope in terms of $\lambda_1$ and $\lambda_2$.

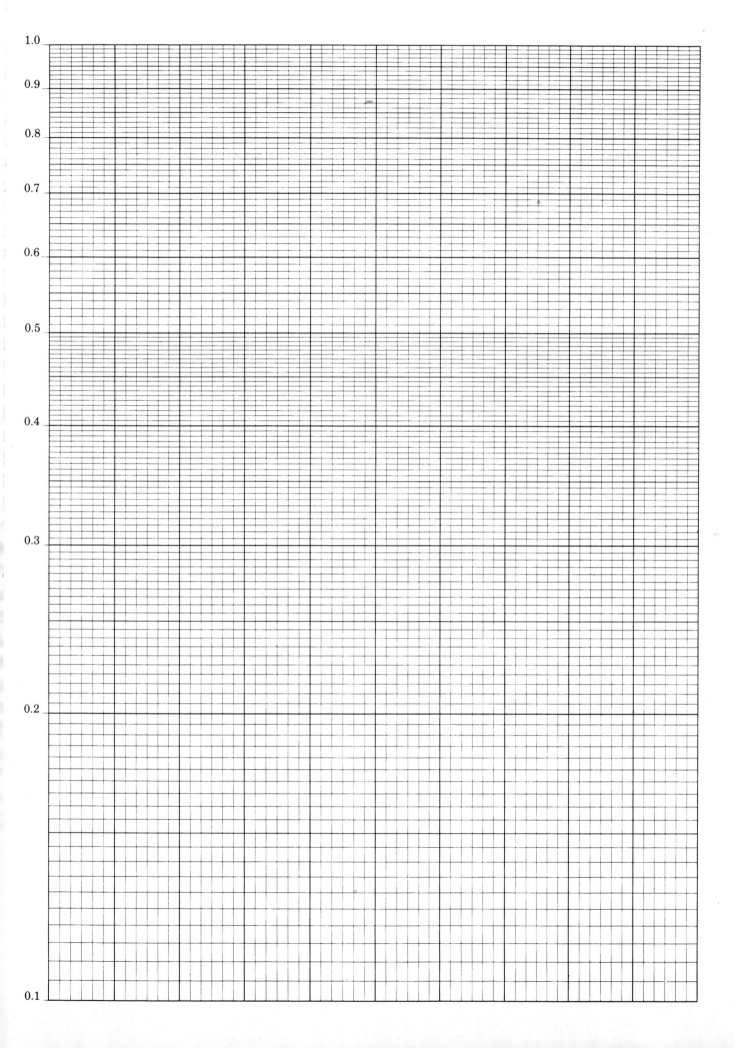

# The Hydrogen Spectrum. 51
# The Rydberg Constant

The research work done in the field of spectroscopy, both experimental and theoretical, has contributed a great deal to our knowledge of the physical nature of things, a knowledge not only of the earth, but of the sun, of the stars, and of interstellar space. This experiment contains a discussion of the Bohr theory of the hydrogen atom, which marked the beginning of a new era in spectroscopy and atomic structure associ-ated with the original introduction of quantum theory. The object of this experiment is to measure the wavelengths of the first four lines of the Balmer series of the hydrogen spectrum by means of a diffraction grating spectrometer, and from these measurements to substantiate Bohr's early theory and to calculate the value of the Rydberg constant.

## THEORY

The *bright-line* spectrum of an element is produced by exciting that element in the gaseous state with an electric discharge or by heating. Such a spectrum consists of bright lines, each being an image of the spectrometer slit formed by light of a particular color and hence wavelength. The number and relative positions of the lines depend on the element and on the method of excitation employed. Wavelengths of visible light are usually expressed in angstrom units, although in recent years there has been an attempt to make the nanometer the standard unit of wavelength for infrared, visible, and ultraviolet electromagnetic radiation. One angstrom unit (abbreviated Å) is defined as $10^{-10}$ m and is therefore $10^{-8}$ cm. A nanometer is $10^{-9}$ m and hence $10^{-7}$ cm. Clearly there are 10 Å in a nanometer.

The frequencies of the light emitted by an excited atom of a gas or vapor at low pressure are characteristic of that particular atom, and the spectrum produced is the bright-line spectrum of that element. When an atom absorbs radiant energy, it absorbs the same frequencies of light it would emit if it were excited. The sun's spectrum is an absorption spectrum. The main body of the sun emits a continuous spectrum, but the relatively cooler gases surrounding the sun absorb the energy corresponding to the line spectra of the elements present in the sun's atmosphere. These absorption lines appear as dark lines or missing lines in the continuous spectrum of the sun. They were first observed by Joseph von Fraunhofer (1787–1826) and are called *Fraunhofer lines*.

It is found that among the many lines appearing in the spectrum of an element, certain ones may be picked out that are clearly related to each other. Such a group forms a series following a simple law, and there may be many different series for the same element. Over a period of years, many attempts were made to explain the origin of spectral lines and the various relationships found. It was thought at first that the frequencies of the light emitted by a particular element might be related like the fundamental and harmonics of a vibrating system. But this explanation turned out to be wrong. Even the one visible series in hydrogen, the simplest element and therefore the element with the simplest spectrum, defied explanation until finally, in 1885, Johann Jakob Balmer (1825–1898), a Swiss school teacher and lecturer, found an empirical formula that correctly gave the wavelengths of the lines of this series. The Balmer formula may be written

$$\frac{1}{\lambda} = R_H\left(\frac{1}{2^2} - \frac{1}{n^2}\right) \qquad (51.1)$$

where $\lambda$ is the wavelength of a visible line in the hydrogen spectrum, $R_H$ is a constant called the Rydberg constant, and $n$ takes on the integral values 3, 4, 5, etc., corresponding to the various lines. Thus setting $n = 3$ gives the wavelength of the longest wavelength (red) line $H_\alpha$ in the visible spectrum of hydrogen; $n = 4$ the next longest ($H_\beta$); $n = 5$ the next ($H_\gamma$); etc. For very large $n$, the wavelength approaches a limiting short value $4/R_H$ called the *series limit*. Note that in Equation 51.1 the Rydberg constant $R_H$ must have the dimensions of reciprocal length and that the unit of length chosen dictates the units in which the wavelength $\lambda$ will come out. In particular, if $R_H$ is given in centimeters$^{-1}$, $1/\lambda$ will be in *wave numbers*, a quantity proportional to the frequency of the emitted light. Wave numbers are popular with spectroscopists who want to talk about the frequency of the radiation with which they are dealing but dislike handling the large numbers involved, the frequency of light waves being of the order of $10^{14}$ Hz. If the wave number of a particular line is known, the frequency can be found simply by multiplying by the velocity of light *in centimeters per sec-*

*ond*, this choice of units being left over from days before the SI system was generally accepted.

A theoretical derivation of the Balmer formula was introduced in 1913 by Niels Bohr (1885–1962) and proved to be the beginning of a vast development in atomic theory. Bohr took as his starting point the nuclear atomic model already proposed by Ernest Rutherford (1871–1937). This model of atomic structure postulates a heavy nucleus containing most of the atomic mass and all of the atom's positive charge and around which one or more negative electrons circulate in orbits under the influence of the electrostatic attraction between the unlike charges. The situation is very simple for the case of one-electron atoms such as hydrogen, singly ionized helium, doubly ionized lithium, etc. The attractive force between the electron and the nucleus is then simply

$$F = \frac{1}{4\pi\epsilon_0} \cdot \frac{(Ze)e}{r^2} = \frac{Ze^2}{4\pi\epsilon_0 r^2}$$

where $F$ is the force in newtons, $Z$ is the atomic number (1 for hydrogen, 2 for helium, etc.), $e$ is the charge on the electron in coulombs, $\epsilon_0$ is the permittivity of free space, and $r$ is the distance between the electron and the nucleus in meters. Assuming a circular orbit, we note that $r$ is the radius of this orbit and $F$ provides the centripetal force $mv^2/r$ necessary to maintain it. We therefore require

$$\frac{Ze^2}{4\pi\epsilon_0 r^2} = \frac{mv^2}{r}$$

or

$$mv^2 = \frac{Ze^2}{4\pi\epsilon_0 r} \quad (51.2)$$

where $m$ is the mass of the electron in kilograms and $v$ its speed in meters per second.

Equation 51.2 is a necessary relation between the speed of the electron and the radius of its orbit in the Rutherford model, but it does not predict that any particular orbit will be preferred. Moreover, the model itself contains a serious flaw in that, according to classical electromagnetic theory, the radial acceleration of the charged electron should cause electromagnetic energy to be radiated continuously until the electron falls into the nucleus and the atom collapses.

Since actual hydrogen atoms neither radiate continuously nor collapse, Bohr postulated the existence of certain *stationary orbits* in which the electron did not radiate classically. These orbits are specified by the *quantum condition*

$$mvr = n\frac{h}{2\pi} \quad (51.3)$$

which states that the electron's angular momentum $mvr$ cannot have any value but only certain allowed ones. These values must all be integral multiples of a basic package or *quantum* $h/2\pi$ of angular momentum that cannot be subdivided. Here $h$ is Planck's constant, and the fact that the angular momentum must be an integral multiple of $h/2\pi$ is expressed in Equation 51.3 by requiring that $n$ be an integer (1, 2, 3, etc.). An integer that tells how many quanta are to be taken is called a *quantum number*, and $n$, the first such number in atomic theory, is called the *principal quantum number*. Notice that $h$ must have the same dimensions as angular momentum, namely kilogram-meter²-seconds⁻¹ or joule-seconds.

Equations 51.2 and 51.3 are two simultaneous equations in $v$ and $r$ and may be solved simultaneously for these quantities. The result is

$$r = n^2 \frac{\epsilon_0 h^2}{\pi m Z e^2} \quad \text{and} \quad v = \frac{1}{n} \cdot \frac{Ze^2}{2h\epsilon_0} \quad (51.4)$$

These relations show that only certain values of $r$ and $v$ are allowed, namely those obtained when the positive integers are successively substituted for $n$.

Following in the footsteps of the German physicist Max Planck (1858–1947), Bohr further assumed that in transferring from one allowed orbit to another, an electron had to make the change instantaneously (in order not to be "caught" in a forbidden position) and accomplished this by either absorbing or giving up the energy difference between these two states (orbits) in a single package, that is, an energy quantum. To determine the size of this package, we must investigate the energy the electron has when moving in any given orbit. Its kinetic energy is $mv^2/2$, and Equation 51.2 shows that this is just $Ze^2/8\pi\epsilon_0 r$. The electron also has potential energy $-Ze^2/4\pi\epsilon_0 r$ by reason of its position in the electric field of the nucleus considered as a point charge $Ze$. Note that the negative sign comes from assuming that the zero of potential energy is at infinity and that since the electron and the nucleus attract each other, work is done *by* the electric field on the electron rather than *on* the field when the electron is brought from infinity into its orbit.

The total energy of the electron is the sum of its kinetic and potential energies and is thus given by

$$E = \frac{Ze^2}{8\pi\epsilon_0 r} - \frac{Ze^2}{4\pi\epsilon_0 r} = -\frac{Ze^2}{8\pi\epsilon_0 r}$$

Notice that it is zero when the electron is stopped at infinity, that is, when it has been just removed from the atom to form a hydrogen ion, a doubly ionized helium ion, or a triply ionized lithium ion.

Insertion of the allowed values of $r$ from Equation 51.4 yields allowed values of the energy, one corresponding to each allowed orbit. These are given by

$$E_n = -\frac{1}{n^2} \cdot \frac{mZ^2e^4}{8\epsilon_0^2 h^2} \quad (51.5)$$

The difference between any two of these energies is

$$\Delta E = E_{n_2} - E_{n_1} = \frac{mZ^2e^4}{8\epsilon_0^2h^2}\left(\frac{1}{n_1^2} - \frac{1}{n_2^2}\right) \quad (51.6)$$

and if in a change between the orbits specified by the quantum numbers $n_1$ and $n_2$ this energy difference must be absorbed or emitted in a single package, one way to do this is to have the electron absorb or emit a train of electromagnetic waves whose frequency is given by the Planck formula

$$\Delta E = hf \quad (51.7)$$

where $\Delta E$ is the energy difference between two allowed states (orbits) in joules, $h$ is Planck's constant in joule-seconds, and $f$ is the frequency of the radiation in hertz (seconds$^{-1}$). Substitution of Equation 51.7 in 51.6 yields

$$f = Z^2\frac{me^4}{8\epsilon_0^2h^3}\left(\frac{1}{n_1^2} - \frac{1}{n_2^2}\right)$$

which may be stated in terms of the radiation wavelength as

$$\frac{1}{\lambda} = Z^2\frac{me^4}{8\epsilon_0^2ch^3}\left(\frac{1}{n_1^2} - \frac{1}{n_2^2}\right) \quad (51.8)$$

where $c$ is the velocity of light and use has been made of the usual relation $f = c/\lambda$ between frequency and wavelength.

Equation 51.8 shows that if a hydrogen atom is excited to any high energy state, that is, if it has its electron placed in an orbit of relatively large radius specified by the value $n_2$ of the quantum number $n$, it may make transitions to any lower energy state (smaller radius orbit) specified by the smaller quan-

tum number value $n_1$ by radiating electromagnetic waves whose wavelength is given by this equation. Various series in hydrogen are predicted by taking $Z = 1$ and setting $n_1 = 1$ while $n_2$ successively takes on the values 2, 3, 4, . . . for the first possible series; $n_1 = 2$ and $n_2 = 3, 4, 5, . . .$ for the next; $n_1 = 3$ and $n_2 = 4, 5, 6, . . .$ for the next; etc. The situation is illustrated in Fig. 51.1, which depicts the Bohr circular orbits schematically (not to scale) and shows various possible transitions arranged according to series. Notice that the Balmer formula, Equation 51.1, is identical to Equation 51.8 when $Z = 1$, $n_1 = 2$, and the Rydberg constant is defined as

$$R_{\mathrm{H}} = \frac{me^4}{8\epsilon_0^2ch^3} \quad (51.9)$$

Agreement between accurately measured values of the Rydberg constant and values calculated from this relation using the best available numerical data for the fundamental physical constants is remarkably close. This is particularly true if account is taken of the fact that the nucleus is not truly stationary but rotates along with the electron about the atom's common center of mass. To do this, a so-called reduced mass is calculated for the electron, which compensates for the assumption that the nucleus is a fixed point. The effect decreases for heavier nuclei, since the atom's center of mass then comes closer and closer to coinciding with the nucleus, but there is nevertheless an observable shift in the value of the Rydberg constant as applied to hydrogen, deuterium, singly ionized helium, etc., as shown in Table I at the end of the book. Notice that the change occurs in the fifth significant figure and thus affects only very precise measurements.

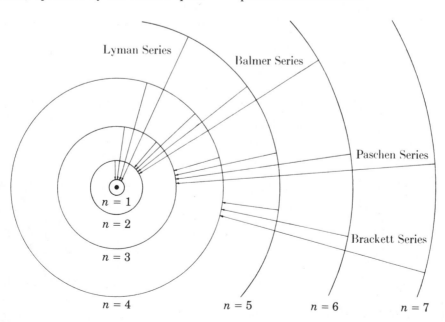

**Figure 51.1** *Bohr circular orbits of hydrogen with transitions for different series of lines*

In addition to the excellent agreement between the theoretical and experimental values of the Rydberg constant, Bohr's prediction of other series besides that of Balmer has been verified. Thus the Lyman series ($n_1 = 1$, $n_2 = 2$, 3, . . . ), the Paschen series ($n_1 = 3$, $n_2 = 4$, 5, . . . ), the Brackett series ($n_1 = 4$, $n_2 = 5$, 6, . . . ), and even the Pfund series ($n_1 = 5$, $n_2 = 6$, 7, . . . ) have all been observed since Balmer's time and are found to follow Equation 51.8 with great precision. Special equipment is required to observe these series, however, since the Lyman lines are all in the ultraviolet region of the spectrum and the other series are all in the infrared. Thus the Balmer series forms the only visible part of the hydrogen spectrum, and even its shorter wavelength lines fall in the ultraviolet. In the present experiment, only the four longest wavelength Balmer lines will be studied. In order of decreasing $\lambda$, these are often designated $H_\alpha$, $H_\beta$, $H_\gamma$, and $H_\delta$.

The presentation of energy states and the transitions between them can be made much more quantitative by using an *energy level diagram* than is possible with a diagram like Fig. 51.1. Figure 51.2 shows the energy level diagram for hydrogen. The allowed energies are plotted on a vertical scale, a horizontal line being drawn at the appropriate level for each such energy and accordingly called an energy level. The levels are labeled both with the appropriate energies computed from Equation 51.5 and with the corresponding quantum numbers. When transitions are indicated by vertical lines drawn between the levels involved, the length of each line is proportional to the frequency of the emitted radiation and hence inversely proportional to the wavelength. Figure 51.2 clearly indicates why the Lyman series should be in the ultraviolet and the Paschen and Brackett series in the infrared. The energy level diagram greatly simplifies the description of the spectrum. It must be remembered that each series is an aggregate of infinitely many lines and that the complete spectrum contains an infinite number of series. An energy level diagram allows the whole spectrum to be simply represented by only one aggregate of infinitely many levels. From these levels, all the lines of all the series can be derived.

In the present experiment, a diffraction grating will be used to measure the hydrogen wavelengths.

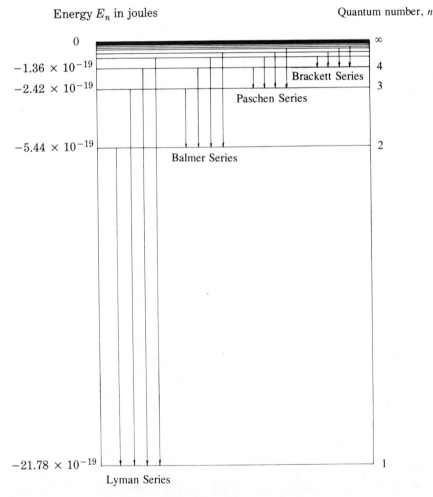

**Figure 51.2** *Energy level diagram of hydrogen*

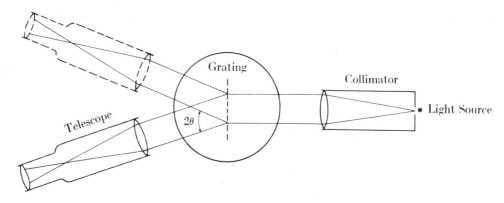

**Figure 51.3** *The diffraction grating spectrometer*

The theory and operation of such a grating is fully covered in Experiment 46, which should be reviewed in preparation for the present work. If you have not done that experiment, read the Theory and Procedure sections, noting Equation 46.1 in particular. You should also note, however, that the setup in Experiment 46 was an extremely elementary one that enabled you to concentrate your attention on the operation of the grating. In particular, the method used for determining the diffraction angle $\theta$ was not capable of great precision. We shall, therefore, now use a spectrometer similar to that studied in Experiment 44 with the transmission grating of Experiment 46 substituted for the prism, as shown in Fig. 51.3. Here again the collimator contains a slit through which light from the source enters the spectrometer and a lens whose function is to render the rays coming from the slit parallel. Compare this arrangement with that used in Experiment 46 to direct approximately parallel rays at the grating. The grating is mounted on the rotating table in the middle of the spectrometer, which is adjusted so that the plane of the grating is at right angles to the incoming rays. The diffracted image of the slit is then viewed with the telescope, which is rotated until the image falls on the cross hairs in the eyepiece. The angle $\theta$ can then be read directly from the spectrometer's divided circle rather than having to be calculated from length measurements as in Experiment 46. Note that with the grating, an absolute measure of the wavelength $\lambda$ forming a particular image of the slit can be obtained if the grating constant is known. No calibration curve is required as it was with the prism in Experiment 44. A grating spectrometer is used in the present work both for this reason and to give the opportunity for experience with this important instrument.

## APPARATUS

1. Spectrometer, for prism or grating
2. Diffraction grating
3. Grating holder
4. Atomic hydrogen discharge tube
5. Discharge-tube power supply
6. Tube holder (unless part of the power supply)

## PROCEDURE

**1.** The spectrometer should be already adjusted by the instructor, with the collimator and telescope properly focused and aligned and the spectrometer table leveled so that its surface is perpendicular to its axis of rotation. CAUTION: Do not handle the spectrometer at all until the instructor has demonstrated its operation to you, as the adjustments are very easily upset. Then, be extremely careful in operating it.

**2.** Place the diffraction grating in its mounting at the center of the spectrometer table. The unruled side of the grating should face the collimator, and the plane of the grating should be made as nearly perpendicular to the collimator's axis as can be judged with the eye. The lines on the grating should be parallel to the collimator slit, and the center of the grating itself should be as nearly as possible in the center of the spectrometer table.

**3.** Place the stand containing the hydrogen discharge tube in front of the spectrometer, with the discharge tube just outside the collimator slit (see Fig. 51.4). Connect the leads from the tube holder terminals to the high-voltage power supply. Adjust the spectrometer so that the collimator points directly at the discharge tube's center. Close the power supply switch. CAUTION: Do not touch the discharge tube or the high-voltage wires while the current is on.

**4.** Observe the central image produced by the diffraction grating. To do this, turn the telescope into

**Figure 51.4**   *Arrangement of apparatus for the diffraction grating spectrometer: (a) telescope, (b) collimator, (c) discharge tube, (d) power supply*

line with the collimator and view this image through the eyepiece. It will be an image of the collimator slit and will have the same color as the hydrogen discharge. The slit should be made as narrow as possible and yet remain visible in the telescope. Make sure the telescope is adjusted so that when the eyepiece is focused on the cross hairs, the image of the slit is also in focus.

**5.** Move the telescope to the right so as to view the first-order hydrogen spectrum. Adjust the position of the telescope until the red hydrogen line falls on the intersection of the cross hairs of the telescope.

**6.** A clamping device that rotates with the telescope and permits it to be clamped by means of a radial screw is located directly under the telescope's supporting arm. A tangent screw on the arm allows a fine adjustment of the setting of the telescope; however, the tangent screw is used only after the telescope has been clamped with the radial screw. Clamp the telescope by means of the radial screw and locate

the position of the red hydrogen line precisely by setting the intersection of the cross hairs exactly on that line by means of the tangent screw. Record the setting of the telescope to the nearest minute of arc, reading the divided circle with the help of the vernier.

**7.** Repeat Procedures 5 and 6 for the blue-green, the blue, and the violet hydrogen lines. Record the setting of the telescope to the nearest minute of arc for each of the lines measured.

**8.** Move the telescope to the left of the central image, so as to view the left-hand, first-order hydrogen spectrum. Adjust the position of the telescope until the red hydrogen line falls exactly on the intersection of the cross hairs. Record the setting of the telescope to the nearest minute of arc.

**9.** Repeat Procedure 8 for the blue-green, the blue, and the violet hydrogen lines. Record the setting of the telescope to the nearest minute of arc for each of the lines measured.

## DATA

Number of lines per inch on the
grating _____

Grating constant in centimeters _____

| Hydrogen lines | | Setting of telescope | | Calculated values | | | | |
|---|---|---|---|---|---|---|---|---|
| Wavelength, Å | Color | Right | Left | $\theta$ | Sin $\theta$ | Computed wavelength | Percent deviation | Computed value of Rydberg constant |
| 6562.8 | red | | | | | | | |
| 4861.3 | blue-green | | | | | | | |
| 4340.5 | blue | | | | | | | |
| 4101.7 | violet | | | | | | | |

Calculated value of Rydberg
constant _____

Accepted value of Rydberg
constant for hydrogen _____

Percent error _____

## CALCULATIONS

**1.** Compute the grating constant, that is, the distance between the lines on the grating in centimeters, from the number of lines per inch stated on the grating.

**2.** Compute the angle of diffraction $\theta$ for each of the four lines measured. The angle of diffraction is one half of the angle between the two settings of the telescope, for each line.

**3.** Compute the value of sin $\theta$ for each of the four lines measured.

**4.** Compute the wavelength of each line, using the value of the grating constant previously calculated and applying Equation 46.1. The wavelength should be expressed in angstrom units, evaluated to four significant figures.

**5.** Calculate the percent deviation of the computed values of the wavelengths from the known values.

**6.** Calculate the value of the Rydberg constant from the computed wavelength of each line by applying the Balmer formula, Equation 51.1, in each case. Pay careful attention to your units.

**7.** Find the average value of the computed Rydberg constant and compare it with the known value from Table I at the end of the book by finding the percent error.

## QUESTIONS _____

**1.** Calculate the angular separation between the D lines of sodium for a diffraction grating of 15,000 lines per inch in the first-order spectrum. Calculate the separation in the second-order spectrum. The wavelengths of the lines are 5890 and 5896 Å, respectively.

**2.** Calculate the frequencies in hertz and the wave numbers in centimeters$^{-1}$ of the first four lines in the Balmer series.

**3.** Using the Balmer formula, calculate the wavelength of the fifth line in the Balmer series of hydrogen lines. What does your answer tell you about the possibility of seeing this line in your spectrometer?

**4.** Calculate the wavelength of the first line in the Balmer series in deuterium (heavy hydrogen), using the Balmer formula with the appropriate value $R_D$ for the Rydberg constant. See Table I at the end of the book. Calculate the wavelength in angstrom units to five significant figures and find the wavelength difference in angstrom units between this line and the corresponding hydrogen line. It was the accurate measurement of these lines, and the interpretation given by the Bohr theory, that led to the discovery of heavy hydrogen by Urey, Brickwedde, and Murphy in 1932.

**5.** Explain the significance of the Fraunhofer lines in the spectrum of the sun.

**6.** A famous constant called the *fine structure constant* may be defined as the ratio of the speed of the electron in the lowest ($n = 1$) Bohr orbit to the speed of light. Derive an expression and calculate the numerical value of the fine structure constant. Also calculate the reciprocal of this value. It is a famous number which you may already have heard of.

**7.** Carry out the derivation of Equation 51.5.

**8.** Suppose an electron makes a transition between two energy states in hydrogen both of which have very high quantum numbers $n_1$ and $n_2$ but which are close together, that is, for which $n_2 - n_1$ is a small integer like 1, 2, or 3. Show that the frequency of the light radiated is approximately an integral multiple or harmonic of the fundamental frequency of revolution of the electron in the associated orbit. This is an example of Bohr's *correspondence principle,* which states that at high quantum numbers the quantum picture must become identical to the classical one.

**9.** (a) What is meant by an atom's *ionization energy*? (b) What is the ionization energy of hydrogen? (c) Show that the *series limit* of the Lyman series involves an energy identical to the hydrogen ionization energy.

**10.** Obtain a numerical value of the Rydberg constant from Equation 51.9 by inserting values of the quantities involved from Table I at the end of the book. Compare your result with the known value.

**11.** What lines in the spectrum of ionized helium lie in the visible range (4000 to 7000 Å)?

**12.** Which, if any, of the first four lines in the visible spectrum of hydrogen could you see in second order with your grating spectrometer? Explain your answer.

# The Ratio *e/m* for Electrons

52

The physical nature of cathode rays is revealed by the deflections they undergo as they pass through electric and magnetic fields. The direction of these deflections shows that these rays consist of particles that have a negative electric charge and a definite mass. In 1897, the investigations of J. J. Thomson (1856–1940) led to a measurement of the ratio of charge to mass for these particles, which have been called electrons. The results of these measurements showed that the atom is not indivisible, that electrons make up a part of any atom, and that this particle is much smaller than the smallest atom. The object of the present experiment is to determine the value of the ratio of charge to mass of an electron. From the known value of the charge on the electron and the value of *e/m* so obtained, the mass of the electron can be computed.

## THEORY

One of the simplest methods of measuring the ratio *e/m*, or the *ratio of charge to mass* of an electron, was first devised around 1900 by T. Lenard and consists of measuring the deflection of a beam of electrons in a known magnetic field. The beam of electrons is bent into a circular path. The value of *e/m* can then be calculated from the known values of the accelerating potential, the strength of the magnetic field, and the radius of curvature of the path.

When an electron or any other charged particle moves in a magnetic field in a direction at right angles to the field, it is acted upon by a force perpendicular to the direction of the field and the direction of motion of the particle. The force on an electron traveling perpendicular to a magnetic field is given by

$$F = Bev$$

where $B$ is the strength of the magnetic field, $e$ is the charge on the electron, and $v$ is the electron's velocity.

Since the force is always perpendicular to the electron's direction of motion, it acts as a centripetal force to make the electron move in a circular path whose plane is perpendicular to the field direction. The centripetal force required to keep a body moving in a circle is

$$F = m\frac{v^2}{r}$$

where $F$ is the force in newtons, $m$ is the body's mass in kilograms, $v$ is the speed in meters per second, and $r$ is the radius of the circle in meters. This force is provided on the electron by the magnetic field so that we have $mv^2/r = Bev$ or

$$\frac{mv}{r} = eB \qquad (52.1)$$

Note that the *momentum mv* of the electron determines the radius of its path in a given magnetic field. Note also that in our SI system of units the field strength $B$ must be measured in webers/meter$^2$, or (as they are now called) teslas.

When an electron is accelerated by an electric potential difference $V$, it gains a kinetic energy equal to the work done on it by the electric field. Because the potential difference is by definition the work done per unit charge, the work done on the electron is $Ve$. Therefore, the kinetic energy $\frac{1}{2}mv^2$ of an electron starting from rest and accelerated by voltage $V$ is given by

$$\frac{mv^2}{2} = Ve$$

where $m$ is the electron's mass in kilograms, $v$ is its velocity in meters per second, $V$ is the accelerating potential in volts, and $e$ is the electron's charge in coulombs. Solving, we find that the velocity acquired by the electron is

$$v = \sqrt{2Ve/m}$$

Substitution in Equation 52.1 then yields

$$\frac{e}{m} = \frac{2V}{B^2r^2} \qquad (52.2)$$

This expression gives the ratio of charge to mass of an electron in terms of the accelerating potential, the strength of the magnetic field, and the radius of the circular path of the beam of electrons. When these quantities are known, the value of *e/m* can be computed.

The apparatus to be used for measuring the ratio of *e/m* is based on a design originally proposed by K. T. Bainbridge. It consists of a large vacuum tube mounted at the center of a pair of helmholtz coils that provide the uniform magnetic field needed for the operation of the tube. The arrangement is shown in Fig. 52.1. The beam of electrons is produced by an electron gun composed of a straight filament parallel to the axis of the coils and surrounded by a cylindrical anode containing a single slit parallel to the filament. Electrons emitted from the heated filament are accelerated by the potential difference applied between the filament and the anode. Some of the electrons come out as a narrow beam through the anode slit. The tube contains a trace of mercury vapor that serves to render the path of the electrons visible. When the electrons collide with the mercury atoms, some of the atoms will be ionized; as these ions recombine with stray electrons, the characteristic mercury light is emitted. Since the recombination and emission of light occur very near the point where ionization took place, the path of the beam of electrons is visible as a bluish white streak.

A pair of helmholtz coils is a particular arrangement of two coaxial, circular coils of radius *a* with their planes parallel and separated by a distance *a* equal to their radius. This arrangement is very useful because it produces an almost uniform magnetic field over a fairly large region near the center of the coils. The magnetic field at the center is parallel to the coil axis and has a magnitude given by

$$B = \frac{8\mu_0 NI}{a\sqrt{125}} \tag{52.3}$$

where $B$ is the magnetic field strength in teslas, $N$ is the number of turns of wire on each coil, $I$ is the current through the coils in amperes, and $a$ is the mean radius of each coil in meters.

Each coil of the pair used in this experiment has 72 turns of copper wire with a resistance of approximately 1 $\Omega$. The mean radius $a$ is 0.33 m. The coils are supported in a frame that can be adjusted with reference to the angle of dip of the earth's magnetic field so that the field produced by the coils will be parallel to that of the earth but in the opposite direction. A graduated scale indicated the angle of tilt.

The magnetic field of the helmholtz coils causes the beam of electrons to follow a circular path whose radius decreases as the magnetic field increases. The field can be adjusted until the sharp outer edge of the beam coincides with the outer edge of one of the five bars spaced at different distances from the filament. The vacuum tube used in this experiment has been so designed that the radius of the circular path can be conveniently measured. In its manufacture, five crossbars have been attached to the staff wire, and this assembly is attached to the filament assembly in such a way that the distance between each crossbar and the

**Figure 52.1**   *e/m vacuum tube and helmholtz coils*

filament is accurately known. The distance in centimeters between the filament and the outer edge of each bar is given in the following table. These distances are the diameters of the circles which the electron beam will be made to describe.

| Crossbar Number | Distance to Filament |
|---|---|
| 1 | 6.48 cm |
| 2 | 7.75 cm |
| 3 | 9.02 cm |
| 4 | 10.30 cm |
| 5 | 11.54 cm |

Figure 52.2 is a sectional view of the tube and filament assembly. $A$ represents the five crossbars attached to staff wire $W$, $B$ shows a typical path of the beam of electrons, $C$ represents the cylindrical anode with slit $S$, and $F$ is an end view of the filament.

The circuit for the tube and its associated helmholtz coils is shown schematically in Fig. 52.3. The coils are powered from a variable dc power supply capable of delivering 5 A of well-filtered direct current continuously. Because the total resistance of the two coils in series is about 2 $\Omega$, this supply will not be called upon to produce over 10 V, but it should have a smooth output control to enable you to

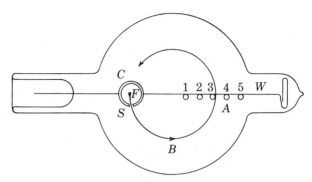

**Figure 52.2** *Sectional view of e/m tube and filament assembly*

set the coil current precisely. A similar supply is used to heat the tube's filament. Note that this filament must not be run at currents greater than 4.5 A, and a fuse is included in the apparatus to prevent operation at currents in excess of this value. A regulated, low-current 0–30-V dc supply is used to establish the accelerating voltage between the filament and the anode.

It is of interest to note that the apparatus used in this experiment demonstrates the principle of the *mass spectrometer*, an instrument that can determine the mass of individual atoms. Thus if ionized atoms could be projected from a source as the electrons are from the filament of your *e/m* tube, Equation 52.2

**Figure 52.3** *Circuit for the e/m tube and its helmholtz coils*

shows that, for a given accelerating voltage *V* and magnetic field *B*, all particles having a particular charge-to-mass ratio will follow a circular path of radius $(1/B)\sqrt{2mV/e}$. But if the ions in the beam all have charge *e* but different masses, they will follow separate paths, one for each mass, and can thus be separated. Such an arrangement is one form of mass spectrometer, but we should note that the determination of mass depends on our prior knowledge of the charge on the particle. It is most important to realize that *no deflection experiment with charged particles (such as the present one) can ever determine the charge or mass separately; it can only determine their ratio.* This follows from Newton's second law, *F = ma*, in which we note that a deflection experiment basically measures some sort of acceleration produced by a force due to an electric or magnetic field. Such a force is proportional to the particle's charge, so that we are always measuring $a = F/m \sim e/m$. Thus either *e* or *m* must be obtained from some independent measurement.

During the years 1909 to 1913, Robert A. Millikan (1868–1953) devised a method of measuring *e* by essentially determining the charge-to-mass ratio of oil drops small enough to carry only a few excess electrons. He could determine the masses of his drops by measuring their velocities when falling under gravity, and he knew that their charges had to be small integer numbers of electron charges. Thus having found the charges on a large number of drops, he could conclude that the electronic charge would be the lowest common multiple, that is, a value that would divide evenly into all his results. Strange as it may seem, this method, although admittedly after much repetition and refinement, remains today as the basic source of our knowledge of *e*.

## APPARATUS

1. *e/m* tube mounted between helmholtz coils
2. Two power supplies capable of delivering at least 5 A at 10 V dc
3. Two variable line transformers, about 150-W rating ("Variac" or "Powerstat") if not included in above power supplies
4. 30-V dc power supply
5. dc voltmeter (0–50 V)
6. Two dc ammeters (0–5 A)
7. dc milliammeter (0–10 mA)
8. Compass
9. Dip needle

## PROCEDURE

**1.** Wire the tube and helmholtz coils in the circuit shown in Fig. 52.3. Make sure all power supply controls are turned all the way down and that their power switches are in the off position. Have your wiring approved by the instructor before plugging in any of the supplies.

**2.** The helmholtz coils have been designed to provide the uniform magnetic field at the location of the tube given by Equation 52.3. For quantitative measurements, the axis of the coils should be made parallel to the earth's magnetic field. This is done by orienting the apparatus so that the *e/m* tube will have its long axis in a magnetic north-south direction, as determined by a compass. Then measure the magnetic dip at this location by using a good dip needle. Finally tip the coils up until the plane of the coils makes an angle with the horizontal equal to the complement of the angle of dip. The axis of the coils will now be parallel to the earth's magnetic field. For example, in New York State the magnetic dip is about 74° and thus the plane of the coils should make an angle of about 16° with the horizontal. A graduated scale connected to the apparatus indicates the angle of tilt of the plane of the coils. If the dip needle is now placed in a north-south direction on the inclined board supporting the coils, the needle should point along the axis of the coils.

**3.** The earth's magnetic field should be directed along the axis of the helmholtz coils in a direction opposite to that of the field you will apply. Because in the Northern Hemisphere the vertical component of the earth's field is down, the vertical component of the field due to the coils should be up when the apparatus is used in this hemisphere. The field of the coils will be so directed if the *e/m* tube is oriented so that the electrons travel in a counterclockwise direction, which means that the slit in the anode should face to your right when you are standing looking at the base end of the tube (the end where the leads to the filament and anode come out). Make sure the tube is mounted in the apparatus with this orientation.

**4.** Apply the accelerating potential to the tube's anode by plugging in and turning on the 30-V power supply. Turn up its output control to give an accelerating voltage of about 25 V as read on the dc voltmeter.

**5.** Make sure that the filament power supply's output control is turned all the way down; then plug it in and turn it on. Slowly increase the filament current by turning up the output control while watching both the milliammeter in the anode circuit and the light from the filament wire viewed through the slit in the anode. *Be very careful in this step.* You will not see any emission current until the filament current gets up to about 3 A, but further increases will cause the emission to rise rapidly. As soon as you see the filament lighting up, proceed with great caution, while continually monitoring the value of the anode current. Set it at 8 mA, after which the filament current should not be increased any further. This will avoid overheating the filament and possibly burning it out. Remember that under no circumstances is the filament to be run at a current higher than 4.5 A. Record the values of the anode and filament currents you have just set.

**6.** The thin ray of light produced by the electrons as they pass through the mercury vapor should now be clearly visible, especially in a darkened room. We will refer to this ray as the electron beam. If the tube has been oriented as described in Procedure 3, the crossbars will extend upward from the staff wire and the electron beam will emerge from the anode slit to your right as you stand facing the tube from the base end. But the beam will be deflected slightly toward the base by the earth's magnetic field. To balance out this effect, a small current will be sent through the helmholtz coils, but first you must determine whether these coils are properly connected. Plug in their power supply; first make sure that its output control is turned all the way down, then turn it on, and increase its output slightly from zero. If the magnetic field of the coils straightens the beam, and, upon further increase, bends it around toward the crossbars, the coils are properly connected. If the field increases the deflection toward the base, the field current is flowing in the wrong direction and the connections from the power supply and meter should be interchanged.

**7.** Adjust the value of the field current until the electron beam is perfectly straight and perpendicular to the long axis of the tube. This adjustment should be made very carefully and accurately. When the beam is straight, the magnetic field of the coils is just equal to the earth's magnetic field, so that the resultant field at the beam is zero. Record this value of the current as $I_1$. It will be constant throughout the experiment, since the accelerating voltage is kept constant.

**8.** Increase the field current slowly until the electron beam describes a circle. Adjust the value of this current carefully until the bright, sharp outside edge of the beam just clears the outside edge of crossbar number 5, the one that is farthest away from the filament of the tube. Record this value as the $I_2$ for crossbar number 5 in the table. Also record the accelerating voltage indicated by the voltmeter.

The outside edge of the beam is used because it contains the electrons emitted from the negative end of the filament. These electrons fall through the greatest potential difference between filament and anode and thus have the greatest velocity. This is the potential difference that the voltmeter measures, and it is the one to be used in the calculations. The electron beam spreads as it goes around the tube, but all of the spreading is toward the inside of the circle. It is caused by the fact that electrons coming from the positive end of the filament are accelerated by a potential less than the voltage applied to the anode by the amount of the voltage drop across the filament. This can be seen in Fig. 52.3, which shows the accelerating voltage supply connected between the negative end of the filament and the anode and the filament supply connected between the negative end of the filament and the positive end. Thus the potential difference between the positive end and the anode is the difference between the two supply voltages, the potential difference between the negative end and the anode is the full anode supply voltage, and intermediate points along the filament are at intermediate potentials.

**9.** Keep on increasing the field current slowly and adjust its value carefully until the bright, sharp outside edge of the electron beam just clears the outside edge of crossbar number 4. Record this value of the current as the $I_2$ for crossbar number 4 in the table. Record the accelerating voltage to be sure there has been no change. Repeat the same procedure for crossbars 3, 2, and 1 and record the value of the current $I_2$ and the accelerating voltage in the boxes appropriate for each case.

# DATA _____

Number of turns on each coil     _____     *e/m* calculated value     _____

Mean radius of each coil     _____     *e/m* accepted value     _____

Filament current     _____     Percent error     _____

Anode current     _____

| Crossbar Number | Radius of circle | Anode voltage | $I_1$ | $I_2$ | $I$ | $B$ | $e/m$ |
|---|---|---|---|---|---|---|---|
| 1 | | | | | | | |
| 2 | | | | | | | |
| 3 | | | | | | | |
| 4 | | | | | | | |
| 5 | | | | | | | |

# CALCULATIONS _____

**1.** Calculate the *radius in meters* of the circle of the electron beam for each crossbar distance.

**2.** For each value of the field current $I_2$, subtract the current $I_1$ to obtain the value of $I$, the current needed to produce the magnetic field that deflects the electrons in the circle of the corresponding radius.

**3.** For each value of the field current $I$, calculate from Equation 52.3 the field strength $B$ of the magnetic field produced within the helmholtz coils.

**4.** For each circular path described by the electron beam, compute the value of $e/m$ by substituting in Equation 52.2 the appropriate value of the accelerating potential, the magnetic field strength obtained in Calculation 3, and the radius of the circle in question.

**5.** Find the average of the values of $e/m$ obtained in Calculation 4 and record this as the calculated value of $e/m$.

**6.** Compare your experimental result with the accepted value given in Table I at the end of the book by finding the percent error.

## QUESTIONS _____

**1.** Name two sources of error in your measurements that are most likely to affect the accuracy of your results.

**2.** Compute the velocity of an electron that has been accelerated through a difference of potential of 100 V. Express your result in meters per second. Do you think relativistic effects will have to be considered in this case?

**3.** Calculate the strength of the magnetic field required to bend electrons that have been accelerated through a potential difference of 100 V into a circle 10 cm in diameter. Express your result in gauss, a commonly used unit of magnetic field strength. $10^4$ G = 1 T.

**4.** Find the mass of the electron from your result for $e/m$ and the value of the electronic charge $e$ given in Table I at the end of the book.

**5.** Calculate the expected spread in the beam at cross-bar number 5 due to the voltage drop across the filament. Do this by noting that the accelerating voltage between the anode and the negative end of the filament is the voltage read on the voltmeter, whereas the voltage between the anode and the positive end is the voltmeter reading minus the filament voltage. This latter may be taken to be 3.0 V unless your instructor gives you a better value.

**6.** Show that if the magnetic field is held constant, the time $T$ required for an electron to make a complete circle in your *e/m* tube and return to the anode is independent of the accelerating voltage by deriving an expression for this time. The reciprocal $f = 1/T$ is called the electron's *cyclotron frequency*.

**7.** Substitute Equation 52.3 in Equation 52.2 to obtain an expression for *e/m* in terms of known constants and the quantities you actually measured in this experiment.

**8.** (a) Answer Question 5 of Experiment 35 if you have not already done so. (b) Does the ratio *e/m* appear in the formula you obtained in Part (a) of that Question? (c) If not, explain why not.

**9.** Suppose instead of having a filament emitting electrons you had a source of neon ions so that your experimental arrangement became a mass spectrometer as discussed in the Theory section. Because an atomic ion is so much heavier than an electron, higher voltages and magnetic fields will be needed, and so take the acceleration voltage $V$ to be 1000 V and the magnetic field B to be 0.4 T (4000 G). (a) Find the diameter of the circle in which ions of the principal isotope of neon (Ne$^{20}$, atomic mass number 20) would move in this magnetic field. (b) Another reasonably abundant neon isotope is Ne$^{22}$ (atomic mass number 22), and so a source of natural neon ions should emit a beam of Ne$^{22}$ ions as well as Ne$^{20}$. What would be the separation of these two beams when they had gotten halfway around their respective circular paths? Your answer demonstrates how a mass spectrometer can separate different isotopes of the same element—something that cannot be done chemically. In 1913, J. J. Thomson was the first to show that atoms of the same element could have different masses by separating the Ne$^{20}$ and Ne$^{22}$ isotopes in this manner.